... ASSOCIATES INC.
177 Jenny Wrenway
Willowdale Ont M2H 2Z3
Phone: (416) 494 2816   Fax (416) 494 - 0303

Human Reliability Analysis

RHODES & ASSOCIATES INC.
177 Jenny Wrenway
Willowdale, Ont. M2H 2Z3
Phone: (416) 494-2816 Fax: (416) 494-0303

Computers and People Series

Edited by

B. R. GAINES and A. MONK

Monographs

Communicating with Microcomputers. An introduction to the technology of man–computer communication, *Ian H. Witten* 1980
The Computer in Experimental Psychology, *R. Bird* 1981
Principles of Computer Speech, *I. H. Witten* 1982
Cognitive Psychology of Planning, *J-M. Hoc* 1988
Formal Methods for Interactive Systems, *A. Dix* 1991
Human Reliability Analysis: Context and Control, *E. Hollnagel* 1993

Edited Works

Computing Skills and the User Interface, *M. J. Coombs and J. L. Alty (eds)* 1981
Fuzzy Reasoning and Its Applications, *E. H. Mamdani and B. R. Gaines (eds)* 1981
Intelligent Tutoring Systems, *D. Sleeman and J. S. Brown (eds)* 1982 (1986 paperback)
Designing for Human–Computer Communication, *M. E. Sime and M. J. Coombs (eds)* 1983
The Psychology of Computer Use, *T. R. G. Green, S. J. Payne and G. C. van der Veer (eds)* 1983
Fundamentals of Human–Computer Interaction, *Andrew Monk (ed)* 1984, 1985
Working with Computers: Theory versus Outcome, *G. C. van der Veer, T. R. G. Green, J-M. Hoc and D. Murray (eds)* 1988
Cognitive Engineering in Complex Dynamic Worlds, *E. Hollnagel, G. Mancini and D. D. Woods (eds)* 1988
Computers and Conversation, *P. Luff, N. Gilbert and D. Frohlich (eds)* 1990
Adaptive User Interfaces, *D. Browne, P. Totterdell and M. Norman (eds)* 1990
Human–Computer Interaction and Complex Systems, *G. R. S. Weir and J. L. Alty (eds)* 1991
Computer-supported Cooperative Work and Groupware, *Saul Greenberg (ed)* 1991
The Separable User Interface, *E. A. Edmonds (ed)* 1992

Practical Texts

Effective Color Displays: Theory and Practice, *D. Travis* 1991

EACE Publications
(Consulting Editors: *Y. WAERN and J-M. HOC)*

Cognitive Ergonomics, *P. Falzon (ed)* 1990
Psychology of Programming, *J-M. Hoc, T. R. G. Green, R. Samurcay and D. Gilmore (eds)* 1990

Human Reliability Analysis
Context and Control

Erik Hollnagel
*Computer Resources International,
Birkerod,
Denmark*

ACADEMIC PRESS
Harcourt Brace & Company, Publishers
London San Diego New York
Boston Sydney Tokyo Toronto

ACADEMIC PRESS LIMITED
24—28 Oval Road
London, NW1 7DX

US edition published by
ACADEMIC PRESS INC.
San Diego, CA 92101

Copyright © 1993 by
ACADEMIC PRESS

All Rights Reserved
No part of this book may be reproduced in any form, by photostat, microfilm, or any other means, without written permission from the publishers

This book is printed on acid-free paper

A catalogue for this book is available from the British Library

ISBN 0-12-352658-2

Printed in Great Britain by Hartnolls Limited, Bodmin, Cornwall

Contents

Foreword

1. **Reader's Guide** xiii
2. **Rationale** xiii
   - 2.1 Credo xv
   - 2.2 The Root Cause xvi
   - 2.3 Who Should Read This Book xvii
   - 2.4 ... And Who Should Not! xviii
3. **Chapter by Chapter** xix
4. **Miscellanea** xxii
   - 4.1 Model Multiplicity xxii
   - 4.2 Scope of the Model and Method xxiii
   - 4.3 Terminology: An Apology to the Sensitive Reader xxiv
   - 4.4 Acknowledgements xxv

Chapter 1. Performance, Reliability, and Unwanted Consequences

1. **Emergence of the Human Factor** 1
   - 1.1 The New Environment 2
   - 1.2 The Rise of "Human Errors" 3
2. **Coupling between Complexity and Unwanted Consequences** 6
   - 2.1 Risk Homeostasis 8
3. **The Anatomy of an Accident** 12
   - 3.1 Accident Signatures 12
   - 3.2 Software Reliability 15
   - 3.3 Reactions to Failures and Accidents 16

4.	**Human Reliability** 20	
	4.1 The Cognitive Viewpoint 21	
	4.1.1 Knowledge Mismatch 22	
	4.2.1 Inaccurate Execution of Plans 23	
	4.2 System Induced and Residual Erroneous Actions 23	
	4.3 Qualitative and Quantitative Analyses 24	
5.	**The President's Heart Attack** 26	
6.	**"Human Error" and Erroneous Actions** 29	
	6.1 Human Erroneous Actions *versus* Unwanted Consequences 31	
7.	**Complexity and Cognition** 33	
	7.1 The Ant Analogy 33	
	7.2 The Origin of Complex Performance 34	
	7.3 The Complexity of Human Performance 36	
8.	**The Artifacts of Decomposition** 37	
	8.1 The Individual Action 38	
	8.2 The Performance Shaping Factor 39	
	8.2.1 Use of Performance Shaping Factors 41	
	8.2.2 Innocence of PSF Calculations 42	
	8.3 Common Modes: Rule or Exception? 44	
9.	**Summary** 45	

Chapter 2. The Need for Models of the Reliability of Cognition

1.	**Introduction** 49	
	1.1 Definitions of Human Reliability 51	
2.	**The Decomposition Principle** 54	
	2.1 Systematic Human Action Reliability Procedure (SHARP) 54	
3.	**Reliability Analysis and Event Estimation** 56	
	3.1 Granularity of Decomposition 58	

4.	**Repeatability and Similarity** 60
	4.1 Reliability and the Cumulation of Effects.................... 61
	4.2 Reliability and Situation Equivalence......................... 62
	4.3 The Person as a System Component......................... 63
	4.4 Man as a Fallible Machine...................................... 64
5.	**Human Reliability and the Analysis of Erroneous Actions** 65
	5.1 The Duality of "Human Error"................................. 66
	5.2 The Systematic Study of Erroneous Actions................. 67
	5.3 The Problem of Privileged Knowledge....................... 68
	5.4 Phenotypes and Genotypes...................................... 69
	5.4.1 Slips and Lapses 70
	5.5 A Logical Classification of Phenotypes 71
	5.6 Simple and Complex Phenotypes 73
	5.6.1 Errors of Omission and Commission 78
6.	**Human Reliability and the Reliability of Cognition** 79
	6.1 The Changing Nature of Tasks................................. 80
	6.2 Task Change and Function Amplification 84
	6.3 The Social System Analogy..................................... 87
7.	**Summary**.. 90

Chapter 3. The Nature of Human Reliability Assessment

1.	**Engineering Quantification** 93
	1.1 The Atomistic Assumption..................................... 95
	1.2 The Mechanistic Assumption 96
2.	**Differences Between Humans and Machines** 98
3.	**Identifiable Models *versus* Curve Fitting**........................ 102
4.	**The Need for Better Models** 105
	4.1 Parameter Uncertainty and Model Imprecision 106
5.	**The Art of Human Reliability Analysis**........................ 109
	5.1 The Problem of Quantification 110

6.	**Obstacles to the Study of Human Reliability**		111
	6.1	Observation	112
	6.2	Registration of Data	113
	6.3	Specification of Data	113
	6.4	Data Collection and Data Analysis	114
	6.5	Experimentation: The Use of Micro-Worlds	118
7.	**Assessment of Human Reliability**		122
	7.1	Empirical Data	122
	7.2	Data from Simulators and Simulations	123
		7.2.1 Cognitive Modelling	124
	7.3	Expert Judgment	125
	7.4	A Procedure for Using Expert Judgment Data	127
	7.5	The Value of Data	129
8.	**Human Factors Reliability Benchmark Exercise**		129
9.	**Short Survey of Human Reliability Methods**		133
	9.1	An "Ideal" Method for Human Reliability Analysis	138
10.	**Summary**		141

Chapter 4. Fundamentals of the Model

1.	**Metaphors and Models of Cognition**		145
	1.1	Stimulus-Organism-Response	145
	1.2	The Human as an Information Processing Mechanism	147
	1.3	The Cognitive Viewpoint	149
2.	**Procedural Prototype Models of Cognition**		152
	2.1	The Step-Ladder Model	154
	2.2	Predominance of Procedural Prototype Models	156
	2.3	Loose Ordering (TOTE)	157
3.	**Contextual Control Models of Cognition**		159
	3.1	Competence and Control	162
	3.2	The Model of Competence	164
		3.2.1 The Activity Set	164
		3.2.2 The Template Set	166

4.	**Control Modes**	168
	4.1 Scrambled Control	168
	4.2 Opportunistic Control	169
	4.3 Tactical Control	170
	4.4 Strategic Control	170
	4.5 Control Mode and Subjectively Available Time	171
	4.6 Interaction between Competence and Control	173
5.	**The Contextual Control Model (COCOM)**	175
	5.1 Main Control Parameters	177
	5.2 Other Dimensions of Control	178
	5.2.1 Number of Simultaneous Goals	178
	5.2.2 Availability of Plans	179
	5.2.3 Event Horizon	180
	5.2.4 Mode of Execution	182
6.	**Control Modes and Performance Characteristics**	184
	6.1 Scrambled Control	185
	6.2 Opportunistic Control	187
	6.3 Tactical Control	189
	6.4 Strategic Control	190
	6.5 Changes between Control Modes	193
	6.6 Relations to Other Descriptions	194
	6.7 Control Modes and Reliability of Cognition	196
	6.8 Control Modes and User Modelling	198
7.	**Summary**	200

Chapter 5. The Dependent Differentiation Method

1.	**Framework for Assessing Human Reliability**	203
	1.1 The Decomposition Principle	204
	1.2 The Need for Task Analysis	206
2.	**Task Analysis Principles**	207
	2.1 Task Analysis	207
	2.2 Task Description	208
	2.3 Task Representation	209
	2.4 Specialised Analyses	209

3.	**The Logic of Task Analysis**		211
	3.1	Pure Tasks	212
	3.2	Tasks with Pre-conditions	213
		3.2.1 Timing Conditions	215
		3.2.2 Execution Conditions	216
	3.3	Post-conditions	218
4.	**Goals-Means Task Analysis Method**		220
	4.1	Example: "Feed and Bleed"	220
	4.2	Formalisation of the GMTA Method	226
5.	**Task Description Requirements**		229
	5.1	Continuity	230
	5.2	Performance Variability	230
	5.3	Communication and Interaction	231
	5.4	Requirements for Human Reliability Analysis Methods	231
6.	**Common Performance Modes**		233
	6.1	Available Time	234
	6.2	Availability of Plans	234
	6.3	Number of Simultaneous Goals	235
	6.4	Mode of Execution	236
	6.5	Process State	237
	6.6	Adequacy of Man-Machine Interface and Operational Support	238
	6.7	Adequacy of Organisation	238
	6.8	Other Common Performance Modes	239
7.	**From Task Analysis to Common Performance Modes**		240
	7.1	Stage 1: Task Analysis	241
	7.2	Stage 2: Assessment of CPMs	242
	7.3	Stage 3: Identification of Critical Events	243
	7.4	Stage 4: Quantification	244
	7.5	Utilisation of the Dependent Differentiation Method	245
8.	**Using the Dependent Differentiation Method**		246
	8.1	Stage 1: Task Analysis	246
	8.2	Stage 2: Common Performance Modes	249
		8.2.1 Available Time	249

		8.2.2 Availability of Plans 250

        8.2.2 Availability of Plans 250
        8.2.3 Number of Simultaneous Goals 250
        8.2.4 Mode of Execution 250
        8.2.5 Process State 250
        8.2.6 Adequacy of Man-Machine Interface and
              Operational Support......................................251
        8.2.7 Adequacy of Organisation..............................251
        8.2.8 Summary of Common Performance Modes for
              FAB 251
   8.3 Stage 3: Identification of Critical Task Steps 252
   8.4 Stage 4: Quantification of Analysis Results 256
        8.4.1 Quantification Issues 260

9. **The Dependent Differentiation Method and COCOM** 262
    9.1 Quantification of the Reliability of Cognition..............264

10. **Summary**..265

Chapter 6. Discussion

1. **Analysing Human Performance** 269
   1.1 Accident Analysis and Performance Prediction............271

2. **Consequences of the Contextual Control View** 273
   2.1 Interface Design 274
   2.2 Other Issues 275

3. **Hypernatural Environments** 276
   3.1 Adaptation through Design....................................277
   3.2 Adaptation during Performance 278
   3.3 Adaptation through Management............................279
   3.4 Adaptation and Reliability....................................279

4. **Performing a Human Reliability Analysis** 280
   4.1 Model Verification 282
   4.2 Computerisation of the Dependent Differentiation
        Method..282
   4.3 Application of DDM Outcomes ,.........................283

5. **Comparing Analysis Methods** 284

		5.1 Completeness, Consistency, and Decidability 286

6. **Attention and the Reliability of Cognition** 288
   - 6.1 The Limits of Attention 290
     - 6.1.1 Attention Capacity 292
     - 6.1.2 Attention Demand....................................... 293
   - 6.2 Consequences for Design 294
     - 6.2.1 Coping with Multiple Tasks 294

7. **COCOM and the Reliability of Cognition**..................... 296
   - 7.1 Attention and Performance Reliability..................... 296
   - 7.2 Attention, Specificity, and Control......................... 298
     - 7.2.1 An Example... 299
   - 7.3 Attention and Control 300

8. **The Last Word** 302

Appendix: Introduction to the System Response Generator

1. **The Practice of Safety and Reliability Analyses**............... 305
   - 1.1 Point-to-Point Analyses 306
   - 1.2 Static and Dynamic Analyses 307

2. **The System Response Generator** 308
   - 2.1 The Generation of System Responses...................... 307
   - 2.2 SRG Modules.. 308
   - 2.3 Operator and Process Modelling............................ 309

References 315

Index 327

Foreword

1. Reader's Guide

The purpose of this foreword is to provide the reader with a survey of the topics that are covered in the book, as well as some supplementary information about the book itself. The purpose of these first paragraphs is to provide the reader with a guide to the foreword itself:

(1) The first section outlines the purpose of the book as well as the **rationale** for writing it. It also provides some advice about who should read the book and who should not. This section should therefore be read by all, even the casual browser in the bookstore.

(2) The second section briefly goes through the book in a **chapter-by-chapter** fashion. Readers who have not been put off completely by the first section are encouraged to read the second section. It will enable them to decide which chapters of the book they should concentrate on and in which order.

(3) The third and last section provides **miscellaneous information** and comments. Readers, whose curiosity is aroused by the headings, should read the associated text at some time, although not necessarily before starting on the main chapters of the book.

2. Rationale

In the beginning of the 1990s the field of human reliability analysis (HRA) was in a state where there was pronounced dissatisfaction with the available methods, theories, and models, but where there as yet were no clear alternatives (Dougherty, 1990). The intention of this book is to present such an alternative, based on the principles of cognitive systems engineering.

Throughout the 1980s there was a growing recognition in the engineering world of the role of human cognition in shaping human action - both when it led to accidents and when it prevented them. This recognition was not felt in human reliability analysis alone, but also in the concern with man-machine systems in general, with decision support systems, with human-computer interaction, etc. One consequence was that "cognitive" and "cognition" became fashionable terms for almost all aspects of man-machine interaction. As an example, the book on "Accident Sequence Modelling" by Apostolakis *et al.* (1988) has the following main entries:

(1) cognitive activity,
(2) cognitive competencies,
(3) cognitive environment simulation,
(4) cognitive modelling,
(5) cognitive primitives,
(6) cognitive processing,
(7) cognitive reliability analysis technique,
(8) cognitive structures,
(9) cognitive sub-elements, and
(10) cognitive under-specification.

In many cases, however, the allusion to cognition was a matter of convenience rather than a real change in orientation. Cognition, however, **is** of fundamental importance and it is consequently necessary to have adequate methods, theories, and models to address properly the role of cognition in human action - and particularly specific issues such as the Reliability of Cognition.

The study of human cognition has developed from experimental psychology in the 1960s and has gradually grown in several distinct directions (it would probably be going too far to call them scientific disciplines). Some of these focus on basic research issues while others venture into what for academia is the *terra incognita* of applications; among the latter are cognitive science, cognitive systems engineering, and cognitive ergonomics.

Cognitive systems engineering (Hollnagel & Woods, 1983) is based on the principle that human behaviour - in work contexts and otherwise -

should be described in terms of joint or interacting cognitive systems.[1] A joint system where one of the parts is a cognitive system is also in itself a cognitive system. Hence all man-machine systems are by definition cognitive systems. In the classical view on man-machine systems, one could consider the man (= the operator) by himself, the machine (= the process) by itself, and add the interaction between the two. This view, however, misses the notion of integration and dependency and, in particular, that all activities take place in a context.

Cognitive systems engineering is obviously not the only way to look at human cognition and it cannot be proved that it is **the** correct way. It is, however, a usable basis for describing human cognition in the context of human work, i.e., it is pragmatically correct. The specific developments described in this book are focussed on the notion of how actions are controlled and on how control and reliability are related.

2.1 Credo

Better analyses of the reliability of cognition are needed for practical reasons alone. Current approaches to HRA are based on the principle of describing situations in terms of appropriate components or elementary events, e.g. as single actions. This principle of decomposition is basically a consequence of the underlying view of the human operator as a machine - possibly a complex, cognitive machine, but a machine nevertheless.

Such approaches are, however, inadequate as a way of describing human cognition because they are not based on a clear theory of human cognition - or even on a clearly formulated description of what human cognition is. A proper analysis or assessment of human reliability must not only acknowledge the role of cognition, but also include a theory or description of human cognition and of the reliability of cognition.

Any such model - even a very simple model of cognition - will show that cognition must be considered as a whole and as an integrated activity that reveals itself in a context, rather than as a decomposable ordering of elementary functions and bits of knowledge. Any assessment method must

---

[1] A cognitive system (1) is goal oriented, and based on symbol manipulation, (2) is adaptive and able to view a problem in more than one way, and (3) operates using knowledge about itself and the environment and is therefore able to plan and modify its actions on the basis of that knowledge. The definition is intended to be equally applicable to men and machines.

start by recognising this fact and strive to derive a description which does not conflict with that.

An alternative approach to human reliability analysis may make it less straightforward - but also less necessary - to provide point estimates or point probabilities of individual actions. It will, however, improve the qualitative basis for developing solutions that consider the system as a whole and which therefore contribute to the overall goal of reducing the number of unwanted consequences. An alternative approach will also make it easier to assess the overall risk or reliability of a work situation in a meaningful way.

On the other hand, it will also reduce the need to collect data (estimates) for minute aspects of human performance, since such data will no longer be very important. Instead data must be sought on the level of cognitive ensembles, i.e., the practically meaningful segments of work.

2.2 The Root Cause

Risk and reliability analyses are often made on the basis of descriptions that use trees as an underlying structure: operator action trees, event trees, cause-consequence trees, etc. Since every tree has one root - at least in the simplified graphical representations that commonly are used - the notion of a root cause has become widespread. The root cause, of course, means the single, identifiable cause for an observed consequence, even though most practical cases show that there rarely is only one cause.

In the case of this book the root cause was a special issue of the *Journal of Reliability Engineering and System Safety* that dealt with the problems of HRA and the unhappy state of the art. The basis for the special issue was a position paper by Ed Dougherty (1990), which was followed by a number of comments (some short, some long, some agreeing and some disagreeing) from people who, in one way or another, either had experienced the problem or had an opinion on it.

I am sure that there are even more opinions than were expressed in the special issue. In fact, I was asked to contribute a comment and started to write down my views but did not finish them in time for the special issue. As luck would have it, another opportunity came at the International Conference on Probabilistic Safety Assessment and Management (PSAM), which was held in Beverly Hills, February 4-7, 1991. For this occasion I elaborated on my unfinished comments and presented them as a paper entitled "What is a Man that he can be Expressed by a Number?" That paper in turn became the starting point for this book, which can be seen as a elaboration and extension of the main theme of that paper, i.e., a long argument against viewing and describing humans in terms of numbers - whether as reliability measures or something else.

Although the special issue of the *Journal of Reliability Engineering and System Safety* mentioned above can be seen as a root cause for this book, it is certainly not the only cause. The paper by Dougherty (1990) merely expressed the concerns that many HRA practitioners had. In addition, psychologists and others had generally criticised the approach to quantitative modelling that HRA practitioners had taken. In his editorial, Apostolakis (1990) rather bluntly expressed it thus: "... researchers who try to understand human behavior and to develop models for the operators have a very negative view toward the use of such quantitative models, whose foundations they consider to be unacceptable." This critical view can be found in practically all of the books and papers published during the 1980s that looked at "human error" from the behavioural or social sciences point of view (e.g. Perrow, 1984; Rasmussen *et al.*, 1987; Reason, 1990; and Senders & Moray, 1991). It is a criticism which is amplified by the general view of cognitive systems engineering and cognitive ergonomics, as described above. The real "root cause" for this book is therefore an assortment of views and issues that gradually were developed during the 1980s by the international community of people concerned with the study of human cognition.

2.3   Who Should Read This Book ...

I have written this book with a certain audience in mind. The audience is not defined in terms of lines of profession but rather in terms of specific interests or views on man-machine systems and human performance. In other words, there is a certain audience that I hope will find the book congenial. This audience includes:

(1)   The HRA practitioners who have found the current approaches, models, and methods lacking in one way or another.

(2)   The scientists and researchers who adhere to what can generally be called the cognitive viewpoint, i.e., who find that human cognition plays an essential role in analysing and understanding human performance.

Foreword xviii

(3) The specialists and engineers who are practically involved with the design, management, or use of man-machine systems in all fields and who are uneasy about the impact of human performance (the human factor) on system performance.

(4) Those people who have an interest in the practical study of human behaviour and human cognition, and who are genuinely interested in or concerned about human performance in working situations.

2.4 ... And Who Should Not!

Just as there is an intended audience, there are also several groups of people who I expect will find this book rather disagreeable, and who therefore are advised not to read it unless they want to see their views challenged. These people include:

(1) The practitioners and risk analysts who perform human reliability analysis and who are perfectly happy with the current approaches.

(2) The scientists and researchers who firmly believe that the study of human cognition can only be carried out with well-controlled experiments and rigorous quantitative/statistical methods. This also includes those who believe that computational models or information processing descriptions can provide perfectly adequate explanations for human performance.

(3) The specialists and engineers who cannot understand why some people have misgivings about quantifying probabilities for human errors and why these people are therefore reluctant to provide such numbers.

(4) Those people who think that "human error" is a perfectly good root cause, and that the solution to the problem of "human error" basically is to increase the level of automation.

Any readers who feel that they do not belong to either of these groups, for instance because they are not interested in this field at all, should probably decide for themselves whether they want to go on reading. I expect, however, that they will find this book rather boring.

3. Chapter by Chapter

The chapters in this book have been organised to express a certain flow of thought or line of argument. It may, however, not be as obvious to the reader as it is to the author. Furthermore, different readers may be looking for different things, and therefore need not read the chapters in the same order - or indeed read all the chapters.

Chapter 1 provides a broad account of the background for the concern with the Reliability of Cognition. It describes how the need to consider the human factor or human operator arose, and how the technological development apparently has caused a greater susceptibility to incorrect or erroneous actions. This is followed by a discussion of how accidents usually are described and what the typical responses or reactions are.

Next, the chapter opens the discussion of human reliability analysis and how it can be understood from the cognitive viewpoint. The predominant approach is to look for a specific and quantifiable cause, as exemplified by the case of the President's heart attack. The notion of "human error" is examined and the suggestion is made that it should be replaced with the concept of an erroneous action. The point is made that the concern should be to prevent or avoid unwanted consequences rather than to study human reliability and erroneous actions as separate topics.

Chapter 1 ends with a discussion on the nature of human cognition and in particular the debate about whether human cognition is inherently simple or complex. The simple view is consistent with the predominant decomposition approach in human reliability analysis. It is argued that this approach has produced two artifacts: ideas about the individual action and the performance shaping factor. Both artifacts have contributed to the problems of current HRA practice.

Chapter 2 argues for the need to have better models of the Reliability of Cognition. It begins by developing a definition of human reliability, and continues by describing the current decomposition principle. It is argued that the current approaches are based on two assumptions about repeatability of events and similarity between situations. It is pointed out that whereas these assumptions are correct for technical systems, they are not tenable for humans. The assumptions are the result of transferring the notion of a machine to the description of humans, but this is not appropriate - not even as the notion of a fallible machine. A human being should fundamentally be described as a cognitive system, and this has consequences for the methods that can be used.

Chapter 2 continues the discussion of "human error" and erroneous actions by proposing a clear distinction between phenotypes (manifestations) and genotypes (causes) of erroneous actions. This is supplemented by a complete taxonomy for the phenotypes of erroneous actions. Finally, the nature of the Reliability of Cognition is discussed in relation to the ways in which tasks and work contexts have changed. This has led to an increased dependence on tasks that involve "thinking" rather than "doing", hence on human cognition. Human performance assessments must accordingly take this dependence into account, and put greater emphasis on the context of human actions. This requires an adequate model of human cognition.

Chapter 3 gives a critical account of human reliability assessment as it is currently practiced. The consequences of the decomposition principle are further elaborated by characterising the atomistic and the mechanistic assumptions. The predominantly quantitative approaches are exemplified by discussing the difference between curve-fitting and model identification. There is a need for better models to support the assessment of human reliability. However, the effort to quantify such assessments define a paradox: in order to have quantification it is necessary first to have a proper qualitative description or model. In other words, it is necessary to specify the data that are needed before they can be sought.

The practical problems of analysing the reliability of performance are discussed by presenting a comprehensive view on the nature of data. This explains the coupling between data and the underlying concepts, and how the notion of objective raw data is an illusion. It is followed by an overview of the different types of data and associated methods that are used in human reliability analysis: empirical data, data from simulations, and expert judgments. The chapter ends by summarising a major human factors reliability benchmark exercise and by comparing the existing methods to a so-called "ideal" method.

Chapter 4 begins the description of the model of cognition that will be used as a basis for the method. It starts by recapitulating the three main approaches to the modelling of cognition: the S-O-R, the information processing approach, and the cognitive viewpoint. This is followed by a characterisation of two major classes of models, called procedural prototypes and contextual control models. The former express the view that performance can be seen as variations of a pre-defined sequence (the prototype); an example of that is found in the typical decision making model. In contrast to that the contextual control models emphasise that the sequence of actions is the result of an active choice. This choice depends on the current context, and the emphasis therefore should be put on how this choice is made.

A contextual control model has two parts: the competence model which describes which actions and plans are possible, and the control model which describes how the choice of the next action is controlled. A distinction is made between several levels of control, exemplified by four distinct control modes called scrambled, opportunistic, tactical, and strategic. This is further developed in terms of a specific instance of the contextual control model called COCOM. COCOM is described in terms of the main parameters that determine the performance characteristics on each level of control and the ways in which control can change from level to level.

Chapter 5 describes the new approach to human reliability assessment, called the Dependent Differentiation Method (DDM). The basis for the method is a systematic description of the tasks, derived by a Goals-Means Task Analysis (GMTA). This task analysis method is explained in detail and the procedure is illustrated by an example. The DDM uses a characterisation of the common features of the task, named the Common Performance Modes (CPM). The CPMs are a convenient way of describing the impact of the context on the control of actions. The CPMs can be determined from the outcome of the Goals-means Task Analysis. Through an iteration procedure the DDM establishes the likely levels of the CPMs and thereby also the probable control modes. The further characterisation of the performance is based on refining the description of the control modes and how they influence the choice of actions. In cases where specific actions are known to be critical for the system, they can be analysed in detail using the same principles.

The conclusion is that it is not the Reliability of Cognition which is important *per se*, but rather how it influences performance. The DDM therefore does not strive to produce a measure of the Reliability of Cognition, but rather of the reliability of performance as a whole. This can be done in a qualitative fashion and improved, e.g. by using fuzzy set descriptions. It may also ultimately be turned into a quantitative description, but this should only be done if the numbers can be given a meaningful interpretation.

Chapter 6, finally, discusses a number of issues that are affected by the model and method developed in Chapters 4 and 5. It is pointed out that accident analysis is possible because the context is known and that performance prediction consequently should serve to describe the likely context as a prerequisite to describing individual actions. The consequences of the contextual control view are discussed as they apply to the design of man-machine systems and human-computer interaction. The practical problems in carrying out a human reliability analysis are addressed, and the prospects of providing computer support for the method are outline. Following that, a framework is proposed to compare various methods for human reliability analysis.

The chapter ends by bringing forward an important concept of human cognition: attention. Attention is considered in relation to the Reliability of Cognition and in relation to the COCOM. It is argued that attention is a concomitant rather than a direct aspect of the contextual control view, and that it comprises several of the parameters that were described for the model. The effect of (a lack of) attention can best be seen by describing how it affects the choice of actions. The possible effects of a lack of attention depend on the relative task demands and on the possibilities for unwanted consequences to manifest themselves - both of which can be understood in terms of the contextual control model and determined by the DDM.

4. Miscellanea

4.1 Model Multiplicity

The notion of models of cognition is widespread and is used in many different ways. The need for models can, however, be made clearer if a distinction is made between different instances of models:

(1) **Scientific:** the primary purpose here is to aid understanding of something (a phenomena, a system). A scientific model explains the phenomenon in question and provides an account of the mechanisms or functions (the causal or functional architecture) that underlie the phenomenon.

(2) **Engineering:** the primary purpose is to develop a representation of a system which can be used to calculate or predict future developments. The model is a translation of essential functional

relationships and dependencies into a form which enables controlled manipulation of the independent parameters (including the environment). An engineering model can serve its purpose without necessarily constituting an explanation.

(3) **Cybernetic:** the primary purpose is to provide the representation necessary to control a system. This usage is based on the Law of Requisite Variety, which can be interpreted as saying that a regulator of a system must be a model of that system. Control implies a certain amount of prediction, but the needs for precision and details are quite different from the engineering use of models. Similarly, a cybernetic model is not always useful as an explanation.

In the field of human reliability analysis a distinction is often made between engineering models and rigorous models. This book takes neither route, but instead proposes a pragmatic (read: cybernetic) model. This model was originally developed to help in controlling a simulation of man-machine interaction, and can therefore easily be used to describe how actions are controlled. In this way it can serve as the basis for developing a method to analyse the reliability of human performance. It does not try to fulfil the need for engineering or rigorous models that is expressed by current HRA; the view is rather that this need is an artifact of the dominating approaches, hence that it disappears if an alternative solution can be developed.

4.2 Scope of the Model and Method

The book develops both a specific model and a specific method. The obvious question is how general these are. The detailed example is taken from the field of nuclear power plants; since this field has had an influence on cognitive engineering which is disproportionately large - due mainly to a limited number of widely published accidents - it is not unreasonable to ask whether the model and the method, unintentionally, is limited to this area.

The answer is that both the model and the method have been developed to be applicable to a wide range of fields. The model is about how actions are chosen and controlled; there is nothing in the model itself which favours one particular field of application. The restriction is rather that the model is concerned with human actions in the context of work with dynamic processes; this may possibly exclude other areas, such as information retrieval or text processing, although this is far from certain. Anyway, it is a limitation that is not unacceptable.

The method is designed to identify the influences from the context where the actions take place and to find the possible ways in which unwanted consequences can occur. This is predicated on a view of human action as purposeful activities carried out in a complex environment which is only partly known - and only partly knowable. It will therefore not be surprising if this particular method of analysis is inappropriate or even inadequate for other purposes. In order to be useful a method must be of limited scope - it must trade breadth for depth. However, within the field of work with dynamic processes I believe that the method can be of general use to determine the possible effects of limited human reliability. Neither the model nor the method are limited to specific fields such as nuclear power plants or aviation.

4.3 Terminology: An Apology to the Sensitive Reader

A small, but important, issue is which term should be used to describe the combination of people and machines that provide the context for the contents of this book. Until the mid-1970s the preferred term was man-machine system (e.g. Singleton, 1974) and no one seemed to have any problems with that. Due to the growing tendency to avoid sexist language, the term man-machine system fell somewhat into disrepute and was replaced with terms like person-machine system or human-machine system. In the 1980s the developments in the study of how people interact with computers produced two new candidate terms: human-computer interaction (HCI; in Europe) and computer-human interaction (CHI; in the US). HCI/CHI, however, only deal with a subset of the problems that are addressed by the study of man-machine systems, and can therefore not be used as substitutes.

I shall, in this book, continue to use the term man-machine system, abbreviated as MMS. There are several reasons for that. Firstly, one meaning of the word man (and usually the first entry in dictionaries) is human (being), and the man in MMS it is to be understood in this sense rather than as a synonym for male. Secondly, although the term MMS may offend some academics, it is well entrenched in the applied fields. One of the most prestigious journals is called the *International Journal of Man-Machine Studies* and practitioners routinely refer to MMS and MMI (meaning either man-machine interaction or man-machine interface). Changing the term to e.g. human-machine system would also require that the widely used acronyms were changed to HMS and HMI. Since this would probably cause a lot of unnecessary confusion, I have decided to stick with the usage of man-machine system and MMS. I hope that readers will not be offended by this.

A related, but less contentious issue, is the choice of a term to refer to the people or persons who work with the machines. The more frequently used candidates are "operator", "user", "person", and "agent". I have decided to use the term person throughout the book. In cases where it is necessary to use a personal pronoun, I have opted for "he". This is not for sexist reasons, but purely for convenience and conformity with the tradition. Finally, most of the people who work in industrial settings such as power plants and cockpits are undeniably male. So using the pronoun "he" could also be defended on the grounds of the *a priori* distribution in the population.

4.4 Acknowledgements

It is customary to acknowledge intellectual debts in the writing of a book like this, and I am indeed very happy to do so. I will, however, not produce a long list of names. Rather I will acknowledge my intellectual debts to what is sometimes known as the "cognitive circus" - the group of people from a broad range of countries who for the last 10-15 years regularly have met (in subsets) for various occasions and among whom the cognitive viewpoint gradually has matured. Many of the ideas described in this book have developed during the meetings of the "cognitive circus" - in presentations or through discussions. Since it is impossible to attribute every idea to a specific source, I refrain from doing it altogether. The book is both an expression of the views of the "cognitive circus", as of myself.

I would, however, like to mention a few people without whom this book might not have been realised. Firstly, Ed Dougherty who started the whole thing by his (in)famous paper in 1990. Ed was supportive of the idea of writing this book from the very start, and has been willing to provide me with his view on many things as the chapters gradually emerged; in particular, he suggested the "Feed and Bleed" as a good example to use and provided me with many insights on that particular event.

Secondly, much of the theory presented here has been developed as part of the work in two projects, the Human Reliability Analysis Method, sponsored by the European Space Agency, and the System Response

Generator, sponsored by the CEC. I have learned a lot through many discussions with my colleagues in these projects, as I am sure they can see throughout the book. I have enjoyed many hours of discussion with Robert Taylor and particularly (standing, sitting, walking, and running!) with Carlo Cacciabue. During the later phases of writing I have received many useful comments and criticisms from Lisanne Bainbridge, Paul Booth, Yushi Fujita, John Hammer, Jacques Leplat, and Neville Moray. The latter in particular did his best to correct the worst abuses of the English language; the anonymous copy editor took care of the rest. I am also grateful to Ole Grønvig for having transformed a loose sketch into the front cover design. At last, I have to thank Dave Woods; although he has not been closely involved with the writing of this book we are twin brothers of the mind and our irregular collaboration over the last decade or so has helped in cementing the foundations of cognitive systems engineering - and therefore also the views expressed in this book.

Finally, and most of all, I must thank my wife Agnes for her unwavering patience and support during the many evenings and weekends that I have spent time writing and rewriting chapter upon chapter instead of being with her. In addition, her common sense has often forced me to make clear what I was writing about - expressing it without overly using technical jargon, and not writing for the initiated and converted.

Needless to say, despite my discussions and loans from others (at times incompletely acknowledged), the responsibility for the final result is mine. If there is any merit or value in what I have written, I gladly claim the honour. But neither will I shy away from anything that is incorrectly or wrongly put. I have tried to avoid mistakes, but if there are any the blame is certainly mine.

1.

Performance, Reliability, and Unwanted Consequences

1. Emergence of the Human Factor

Until the time of the Second World War it had been tacitly assumed that human beings were able to adapt to the requirements posed by technological systems. Practical experience had on the whole confirmed this assumption, both because the technological systems mostly had been related to assembly and production, i.e., the handling of materials rather than of information, and because the level of automation and the use of electronics had been limited. But the rapid technical developments of military equipment required a revision of this experience:

> "...during the Second World War, the approach of designing the task to fit the operator was added to the more traditional psychological procedures of selecting and training operators to fit their jobs. This was necessitated by the variety and complexity of military equipment. Machinery had finally outrun the man's ability to adapt."
> (Taylor, 1960, p. 643)

This led to the fields of engineering psychology, human factors (engineering) or ergonomics. (Although problems of this nature had been recognised even earlier, there was no specialised academic discipline to deal with them.) The technological development was not for long confined to the military world, but quickly spread into civilian applications. This brought about a change in the requirements to the person: the work moved from being predominantly production (**making** or **doing** things) to being predominantly control and supervision (**thinking** about how things should be done and **planning** how to do them). The important input-output was no longer energy or material but information. The typical task changed from being the direct involvement with a process to being the control of a process, and even further to being the control of machines which, in turn, controlled other machines; the person was in this way removed several steps from the actual process.

1.1 The New Environment

These changes were to a very large extent a consequence of the increased speed and complexity of the machines that came into use (which again were required to improve the efficiency of the processes. The term "process" is used as a general reference to that which is being controlled. It can be anything from using an anti-aircraft gun to producing an automobile or controlling an oil refinery). The pace of the technological systems had become so fast that the unaided human no longer could follow. Typically, reaction times were required which were well below what a human normally could accomplish. Or there was a need for calculations and evaluations of measurements and signals that were beyond normal mental capacity - or human information processing capacity as it gradually started to be called. There is an unmistakable relation between the need for ergonomic knowledge and the appearance during those years of meta-technological sciences such as information theory (the mathematical theory of communication), cybernetics, computer science, signal detection theory, etc. These sciences helped to fulfil the need for a better understanding of the human factor, as well as to the development of concepts and basic tools required for making such descriptions.

The new need spawned a collaboration between engineering sciences and psychology, which led to the appearance of ergonomics (human factors engineering). The requirement that the human should become a controller of machines created a host of new problems. For instance, it became important to know exactly how much a person could attend to, discriminate between, evaluate, remember, manipulate, etc. either simultaneously or over a period of time. Furthermore, since the work environment often became more demanding (in planes, ships, vehicles, etc.) it was also necessary to know what a person could tolerate in terms of vibration, heat, cold, pressure, noise (and silence!), light, acceleration, etc. In short, there was a growing need to design machines, operations, and complete work environments to match human capabilities and limitations (Swain, 1990).

The underlying problem for ergonomics was the **shortcomings of human performance** in Man-Machine Systems (MMS), which was seen as the cause of all kinds of **unwanted consequences**, i.e., events which are neither intended to occur nor desired. These shortcomings had of course always existed, but they became conspicuous during the change from doing to thinking. Today, 35 years later, the problem still remains, despite all the efforts that have been spent and all the solutions that have been proposed in the time between. The problem may in fact have become worse, due to the increasing complexity of the technological systems.

1.2 The Rise of "Human Errors"

One indicator of that is the size of the so-called human errors. Or, to be more precise, the number of cases in which the cause of a mishap is generally attributed to a human action. A human action is traditionally classified either as an omission (non-occurrence of an action) or as a commission (occurrence of a substitute action), although these categories are not conceptually distinct (cf. Chapter 2). The actions in question are typically part of operation (supervision and control), but may also be part of design, building, assessment, maintenance, management, etc. In the 1960s, when the problem first began to attract attention, the estimated contribution of "human errors" was around 20%. In 1990 the consensus seems to be that the contribution is about 80% (cf. Table 1-1 and Figure 1-1). These numbers cover not only the actual operation of the system, i.e., what happens in the control room, but include design and maintenance as well. Even so, and even if these numbers are only estimates or educated guesses, the difference is astonishing and significant. This difference unavoidably leads to speculations on what the reasons may be; even a little thought will show that there may be several different reasons.

Firstly, the growing complexity of the systems in our technological environment (Perrow, 1984) may be creating more opportunities for malfunction. One pervasive phenomenon is **risk homeostasis** described below, i.e., that advances in technology lead to a reduction in perceived risk, hence to behaviour that is closer to the limits of acceptable performance - thereby effectively reducing the margin for safety. It is quite a paradox that the technological improvements do not seem to lead to a reduction in the overall number of malfunctions, but rather to an increase in their severity. One reason is that when automation fails the system may be closer to a state of error - or already in a state of error; this diminishes the opportunities for corrective action, hence effectively leads to a reduced margin for errors.

Table 1-1
Estimated contribution of "human errors" to system accidents

Source	Single estimate (%)	Double estimate Low (%)	Double estimate High (%)	Domain
Shapero et al. (1960)	-	20	50	Weapon systems
LeVan (1960)	-	23	45	Aerospace
Robinson (1970)	25	-	-	General
Rasmussen (1973)	-	20	30	General
Hagen (1976)	-	10	15	NPP Total failures
Christensen et al., (1981)	-	50	70	Electronic equipment (human initiated)
Christensen et al., (1981)	-	60	70	Aircraft maintenance (total failures)
Christensen et al., (1981)	-	20	53	Missile system maintenance (total failures)
INPO (1984)	44	-	-	Nuclear
INPO (1985)	52	-	-	Nuclear
Trager (1985)	60	-	-	Nuclear
Bellamy et al. (1988)	59	-	-	Process control
Guardian (1990)	75	-	-	Air transport
Swain (1990)	90	-	-	General

Secondly, the increased focus on erroneous actions and the improved methods for analysis may result in finding more cases. The very idea that human erroneous actions can play a role has brought them to the foreground and the methods of analysis have improved, enabling more extensive analyses and more detailed distinctions between possible causes. This result may partly be an artifact: erroneous actions have become more

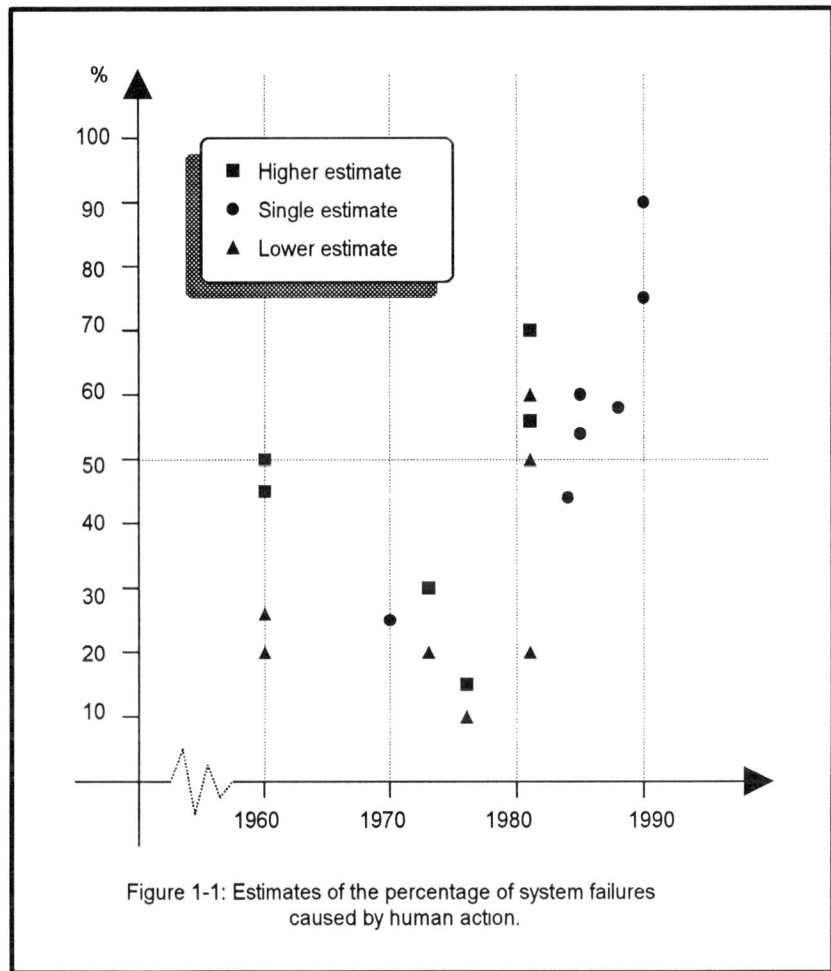

Figure 1-1: Estimates of the percentage of system failures caused by human action.

conspicuous, but since the base rate is unknown, it is impossible to decide whether they actually also have become more frequent.

Further reasons could be:

(1) People have become more prone to make errors or to be sloppy in their work. This could either be due to psychological factors such as lower work morale, complacency, lack of hope for the future, or due more to physical factors such as increased pollution, too much junk food, deteriorating physical condition, etc. Only the imagination sets a limit to what the plausible causes may be.

(2) The physical/mechanical safety mechanisms have actually improved, thereby reducing the category of causes where the technological system fails - at least in a simple way. This has made "human errors" more visible.

(3) Due to the complexity of the systems, and perhaps also due to the complexity of the organisations around them, it is often thought to be "cheaper" or more convenient to solve problems if they can be shown to have a human rather than a technological cause. This is particularly so in cases where the cause is found to be a complicated combination of conditions and events. Since such a combination is unlikely to repeat itself, it may not be worth the effort to do anything about it in terms of redesigning parts of the system. But if something has to be done, the human element is a permanent ingredient and a notably unreliable one, hence an obvious candidate for replacement.

(4) The demands for efficiency and performance have increased, thereby increasing the pressure on the person. In other words, the envelope of safe functioning has been reduced due to external demands, thereby increasing the number of faults that lead to noticeable unwanted consequences.

(5) The level of automation has increased. This leads to a change in the nature of work and in particular to a loss of skills, hence to reduced ability of the person to act appropriately when an unanticipated situation occurs.

There may be even more reasons, but the fact remains that the occurrence of unwanted consequences has not diminished and that the role of human cognition, hence also the question of the reliability of cognition, has become more prominent.

2. Coupling between Complexity and Unwanted Consequences

One intriguing explanation for the observed growth in the occurrence of unwanted consequences, hence also in the number of erroneous actions, is the possible coupling between erroneous actions and system complexity.

Coupling between Complexity and Unwanted Consequences

According to Perrow (1984, pp. 85-88), complex systems have the following characteristics:

(1) Parts or units that are not in a production sequence may have common mode connections or be physically close.
(2) Feedback loops may exist even where they were not intended.
(3) There may be unintended interactions between control parameters.
(4) Subsystems may be interconnected.

Because of all these, complex systems are hard to understand and it may be difficult even to find the right information. The basic scheme underlying the coupling between complexity and unwanted consequences is that growing system complexity leads to growing task complexity, that growing task complexity in turn leads to a growing opportunity for malfunctions, which again leads to a growing number of unwanted consequences; the growing number of unwanted consequences in turn lead to solutions which ultimately increase system complexity, thereby closing the loop. This is sometimes, jokingly, referred to as the "Law of Unintended Consequences": i.e., that the effort to fix things sometimes worsens the damage. More seriously, it corresponds to the notion of risk homeostasis, which means that the level of (perceived) risk remains constant despite technological improvements that could have lowered it. There are, of course, also other factors which may play a role, as illustrated in Figure 1-2.

The growing technological potential constitutes a driving force on its own; the demands for higher efficiency in work also seem to be rising, perhaps because the technology becomes more expensive; and finally, the changes in the methods of analysis may in themselves lead to a growth in the number of cases where human action is seen as the cause of unwanted consequences. All the possible influences shown here contribute, directly or indirectly, to the increase in system complexity. This view on unwanted consequences, hence also on system reliability, is valuable because it also suggests ways in which the situation can be improved. Since the coupling basically is an amplification of deviations (Maruyama, 1963), the only solution is to break the amplification at one or more steps, for instance by trying to reduce task complexity or by designing improved system functionality which does not necessarily lead to increased system complexity - as indicated in Figure 1-2. This, then, becomes a problem of

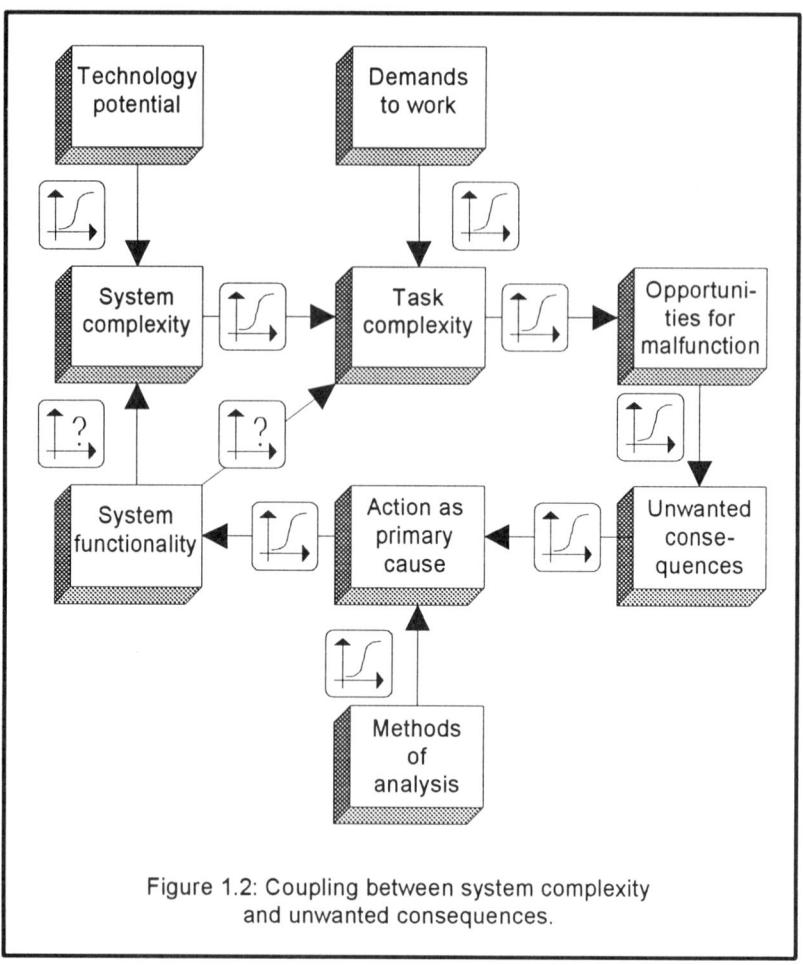

Figure 1.2: Coupling between system complexity and unwanted consequences.

appropriate design of the Man-Machine Interaction (MMI), which in turn requires an improved understanding of human reliability.

2.1 Risk Homeostasis

New technology is often introduced with the aim of improving working conditions and reducing risk. The general experience is, however, that the effect rather is to maintain the same level of risk and instead increase the cost-effectiveness of work (Wilde, 1982). A classical example is the introduction of the Davy lamp in coal mines, which in 1815 was invented to reduce the risk of explosion by isolating the flame from the combustible gases in mines. Mine owners, however, used the lamp as a means to tackle seams that previously had been too dangerous. As a result, the number of mine disasters rose steadily, reaching a peak between 1860 and 1890 (Hamer, 1990, p. 73).

One way of describing risk homeostasis is by considering perceived risk as the outcome of a balance between system dependability/reliability and system utilisation (cf. Figure 1-3). The notion is simply that an increase in the dependability of the system will lead to a decrease in the perceived risk (the balance shifts to the left); similarly, that an increase in the utilisation of the system will lead to an increase in the perceived risk (the balance shifts to the right). In both cases technological improvement is a common underlying factor. The assumption is that people, for one reason or another, want to maintain their perceived risk at a constant level. If, for instance, the dependability of the system increases then the person(s) will be prone to change their use of the system (go closer to the limits, be less concerned about how it is applied) until a balance is reached, so that perceived risk more or less remains the same. If, on the other hand, a higher degree of system utilisation is demanded, then the person(s) will try to improve system dependability until the previous level of perceived risk is regained. It is quite natural that people are unwilling to accept an increase in perceived risk; but the puzzle is that they seem reluctant to accept a decrease in perceived risk!

Consider, for instance, the following example. If a person buys a car with an Anti-lock Braking System (ABS), that person is prone to perceive the risk of not stopping, hence the risk of having an accident, in time as smaller than otherwise (particularly under unfavourable weather conditions, such as rain or snow). The acceptable risk remains the same, i.e., the person wants to avoid a collision. But since the perceived risk is smaller, it follows that the person can increase system utilisation (e.g. reduce the distance to the car in front or increase speed), thereby maintaining the same level of perceived risk. If, on the other hand, the person had maintained the same performance (the same distance), then the increased dependability of system performance would have resulted in a reduction in the actual risk, and thereby also the perceived risk. The technological improvement could have made the person realise that it was not necessary to take the same risk as before, hence the perceived risk would have been lowered. Few people, however, seem inclined to make this kind of adjustment.

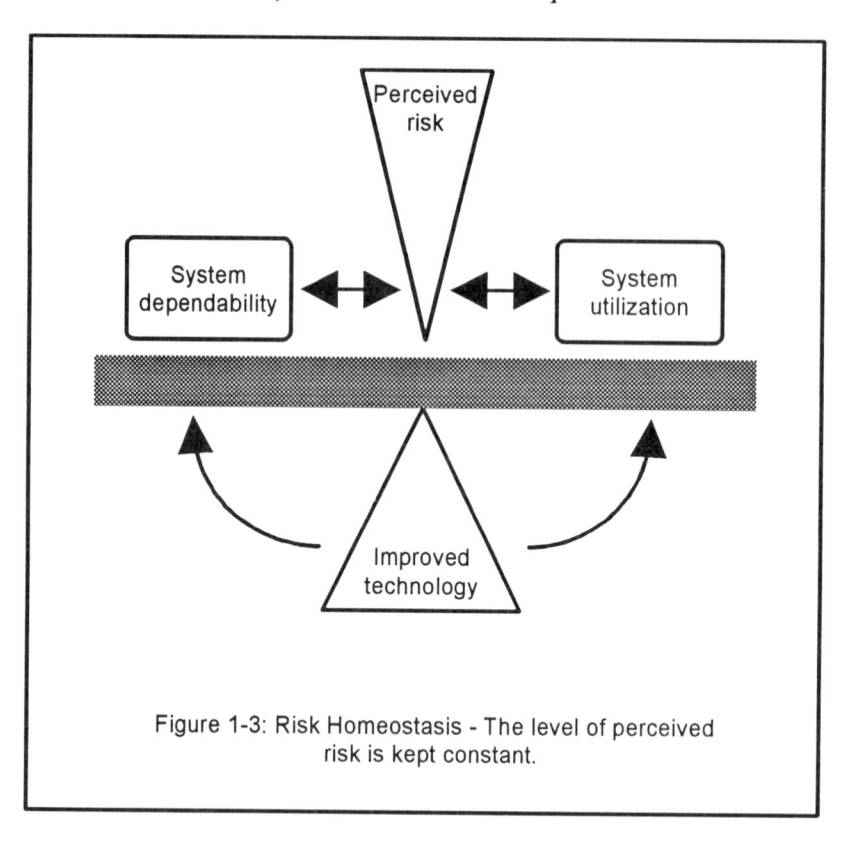

Figure 1-3: Risk Homeostasis - The level of perceived risk is kept constant.

The issue of risk homeostasis (e.g. Wilde, 1982) can be explained by considering two different consequences of improved technology. In the first case the technology is used to increase efficiency while maintaining the same level of risk - or at least that is the usual assumption. The argument is that if a person is safe driving with a distance of X meters to the car in front, then with improved technology the person will be equally safe driving X-dX meters away.

In the second case, the technological improvement is used to increase safety or reduce risk, while maintaining the same performance. The argument is that if a person is safe driving with a distance of X meters to the car in front, then with improved technology the person will be even safer maintaining a distance of X meters. Unfortunately, this line of reasoning is rarely used.

One potent illustration of risk homeostasis in practice can be found in the discussions about the future of Air Traffic Control (ATC) in Europe. Here, as in most other places, the demand for air traffic exceeds system capacity; this situation already exists, and it is expected to grow worse during the 1990s. The solution is, of course, to improve the capacity and this is achieved by such measures as reducing the separation of aircraft, by allowing a more frequent application of separations closer to their minimum values, and by increasing the number of aircraft monitored by a controller. In order not to increase the acceptable level of risk (the perceived risk), the efficiency and dependability of the system must be increased; this involves improved communication facilities, improved capabilities of aircraft to follow flight paths, improvement of the monitoring facilities, and - last but not least - improved support for the air traffic controller. (In addition, the tolerance for pilot errors will be reduced. A seriously proposed solution is simply to automate the cockpit completely, i.e., fly with either a single pilot or no pilot at all!)

In this case the goal is to maintain perceived risk at a constant level, and the means are improved technology and higher demands to the performance of the man-machine systems - on the ground as well as in the air. It does not take a lot of pessimism to fear that this solution is bound to fail in the long run. It also provides a very nice illustration of the coupling between external demands and task/system complexity depicted in Figure 1-2. In terms of risk homeostasis the interesting point is that all these enabling technologies could have been developed and used to reduce the level of risk while keeping system utilisation constant. In other words, although flights would not have become more frequent, it would have become safer to fly.

The importance of risk homeostasis for the reliability of cognition is that if performance takes place closer to the margins a greater demand is put on the reliability of the system - the technological parts as well as the person(s). This is true both in the case of a single person (driving his own car) and a complex organisation (air traffic). It matters little whether the demands are generated internally - the driver's decision to driver closer to the car ahead - or externally - the authorities' specification of reduced separation between planes; the system still has to perform reliably. The consequence is therefore also the same: an increased dependence on reliable performance including "doing" as well as "thinking", hence a need to understand better what human reliability is.

3. The Anatomy of an Accident

The study of human reliability is to a large extent motivated by the need to prevent unwanted consequences from erroneous actions - in other words, to prevent accidents from happening. A recent collection of papers on that can be found in Rasmussen *et al.* (1987). In order to prevent accidents from happening it is necessary to take a closer look at the accidents that have happened, to see if something can be learned from them. Despite the large variations in the types of accidents and their often unique conditions of occurrence, it is nevertheless possible to produce a generally applicable break-down, which can be called "The Anatomy of an Accident" (Green, 1988; also Rasmussen & Jensen, 1973).

According to this description, an accident begins when an unexpected event occurs while the system is working normally. Unless the unexpected event can be immediately neutralised, it will bring the system from a normal into an abnormal state. In the abnormal state attempts will be made, by people or by automatic systems, to control the failure. If this control fails, then the system will enter into a state of loss of control which means that there will be some unwanted occurrences. Usually, even this possibility has been anticipated and specific barriers or obstacles have been provided. It is only in the case where these barriers are missing or when they fail to work according to their purpose that the accident actually occurs. This is described in Figure 1-4.

3.1 Accident Signatures

The description of how an accident unfolds is very general, and there may therefore be differences for different types of systems. There will, in particular, be differences between accidents due to hardware failures (typical of technical systems), accidents due to software failures (typical of information systems), and accidents due to inappropriate human actions. For hardware components (motors, switches, pumps, pipes, etc.) it can be assumed that there will be a basically constant performance combined with a slow but continuous degradation over time - possibly ending with a steep degradation if the component fails. For software faults the main difference is the lack of slow degradation due to wear and tear; software simply does not degrade. Performance is therefore essentially constant but sudden changes in input conditions and/or internal system states may lead to an abrupt breakdown.

Figure 1-4: The anatomy of an accident (simple version).

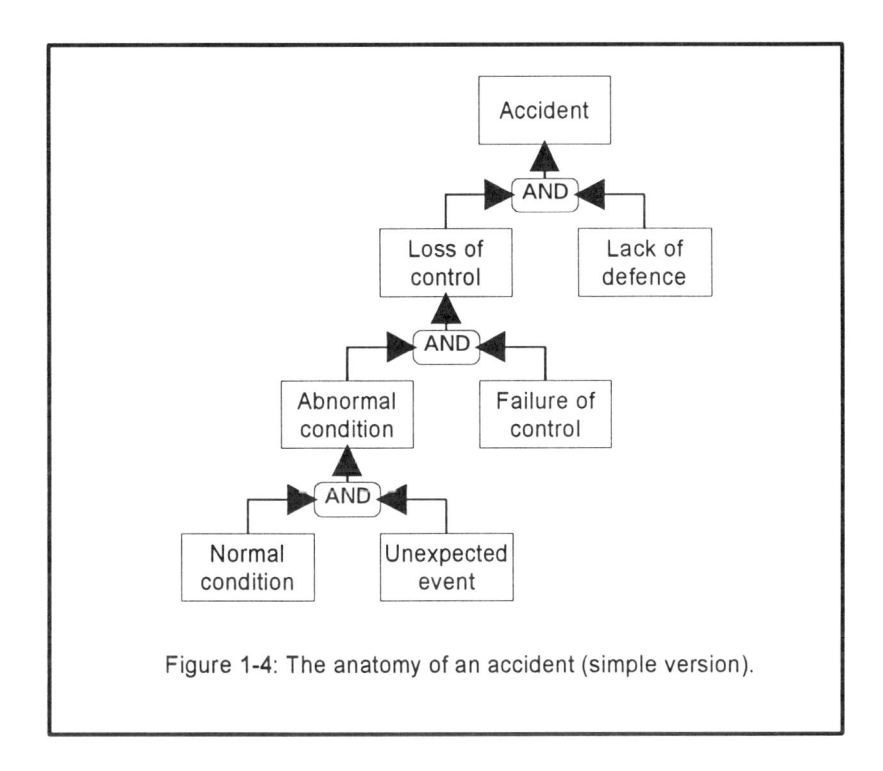

For humans the development is entirely different from either of the above. Performance is, on average, constant although with substantial variations over time. Degradation is not continuous but rather characterised by changes in plateaus - where performance sometimes gets better and sometimes gets worse. Changes between characteristic levels of performance, e.g. from skills, automated perceptual-motor actions, to reflection, use of knowledge and problem solving, are not regular or predictable events and do not appear to a be simple function of, e.g. task load, elapsed time, or any other single parameter. Hardware, software, and human systems each have a different "signature" which shows up in the way accidents develop; in addition, the "signature" of a mixed system may be different from the "signature" of a pure system. Each accident must therefore be treated differently with due regard to the characteristics of the systems involved.

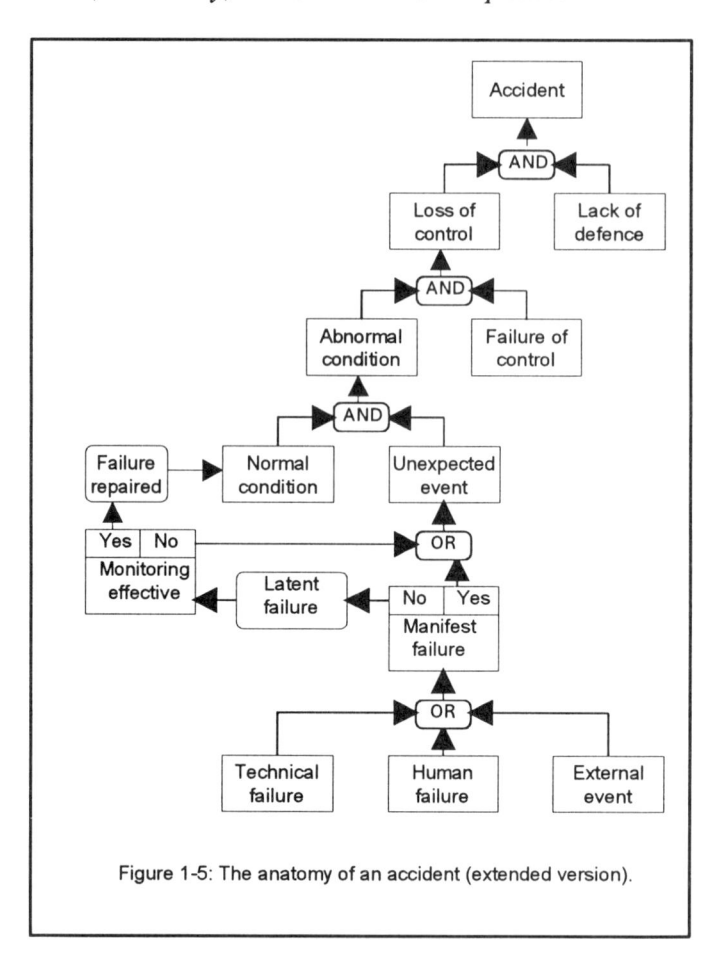

Performance, Reliability, and Unwanted Consequences 14

Figure 1-5: The anatomy of an accident (extended version).

The importance of MMI for how an event may occur or develop can be seen by considering an extended description of the anatomy of an accident (cf. Figure 1-5). One extension is the additional details of how an unexpected event can occur. The three "root" causes are a technical failure (internal to the system), a human failure, or an external event (e.g. a blizzard or a system external technical failure such as loss of communication). These may either give rise to a manifest or a latent failure, which in turn may cause the unexpected event. A second extension is to suggest how a human action can be a contributing cause in practically all of the other conditions that lead to an accident. The "anatomy of an accident" can clearly be extended almost *ad infinitum* - but the extensions will almost all refer to the reliability of human actions in one way or another. Although the focus traditionally has been on human reliability in operation, the contribution from maintenance is actually larger; to that can be added human reliability in design, construction, management, etc.

3.2 Software Reliability

The issue of software reliability, in a broad sense, is similar to the issue of human reliability in the sense that they both defy the decomposition principle. Present day systems can no longer be analysed purely in terms of hardware or physical characteristics, but must take into account the functions of software and humans as well. In both cases the immediate solution has been to carry over the methods and principles that have been applied with notable success in the engineering/physical domain. But in both cases this solution has turned out to be inadequate. Software reliability is in a somewhat better position than human reliability because theories of computation and formal methods of software specification constitute a possible basis for the reliability analysis; no such thing exist for human reliability. Still, software reliability is well suited to demonstrate the limitations of the decomposition principle to reliability analysis (e.g. Leveson, 1991).

Software reliability addresses the problem of quantification of error occurrences and probabilities. Here such methods as hazards and operability analysis, fault trees, etc. can be used as a means to identify the ways in which software failures can contribute to system failures. In distinction to that software safety tries to identify the causes for errors, i.e., the weaknesses in the software that may lead to system hazards and accidents. This is tantamount to looking for software bugs and removing them. But the practical complexity of software systems, measured, for example, in terms of lines of code or number of variables, is enormous and it is therefore often found that the errors are in the specification rather than in the implementation. Rigorous testing of the code is usually not enough and just providing a numerical indication of the reliability as a number or a range may actually be misleading. Software systems differ from hardware systems because they do not fail due to wear and tear, but only because of mistakes and errors in the design. The prevention of failures therefore amounts to an improved design. However, when failures occur they cannot be "repaired" by replacing a component or subsystem, as in the case of hardware failures; instead the repair must lead to a redesign of part of the system. This, however, means that the original

Performance, Reliability, and Unwanted Consequences 16

software reliability analysis has become partly obsolete. The decomposition assumes an unchanging relation between components; but if a component is changed rather than replaced, then the analysis must be done all over again. This particular trait of software means that software reliability (and software safety) is different from engineering/hardware reliability and safety.

The upshot of all this may be that a rethinking of probabilistic safety assessment itself is necessary. The analysis of the mechanical/physical components of the system can certainly continue to use established probabilistic risk analysis methods. But it may no longer be a straightforward matter to combine these with the results from software reliability analysis and human reliability analysis to provide a joint probabilistic safety assessment for the system as a whole. It is, however, difficult to say for certain what the implications of that may be.

3.3 Reactions to Failures and Accidents

Reactions to a failure seem to follow a regular pattern. The first step is to look for an acceptable cause. If no acceptable cause can be found within a reasonable time or with reasonable resource usage, the event is written off as an act of God or a freak event. Even if the cause can be found the occurrence may be classified as an unusual, infrequent event (though possibly a predictable one) and because it may be difficult to do something about this, the response is to do nothing. If that is a deliberate choice, then the event is in fact treated as a stochastic event with a spurious cause that must be tolerated, or as an act of God.

However, once a cause has been found the response may be chosen from several lines of action (cf. Figure 1-6):

(1) **Elimination of the cause by replacement** - either using an identical component or an improved component.

   This is the established solution for hardware and is efficient if the main cause is wear and tear. It is clearly not applicable for software since replacing an incorrectly functioning module with an identical one would be meaningless (cf. above). It may or may not be applicable for humans, depending on a further analysis of the fault; it should, however, not be used as a general response.

   If the function is central to the system, redundancy may be ensured by replacing a single component with a set of components

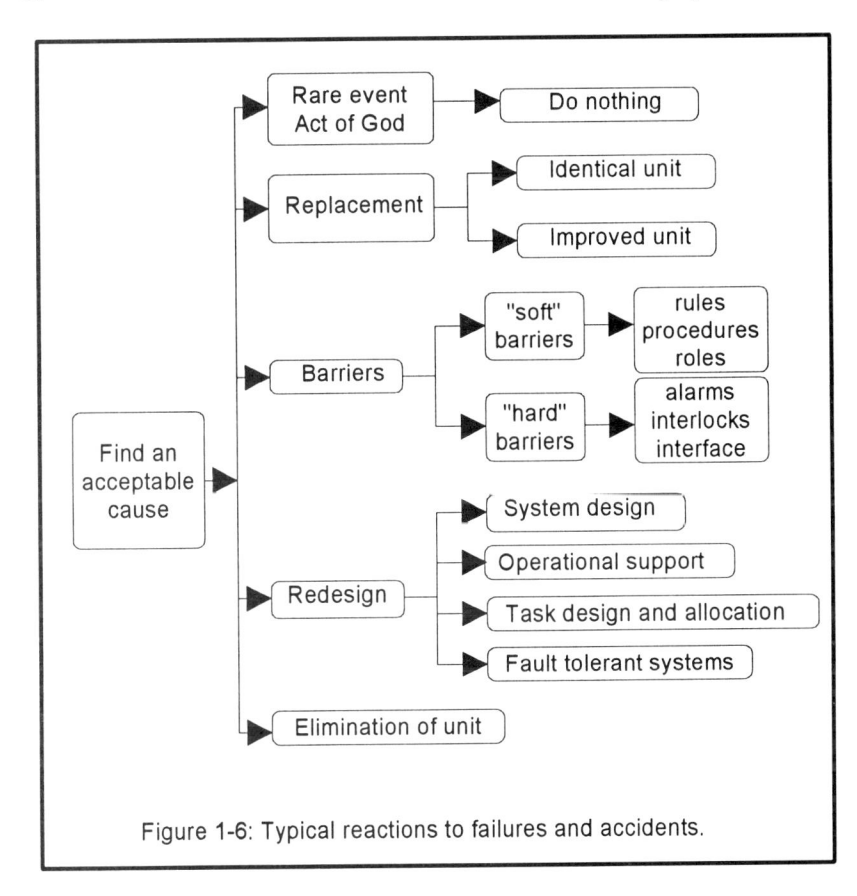

Figure 1-6: Typical reactions to failures and accidents.

where each provides the function in a different way (e.g. *n*-version software); for people that would mean replacing the single person with a crew.

(2) **Eliminate the consequences of the cause by introducing new barriers**. These barriers can be of two types:

(a) **"Hard"** barriers, such as new protection systems, new physical barriers, alarms and annunciators, redesign of the work place (new interlocks, keys, double checks), etc.

(b) **"Soft"** barriers, such as changes in work routine, changes in procedures, introduction of new procedures, changes in the organisation, etc.

The disadvantage of such reactions is that they generally lead to more complex systems, hence contribute to deviation amplification. Another drawback is that such solutions are aimed at avoiding the unwanted consequences if the accident should occur again; but since most accidents are the result of unique conditions this type of solution is not very efficient. Both kinds of solutions can be used for the technological parts of the system, for the software parts, and for human actions.

(As an aside, it is depressing to see that it is not only the engineering community which is partial to this solution. Much of the legislative work in the national parliaments also serves to amend or improve previous legislation, in most cases to compensate for loopholes and oversights.)

(3) **Eliminate the cause (and the consequences) by redesigning the system** in whole or in part. This can be achieved in several ways:

(a) **Improved system design.** This covers all aspects of the system's functioning and is exemplified by risk analyses, probabilistic safety assessment, human factors engineering, staff selection and training, specification of the organisational and communication infrastructure, replacing a module by an improved version, etc. It is often tied to the strong concept of the person as being a limited capacity information processing system. In this view erroneous actions are seen as the result of overloading the human information processor and every effort is therefore made to reduce the demands.

(b) **Improved operational support.** This concentrates on the system in use and is exemplified by automation in various forms, protection systems, "living" probabilistic safety assessment, computerised operator support, and various types of expert systems for "intelligent" information presentation, dialogue management, diagnosis, etc.

(c) **Fault tolerant systems.** One approach is to mitigate the consequences of erroneous actions once they have occurred. Since it is very difficult to bring about a significant reduction in the number of erroneous actions and outright impossible to eliminate them completely, the goal of designing fault tolerant systems becomes very attractive (Hollnagel, 1992d). In most cases a fault tolerant system is forgiving of erroneous actions because it can:

(i) produce automatic counteractions,

(ii) limit consequences through interlocks and automatic shut-down mechanisms,

(iii) provide early detection (improved feedback) and allowing prolonged recovery, and

(iv) provide more possibilities for corrective action.

(d) **Better task design and allocation.** Another perspective is that humans are often asked to perform tasks they are ill suited for rather than asked to do those that they are suited for. The use of improved task design and task allocation, which has a strong cognitive orientation, focuses on a redesign of the task and the provision of tools rather than prostheses for the person.

(4) **Eliminate the consequences by eliminating the system**. Although this is a radical solution it may sometimes be the only one that is available. This could be the closing of a production unit, abandoning a method or a product (e.g. CFCs), firing a group of people or a department, recalling a specific product from the market, etc. This is a brute force solution which is only effective if the function provided is one that can be dispensed with. Otherwise one should expect that the function will reappear in a different guise, which sooner or later may lead to another failure or accident.

This set of typical reactions can be considered in relation to the Reliability of Cognition. In the cases where human reliability is seen as the cause, elimination by replacement usually means increased automation. Improved training could also be used, although the long-term effects of training are uncertain because the impact of the working environment may easily cancel the effect of the training. Yet elimination by replacement is only a viable solution if it can be done in isolation, i.e., if the function to be replaced is isolated and has virtually no interaction with the rest of the system. Elimination through barriers is often used, particularly with the soft barriers (procedures, roles, etc.); the long term efficiency of that is, however, doubtful unless it is complemented by other changes as well.

Elimination by redesign again corresponds to increased automation, for instance through the introduction of expert systems and intelligent operator support. If the established cause is the person as such, rather than something which can be attributed to the interaction with the system, elimination by redesign is not possible since the design of the system - the human - is unknown. The final solution, eliminating the problem by eliminating the system, is clearly not very attractive; although it may solve some problems on the technical level it will easily create other problems, and possibly more difficult ones, on the organisational or managerial level. In any case a prerequisite for an efficient solution is a good knowledge of the problem and an adequate understanding of human reliability, in particular the reliability of cognition.

4. Human Reliability

The problem of human reliability can be posed as the problem of why a person may sometimes succeed and sometimes fail in trying to accomplish a goal, when there otherwise are no noticeable changes in the external conditions. I am here referring to a person carrying out a task or a specific work assignment, typically in the context of industrial process control. The notion of human reliability is, however, valid for all situations that involve or depend upon humans performing part of the system functions. If the conditions of work have changed (e.g. changes in temperature, noise, interfering communication, available time, malfunctioning of components or equipment, etc.), then it would be reasonable to expect performance to change as well. In the absence of that, i.e., in the cases where the task or the assignment as well as the conditions of performance are identical in the sense that they appear to be the same or are described in the same way, it would be reasonable to expect performance to stay the same. Yet human performance may still be different and a cause or explanation for that is naturally sought. This quest for knowledge can be seen as serving three purposes:

(1) To enable specific **system changes** to be made in response to specific unwanted occurrences, i.e., modifications or redesign after the fact (a **pragmatic** purpose);

(2) To be able to make better **predictions** of what will happen under given conditions, as an effort to improve the system design (an

engineering purpose, and also an extension of the first purpose); and

(3) To increase **knowledge** in general about man-machine systems, how they work, and to provide better theories (a **scientific** purpose, again extending the previous purposes). Examples are design guidelines, laws, etc.

All three purposes are to a certain extent intermingled, but all three of them cannot be achieved by the same means or at the same time. For the present discussion their relative importance is the same as the order in which they are stated - corresponding to the above definition that the purposes of studying the Reliability of Cognition and pursuing human reliability analysis as a whole are to reduce the number of unwanted consequences. To pursue any of the three purposes requires a solid conceptual foundation; even though the pragmatic purposes may be more prone to use empirical studies than theoretical analyses, empirical data are not in themselves sufficient. The advantage of the pragmatic purpose is that there are clear criteria which can be used to evaluate the outcome. This is also the case for the engineering purpose. In the case of the scientific purpose, however, the main criterion is the internal consistency of the resulting descriptions and the compliance with the accepted theoretical basis. It is important to realise this before a study is planned and executed, otherwise there is a real risk that the results will be unsatisfactory for any of the purposes.

Since it has been assumed that the observed performance variations are not due to noticeable changes in the environment, other causes or sources of the variation must be found. According to the current view of the human as an information processing system the two most obvious ones are **knowledge mismatch** and the **inaccurate execution of plans**.

4.1 The Cognitive Viewpoint

One view is that performance variations are due to subtle factors in the environment, which have to do with cognitive rather than physical aspects. This is based on the **cognitive viewpoint**, which - in a very simple rendering - says that a person's way of understanding and acting depends on how the context is perceived and interpreted, e.g.:

"...any processing of information, whether perceptual or symbolic, is mediated by a system of categories or concepts which, for the information processing device, are a model of his (its) world."
(De Mey, 1982, p. 84)

It is common to refer to the perceived context as the person's model of the world although the two terms not really are synonymous. It is a consequence of the cognitive viewpoint that the appropriateness of information processing, in particular the appropriateness of the results, depends on the appropriateness of the knowledge (the set of categories or concepts) which is used. If there is sufficient correspondence between the knowledge and the conditions of work, then this cannot reasonably be the source of the unwanted performance variations and the unwanted outcomes that may still occur. Conversely, if the correspondence is inadequate, unsatisfactory performance should be expected as a result. (Cognition is here considered as a whole rather than as just the process - as in the separation between knowledge and knowledge processing.)

4.1.1   Knowledge Mismatch

Planning is an essential part of how humans act (cf. the further discussion in Chapter 5). Planning can be defined as the systematic examination of possible (future) conditions of the environment and the specification of principles for choosing among them. The result of a plan is a scheme for what should be done until certain conditions are achieved (the extension of the plan). Even if the scheme is pre-defined, as in a set of procedures, thereby relieving the person of the actual planning, the need to follow the procedures does in itself require planning and deliberate action (unless one subscribes to the ideal of well-drilled operators mindlessly carrying out a procedure - in which case the operator is not really required). If, therefore, a mismatch exists, or appears to exist, between the conditions of work and the person's knowledge (model or representation) of them, then plans and decisions will be incorrect and may lead to unwanted consequences. Performance will therefore not be reliable or dependable.

4.1.2 *Inaccurate Execution of Plans*

Another view is that performance failures are due to inaccuracies in the carrying out of the plans, rather than in the plans themselves. In other words, it is assumed that the person's knowledge is adequate, but that the execution of the intended actions goes wrong in some way, because of random fluctuations in how the mind works, subtle (or even less subtle) influences from the environment, forgetting, loss of attention, associative jumps, etc. (In a more elaborate view this amounts to a kind of chaos theory for the human mind.) This view does not just postulate a separation between knowledge and knowledge processing, but rather points to an inherent characteristic of human cognition on all levels

4.2 System Induced and Residual Erroneous Actions

A different way of saying that is that erroneous actions can be of two main types: **system induced erroneous actions** which are due to features of the interaction between person and system, e.g. the interface characteristics, and **residual erroneous actions** which are due to the inherent variability of human cognition and performance. One may assume that system induced erroneous actions can be reduced to a given level, provided they are considered important and the consequences are costly enough, but that the residual erroneous actions cannot be significantly reduced. The relation between the two views is that a knowledge mismatch may be due to inadequate background knowledge and insufficient training as well as system or interface characteristics (i.e., it is system induced). Inaccurate processing and execution may, on the other hand, be attributed to both system induced and residual causes (cf. Figure 1-7).

System induced and residual causes will naturally blend in the daily use of the system and may not be easy to separate empirically. It may not even be easy to see whether the performance variation is due to knowledge mismatch or to inaccurate execution, since neither category is operationally pure (cf. the discussion in Chapter 2). The distinction is nevertheless a useful one for analysis as well as for remediation. System induced causes are important **indicators** of weaknesses and inadequacies in the system design which, in principle, can be eliminated. (In practice the elimination will be a question of cost *versus* benefit. There is always a certain number of problems or malfunctions that are accepted, either because they occur very infrequently or because the consequences are too

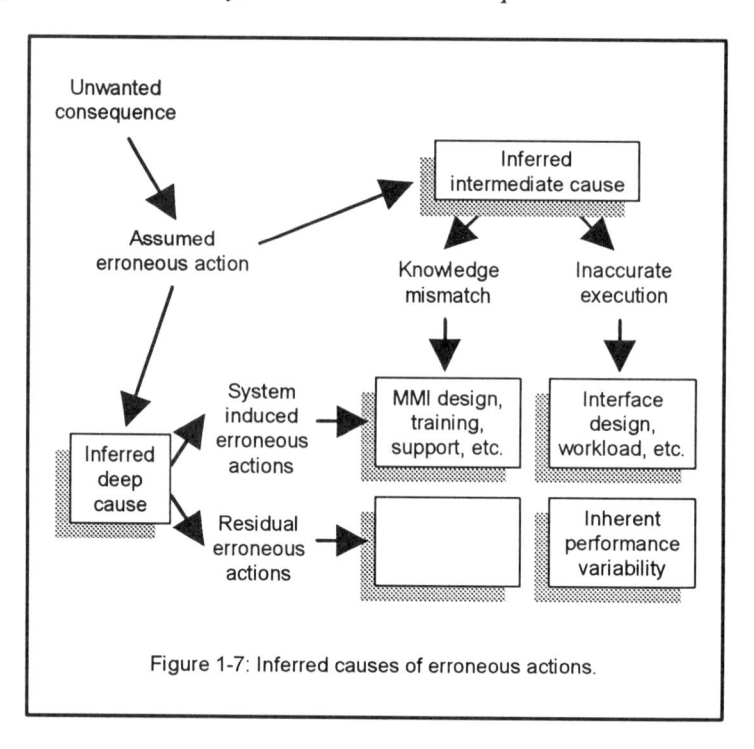

Figure 1-7: Inferred causes of erroneous actions.

small to be considered a real nuisance. The malfunctions may not always be clearly identified, but we nevertheless know - from bitter experience - that they are there.) Residual causes cannot be eliminated but their effects can in many cases be contained by a proper system design. A proper set of concepts for reliability of MMI, human reliability analysis, and the Reliability of Cognition must be able to account for both system induced and residual erroneous actions, as well as for the more detailed explanations that may arise from further scrutiny. Chapters 4 and 5 of this book will present a candidate set of concepts.

4.3 Qualitative and Quantitative Analyses

One issue which can always get people into an argument is the juxtaposition of qualitative and quantitative methods. It seems to be an unquestioned assumption in the current methods for probabilistic safety assessment and human reliability analysis that quantification is the name of the game (cf. Chapter 3). But this is only so on the surface. After working with quantitative techniques for some time it becomes clear that the basis for quantification must be provided by qualitative analyses, and furthermore that the juxtaposition between the two is an artificial problem. Alan Swain, for instance, has clearly acknowledged that "the qualitative aspects of human reliability analysis are at least as important as the quantitative aspects" (Swain, 1990, p. 310). This is another way of saying that even if the quantitative results that are produced by a human reliability analysis are not always very precise or reliable, the (qualitative) analysis that lies behind will probably be of value. The purpose of the qualitative analysis is to identify the potential for human erroneous actions, in particular situations where such actions are very likely. This qualitative analysis is usually some sort of task analysis and/or an ergonomic (human factors) assessment of the system in question. A detailed description of a specific method, called the Dependent Differentiation Method, is provided in Chapter 6.

When numbers (quantitative expressions) are used in well-known contexts it is possible to interpret them without any effort and in a way which makes sense. The ease with which this can be accomplished tends to obscure what actually goes on, hence to hide the underlying process of interpretation. Numbers are, of course, only useful if they are meaningful. A single number without a context, such as "76" is of little use; it could literally mean anything. If the number is used in a context, such as "the weight of the patient was 76" it gains some meaning, but it is still open to interpretation. For instance, what are the units? What are the other characteristics of the patient (age, sex, height, etc.)? It is only when this information is provided that it is possible to begin to use the number "76" in a meaningful way, i.e., to make reasonable guesses about the unspoken conditions, hence understand it.

One way in which to establish a meaningful number is to relate it to a clearly identified structure or a model instantiation. This, in turn, requires that an underlying theory or model exists, and that again means that there must be a prior qualitative description or a conceptual basis. A second way in which a meaningful number can be established is when it expresses a clearly established empirical correlation. Thus, the weight 76 kilos becomes meaningful when it is set in relation to the height of the person, for instance suggesting whether the person is underweight or overweight. In this case the empirical correlation is based mainly on experience and norms. In other cases the correlation is based on a set of operational concepts for the phenomenon in question. For instance, relations between pressure and temperature in a nuclear reactor, or between stall speed and gliding angle in a plane, are meaningful only because they can be related to some operational concepts - which again must be based on a model, hence a qualitative description or conceptual basis.

There is thus an unavoidable link or dependency between qualitative and quantitative descriptions. Properly viewed, they point to two different perspectives of the same phenomenon, rather than to mutually incompatible dimensions. The weakness of human reliability analysis is that it often neglects the importance of the qualitative aspects. The practical goal is to produce well-defined methods which can be used in an almost mechanical way. This may be very effective in the hands of the expert, but may be somewhat misleading if used without the necessary background. Unless sound principles for analysis and description are used there is therefore little to be gained from quantifying the results. Quantification can always be achieved, as psycho-physics has amply shown. But quantification requires an interpretation of the numbers, and that interpretation must in itself be based on a meaningful description of the phenomenon or system (cf. Figure 1-8). The objective should therefore not be to provide numbers *per se*, but rather to provide results in a form which can be treated rigorously according to agreed upon principles. In other words, numbers are only really useful if they have a meaning that can be unambiguously stated.

5. The President's Heart Attack

One day in 1991 (Saturday, May 4th) President Bush of the USA was suddenly taken ill by fatigue and shortness of breath during his routine exercise (jogging). He was immediately flown to a hospital for examination, and it was declared that he had suffered from an attack of atrial fibrillation.

The interesting aspect of this is not the attack itself, nor that it happened to the President. (But the example is useful because it happened to the President, since that caused a lot to be written about it.) The interesting thing is the way in which everybody tried to see it as a phenomenon with a definite cause. Atrial fibrillation can be caused by a general condition of stress, and at that particular time there were a number of political issues which could have contributed to that (for instance, the situation in Northern Iraq, the peace initiative in the Middle East, concerns about the Chief of Staff, and the economy). Despite that

Figure 1-8: The paradox of quantification.

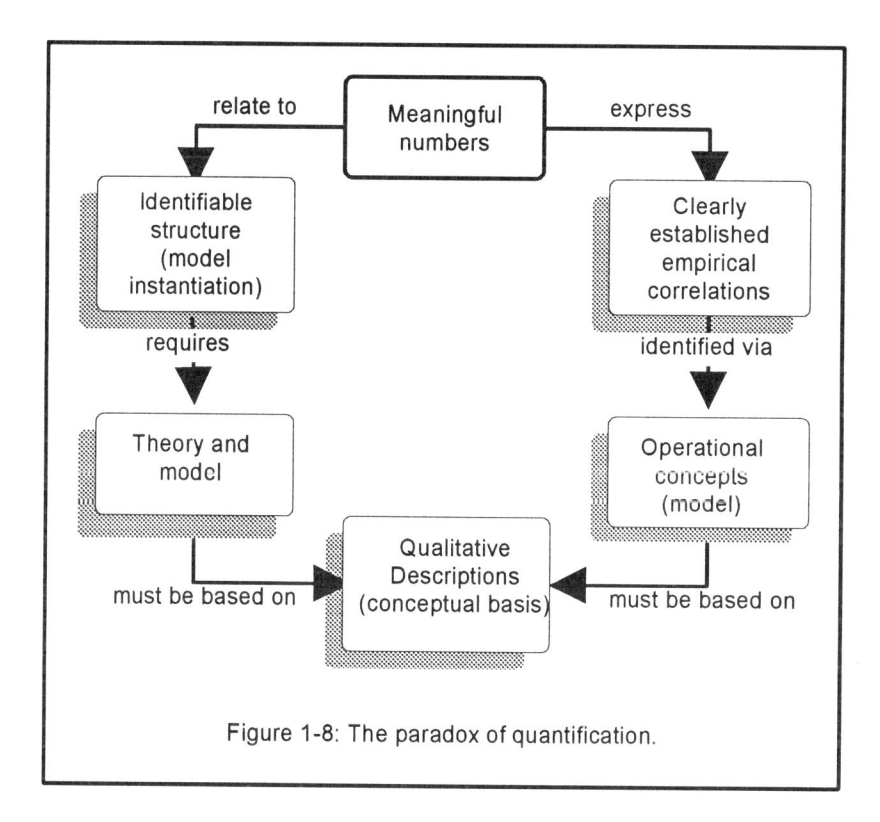

the doctors searched for a specific cause and finally announced that a hyperactive thyroid had produced an excess of hormones which had caused the irregular electrical impulses to the heart, etc.

In other words, the human body was seen as a physical/mechanical system of a deterministic nature; the causal chains were explored and the root cause determined (in this case the goal was, obviously, to find a root cause which could be treated adequately). The reason for mentioning this example here is that it demonstrates, in a very clear way, the predominance of the deterministic mechanical model that lies underneath much of our thinking. It is therefore no surprise to find that this model also lies underneath our conceptualisation of human actions and human reliability. In other words, events which are clearly of a psychological nature are seen as having the same causal structure as non-psychological events. Or, put differently, the psychological mechanism is assumed to exist and be similar to the other mechanisms (physiological, etc.) that are found in humans, and also similar to the many mechanisms found in technical systems. One of the intentions of this book is to show that this is not necessarily so, and that maintaining this assumption leads one to develop methods and theories which are inappropriate or even directly misleading.

To return to the President's case, if it had been realised (as an assumption) that the common performance conditions of being 67, having a high workload and even having specifically stressful problems at the time could have been the "real" cause of the attack, then the remedy would have been an attempt to change those conditions rather than mainly to treat the thyroid - which was an innocent victim of all this. In fact, the problem is **why** the thyroid started to malfunction, not **that** it malfunctioned. But since the thyroid could be treated, the analysis stopped there.

The story, however, continues. The problem of the overactive thyroid (a.k.a. Grave's disease) was suggested to be caused by an agent in the environment, e.g. either iodine or lithium in the water supply (due to old plumbing) or a bacterium called *Yersinia Enterocolitica*. It was noted that the First Lady also had suffered an attack of Grave's disease, and furthermore that the President's dog (Millie) had suffered from lupus (both being autoimmune disorders). This provides a good illustration of how the search for a cause is conducted. The immediate cause (the overactive thyroid) was amenable to treatment. But after a while there appeared a need to extend the search; it was noted that the occurrence of Grave's disease in two cases combined with the case of lupus in the dog constituted a rare event (although it occurred within a span of 16 months); how rare was not really certain - some doctors (!) "... estimated one in several million as being the chance of that happening" (*International Herald Tribune*, May 29th, 1991).[1]

The reasoning seems to be as follows: the overactive thyroid must have an underlying cause (this in itself seems reasonable enough). There had been an apparently rare co-occurrence of similar events (although over a long time), hence there must be a common cause (this conclusion does not really seem reasonable). The common cause would then be some environmental agent, either a fault in the plumbing or a microorganism. This would conclude the search and identify the root cause of the event. It would furthermore be a root cause that could be treated. Notice that in this development any consideration of alternative hypothesis had been

---

[1] One in a million, or so, seems in everyday language to express that something is very rare.

completely abandoned (at least as far as one could read in the press). The notion of a definite physical and treatable cause directed the search, and - "surprise surprise" - the culprit was found. Such outcomes vindicate the original assumption and demonstrate the pervasiveness of the deterministic and mechanistic way of thinking.

6. "Human Error" and Erroneous Actions

Actions carried out by a human can, logically, fail to achieve their goal in two different ways. The actions can go as planned, but the plan can be inadequate. Or the plan can be satisfactory, but the performance can still be deficient. It is the latter situation which is of interest here. Such incorrectly performed actions are often referred to as human errors, although this term is misleading in its connotations. It would be more correct to say that the actions are incorrect *vis-a-vis* the current goal because they produce unwanted effects and thereby make the goal more difficult or even impossible to achieve.

"Human error" has become the established term to describe situations or events where undesirable consequences occur and where the cause can be attributed, in part or in whole, to some aspect of human action. The main problem with "human error" is that the term may denote both a **cause of an event** and a special **class of actions**. The first meaning is found when the term is used as the explanation or cause of accidents. The second meaning is found when the term is used to refer to the internal psychological or cognitive mechanisms of the mind, which are assumed to explain the action.

Both uses of the term are important, but it is even more important that they are not confused. In order to avoid the problems in using a single term to denote two different things, I propose the term **erroneous action** as characterising a certain type of action without implying anything about the cause (cf. Hollnagel, 1991a). An erroneous action can be defined as an action which fails to produce the expected result and/or which produces an unwanted consequence. The expectation is usually that of the person carrying out the action, but can also be the expectations of an external person or an organisation. The cause of an erroneous action can logically lie with either the person, the equipment (the system), and/or the conditions when the action was carried out. Erroneous actions can occur on all system levels, from design to maintenance and from operation to management.

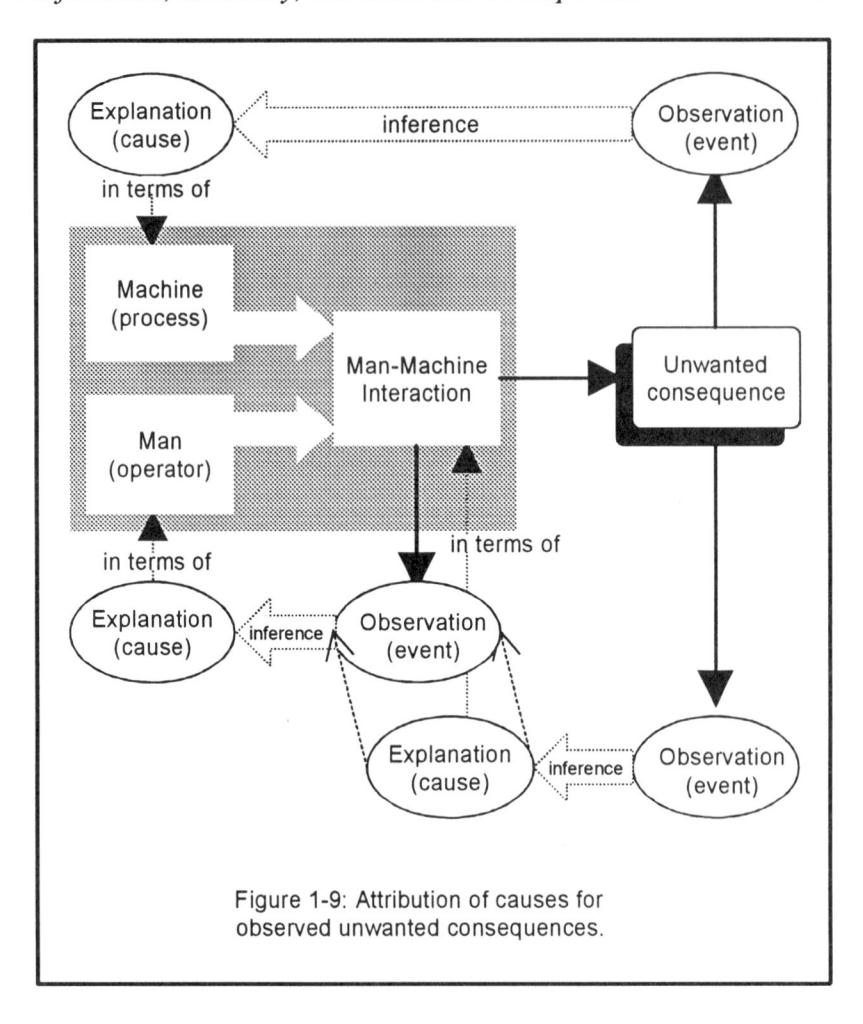

Figure 1-9: Attribution of causes for observed unwanted consequences.

In the context of man-machine systems, the locus of the erroneous action is usually the interface or the man-machine interaction. In other words, this is where the erroneous action occurs. In practice, however, erroneous actions are not usually observed when they occur but are only detected when they have consequences. The observation thus usually refers to some part of system performance, rather than to the erroneous action as such. From the observation of the unwanted consequences, inferences about the cause are made, leading either to the machine or technological system (i.e., the non-human parts) or to the people in the system (cf. Figure 1-9). The locus thereby serves as a starting point for identifying the attributed cause, or even the "root" cause. But if an acceptable cause is found in the first step, the analysis rarely continues beyond that.

6.1 Human Erroneous Actions *versus* Unwanted Consequences

The reliability of Man-Machine Interaction is an important issue because the dependence on MMI is constantly increasing. As described earlier in this chapter, the notion of human erroneous actions gained acceptance in the early 1960s as a probable cause for incidents and accidents. The focal role of individual human performance, hence also of erroneous action, continued to grow until the mid-1980s, reaching almost Gargantuan proportions. Then, gradually, the role of the work environment and the organisation began to be recognised. Simultaneously, it became clear that erroneous actions of system operation often were less important than erroneous actions of design and maintenance. At the present stage of development the focus remains on human performance, but now seen as an outcome of many factors, including training, support, management, etc.

The goals of human reliability research must be seen in relation to this changed perspective. In particular, the reliability of MMI must not focus narrowly on the interaction that takes place in the control room, but must try to maintain a broader view. Swain (1990) has correctly pointed out that the basic problem is to reduce the occurrence of unwanted consequences. Since such occurrences frequently are attributed to human erroneous actions, the problem has often simply been seen as one of minimising erroneous actions. This has created an intense interest in theories and explanations of erroneous actions, usually by extending the practices of engineering analyses to the field of human action. Yet unwanted consequences can occur for a number of reasons, of which erroneous actions - as a directly attributable cause - is only one. Experience has also shown that an erroneous action frequently occurs as a result of other conditions, or to put it differently, that the locus of erroneous actions need not be in the control room or the direct operation; it may equally well be in the design, the managerial structure, the maintenance, etc., as shown in Figure 1-10.

By considering unwanted consequences rather than erroneous actions as the target for the study of reliability of MMI, one avoids the focus on a single possible cause out of many. It also means that there will be more ways in which unwanted consequences can be reduced or

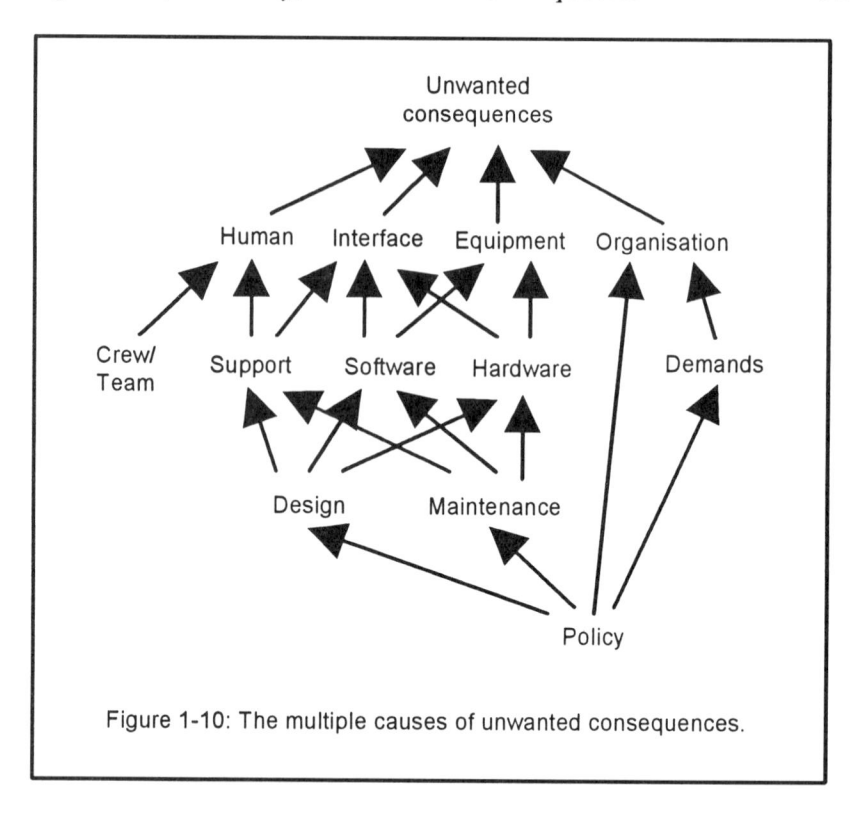

Performance, Reliability, and Unwanted Consequences 32

Figure 1-10: The multiple causes of unwanted consequences.

avoided. Should it, for instance, prove exceedingly difficult to reduce the occurrence of erroneous actions - as it actually is - then the unwanted consequences could be diminished by devising clever automation systems, providing error tolerant interaction systems, changing roles of the persons, etc. When the reliance upon a single all-embracing theory or model is forsaken, the options for intervention increase; and that can only be an advantage. This view has been expressed, in a slightly different way, by Rasmussen, who concluded an overview of human reliability analysis by saying:

> "It also seems to be important to realise that the scientific basis for human reliability considerations will not be the study of human error as a separate topic, but the study of normal human behaviour in real work situations and the mechanisms involved in adaptation and learning. The findings may very well lead to design of more reliable systems, without

improving the basis of quantitative prediction of reliability in the higher-level mental tasks required in new systems."
(Rasmussen, 1985, p. 1194)

(Note, however, that the emphasis (implicitly) remains on developing models of higher-level mental tasks such as the mechanisms involved in adaptation and learning, although the optimism has been reduced.)

It is in accordance with this view that reliability of cognition in this book will be considered not just as the reliability of human performance but as an issue where the MMI constitutes a **context** rather than a focus. The reliability of human cognition plays a pivotal role, but it must not be treated in isolation.

7. Complexity and Cognition

The precondition for purposeful behaviour is that the person knows what he is trying to achieve and in each particular case is able to assess how much remains before the goal has been reached. Furthermore, that this observed discrepancy can be used to determine which direction further actions should take.[2] It follows that if the person has a complete overview of the situation, i.e., if it is possible to see all the way from the present state to the goal, then the person will be able to plan (and execute) the steps needed to get there. If this is the case, and if furthermore the plan can be carried out as specified, then reliability will obviously be very high or even maximal.

7.1 The Ant Analogy

This situation has been captured in the so-called ant analogy introduced by Simon (1969). If a person observes an ant moving around on, say, a beach and tries to make a sketch of its path it will come out as a sequence of irregular, angular segments. Although there is an overall sense of direction, the movements of the ant seen on a smaller scale appear to be random. Simon makes the point that the apparent complexity of the ant's

---

[2] The widespread use of the geographical metaphor comes from viewing problem solving as taking place in a problem space. This carries over to notions such as the distance from the solution, the direction of the strategy, etc. The use of these terms do not imply an underlying metric, but is purely metaphoric.

behaviour over time for the most part is a reflection of the complexity of the environment. This conclusion is carried over to human beings, and formulated thus:

> "A man, viewed as a behaving system, is quite simple. The apparent complexity of his behavior over time is largely a reflection of the complexity of the environment in which he finds himself."
> (Simon, 1969, p. 25)

The reason for mentioning this analogy is quite simply that it lies underneath the whole notion that the human is an information processing system - and furthermore a rather simple one (according to Simon's view). The notion of the human as an information processing system (IPS) is important because it is at the core of the decomposition principle. If a human is an IPS, then it makes sense to analyse and describe the human using information processing (i.e., basically mechanistic) terms. Furthermore, it makes sense to consider the reliability of the human, and in particular the reliability of cognition, in a way similar to how the reliability of an IPS is viewed. According to the strong IPS school of thought the reliability of cognition is not even really an issue, because cognition is very simple: the variability is due to the complexity of the environment rather than the complexity of cognition.

(On the other hand, if cognition in the sense of human information processing really is so simple, then it is somewhat puzzling that Artificial Intelligence is so complex. The glaring inadequacy of computational psychology is, perhaps, an indication that something is not quite right in this view of the world.)

7.2 The Origin of Complex Performance

The main thrust of Simon's argument is that seemingly complex behaviour is determined by the complexity of the environment rather than the inherent complexity of the organism or system itself. Other things being equal, complex behaviour will be less reliable than simple performance because there are more things that can go wrong (more branch points in the event tree, so to speak). According to this view the lack of reliability is therefore mainly due to the complexity of the environment, rather than to inherent factors in the human. If that is the case, then any desired level of performance reliability can be achieved by reducing the complexity of the environment. Things are, however, not always that easy, so perhaps there is something wrong with the argument.

One thing that can have an influence on the reliability of cognition is if the person only has an incomplete view or understanding of the context. In this case it must be expected that reliability will be reduced and that performance will deteriorate. If it is impossible to grasp the whole situation, then it is necessary to use short-ranging tactics rather than long-ranging strategies, and the performance will therefore be less certain (Schützenberger, 1954). In relation to the ant analogy, we (as human observers) have the overview (God's eye, so to speak) while the ant does not. We can therefore see the ant's behaviour as either complex or random. It is complex if we ascribe full intentionality to the ant, and accredit it with cognitive states and cognitive capacities - be they ever so primitive - of planning, having mental maps (as e.g. Tolman's cognitive maps), etc. It is random - but rational in its own sense - if we do not ascribe any cognitive faculties to the ant but realise (as we assume) that the ant responds only to local cues while trying to maintain an overall direction.

It is instructive to consider the situation from the ant's point of view, so to speak.[3] Let us make the bold (and anthropomorphic) assumption that the ant sees the world as a series of (discrete and) partly overlapping views (mini worlds), which are considered one by one as they come into view - literally speaking. In other words, the ant's behaviour is based on what is perceptually present with little or no memory to sustain previous views. As long as there is a certain level of momentum or (conceptual) overlap between neighbouring views, the ant will be able to maintain a steady course (provided, of course, the goal can be related to the situation or even be seen in the situation; in the case of the ant the overall guiding principle is the polarisation of light combined with a tendency to move in a specific angle relative to the polarisation). But if this overlap disappears, or if it is no longer possible to see the goal in relation to the current situation, performance necessarily will break down. (An added factor is the ability to remember a number of steps back and establish the relationship conceptually rather than perceptually.)

---

4   We can, of course, not know what it is like to be an ant since we only know what it is like to be ourselves (cf. Nagel, 1974).

7.3 The Complexity of Human Performance

In the case of a person, the same argument can be used. As long as it is possible to maintain the direction to the goal, so to speak, then performance will be appropriate (except for the inherent performance variation). When the overlap between the mini worlds disappears, then performance will suffer. A continuity in the environment will probably always ensure an overlap between states, hence a conservation of cognitive momentum. Situations may nevertheless occur where the overlap is gone and where there is no relation between neighbouring situations. In that case the person clearly becomes lost and is unable to sustain reliable performance. This can also be seen as a case of a missing match between the world and the model/expectations of the world. It is exactly what happens in, for instance, contingency situations, where a relation between the current situation and the desired goal cannot be established - in a sense, the person cannot see or imagine a path from the present state to the goal. In those situations human behaviour may become as random as that of the ant, but it is random in a different way. Seemingly random behaviour, such as trial and error, precisely serves to find some key point or some link which makes it possible to regain the overview, hence enables purposeful planning (and predictions) to take place. Trial and error are not stochastic but are usually guided by some, possibly only vaguely formulated, high level principles or assumptions. If they fail, the outcome is called an error, but usually the person learns something from that. Control, and therefore also reliability, is in this way slowly regained.

Performance is therefore random only to the extent that one cannot describe the rationale for it or see the pattern in it. In the case of the ant, its behaviour or movements are not random at all; they only appear to be so in relation to a person's notion of how he would go about things. (In a discussion of behaviour in a stochastic environment Schützenberger (1954) pointed out that the optimal strategy would be the simple tactic of attempting to do one's best on a purely local basis. In that sense the ant is actually behaving in a rational way.) If you replace the ant with a hiker in an unknown mountain terrain, and without an adequate map, then the hiker will show the same kind of random movement as the ant does - if seen with God's eye. The point is that the person on the ground does not see the same as does the person high above; any comparison between their points of view is therefore misleading.

Although the ant analogy is seductively simple and intuitively seems to be right, it is actually deceptive because it does not openly reveal the assumptions behind performance and, more importantly, how performance is interpreted. If that is done, all the analogy does is support the argument that random behaviour, which can be seen as akin to low reliability, can be due either to lack of situational overlap (inadequate knowledge) or inherent performance variation. **Man, viewed as a behaving system, is not simple**; on the contrary, the complexity and variety of the environment requires equally complex cognition in order to be able to cope. This is consistent with the notion of requisite variety, as expressed in the cybernetic Law of Requisite Variety (Ashby, 1956). The basic idea is that in order to control a system, the controller needs at least as much variety as the system has. Since we assume that it is human cognition which enable us to cope with the complex environment, it is necessary that cognition is at least as complex as the environment. The reliability of cognition is therefore not a simple matter, but requires an adequate theoretical basis.

8. The Artifacts of Decomposition

The basic principle of analyses of the Reliability of Cognition and the reliability of MMI, as it is currently practiced, is that of decomposition. This follows from the very notion of the human as a passively responding machine (the Stimulus-Organism-Response metaphor) exemplified, for example, by supervisory control models. The reliability of a human is accordingly assessed in essentially the same way as for a piece of equipment (this issue is discussed in greater detail in Chapter 3). The decomposition principle involves the following steps:

(1) first, the person's tasks are identified by means of an event or fault tree modelling of a given (accident) scenario;

(2) next, the tasks are broken down into task elements; in addition, possible Performance Shaping Factors (PSF) are identified;

(3) finally, error probabilities are assigned to each task element, the influence of performance shaping factors is defined and quantified, and the outcome is possibly aggregated over the task.

The main criticisms against the decomposition principle are that it does not fully acknowledge the role of cognitive functions, e.g. reasoning, association, interpretation, and memory, that govern human behaviour. The human is essentially treated as a black box. Human erroneous actions can, of course, only be properly understood if they can be related to a model of cognition as well as to the overall conditions under which the task is carried out. The decomposition principle furthermore produces two artifacts: the concept of the **individual action** and the concept of the **Performance Shaping Factor**. This, therefore, speaks in favour of developing a more sophisticated description of the human, including an approach for modelling of cognition.

The counter-argument is that decisions regarding the safety of major industrial facilities cannot wait until satisfactory operator descriptions become available. There is a practical need for data, which must be fulfilled one way or the other. While the decomposition principle may be theoretically unsatisfactory, it does at least offer a solution which can be used in selected applications with sufficient precautions. Although this counter-argument must be acknowledged as expressing a real need, it does not constitute a license to use whatever method is available.

8.1 The Individual Action

The decomposition principle is focused on single, clearly separable actions which are seen as unique causes - and at times even as root causes. The concern therefore becomes one of determining when and under what circumstances the action can go wrong. The notion of the individual action is borne out by the human reliability analysis event tree, as introduced by THERP (Swain & Guttmann, 1983). Clearly, under the assumption that this representation is correct (and useful) the individual action, as a token for any action at a node of the tree, becomes important. Every effort is therefore made to find out as much as possible about this action, e.g. by estimating the probabilities for when it may go wrong. Yet one might go one step back and question the underlying assumption: that the function of the system can be produced or deduced from an aggregation of the descriptions of the functions of the parts. This assumption requires:

(1) that a description can be given of the function of each element (or that a description can be given of each element itself, e.g. task or task step), and

(2) that the function/characterisation of each element in isolation is not significantly different from the function of each element in the larger context.

In other words, it requires that the effect of the context is quantitative (arithmetic) rather than qualitative.

This assumption may be valid for purely mechanical/technical systems, at least as an idealisation - but except for common mode failures (cf. below). The function of a technical component should obviously not be different *in vitro* and *in vivo*, since that would invalidate most attempts of designing, building and testing components and systems. Neither should the probability of a malfunction be different, since that would invalidate most analytical methods.

The assumption, however, cannot be maintained for human action nor always, for that matter, for complex technological or software systems. Consequently, it does not hold either for systems, such as man-machine systems, where human action plays an essential part. The description of the overall system cannot be produced from a description of the functions of the parts, not even if a description of the interaction is included, because that assumes decomposability - for instance of knowledge and knowledge processing. The whole or the context provides the main influence and the main conditions. The characteristics must therefore be provided on the level of the whole, i.e., as an integrated (but not aggregate) description rather than as a detailed and decomposed one.

8.2 The Performance Shaping Factor

Performance shaping factors have been invented as a useful ingredient of human reliability analyses. Performance shaping factors serve two main purposes.

The first is to compensate for the lack of appropriate empirical data. This problem is addressed again in Chapter 3. The sum and substance of it is that empirical data which describe the probabilities of human actions are derived from a few selected fields of application - mainly the military and nuclear power plants. These data are usually referred to as basic human error probabilities (HEP). It is assumed that the basic HEPs can be used for other fields than the ones where they were collected, as well as for a wider range of situations and tasks. In order to accomplish this a set of performance shaping factors have been defined which serve as correction coefficients for basic HEPs when they are adapted from a nominal to an actual situation. The performance shaping factors thus represent the **specific context** or task while the HEPs represent the general or **universal characteristics** of human performance. This solution is clearly expedient for the need of providing quantitative input to the reliability analyses, but the theoretical foundation is a little sketchy.

The second purpose is to compensate for the lack of context in a decomposition based analysis. Any analysis that concentrates on the individual actions by themselves cannot help losing the context - or must at least temporarily ignore it. To compensate for that the notion of a Performance Shaping Factor (PSF) comes in handy. It is a concentrated representation of the influence of the context, for instance different levels of work load, various degrees of interface appropriateness, etc. A performance shaping factor has been defined as:

> "any factor that influences human behavior. Performance shaping factors may be external to humans or may be part of their internal characteristics."
> (Swain, 1989, p. xxi).

Years of study have produced long lists of possible performance shaping factors and much effort and ingenuity has been spent on finding the numerical (quantitative) values. Despite that, the influence of a PSF is often treated in a very simplistic way. If, for instance, the probability of making a failure, such as missing a step in a procedure, is estimated to be $p = 0.01$, the influence of a PSF, such as moderate stress, is simply assumed to double the value. This really begs the question of whether moderate stress is defined independently, or as that condition which doubles the failure rate!

Performance shaping factors vary considerably. One of the better known collections (Swain & Guttmann, 1983) separates performance shaping factors into external performance shaping factors, stressor performance shaping factors, and internal performance shaping factors. The individual performance shaping factors, however, show little resemblance to each other. They vary, randomly selected, from "shift rotation" and "work methods" over "man-machine interface factors", "task load", and "disruption of circadian rhythm" to "previous training" and "physical condition". In other cases such things as "information interface", "cognitive aspects", "stress/conflicts", and "training" have been proposed as specific performance shaping factors! It is hard to disagree that these factors may have an influence on the person's performance. On the other hand, it is also hard to see how their value can be estimated or measured, or sometimes to see even how they can be distinguished from each other.

8.2.1 Use of Performance Shaping Factors

A clear example of the problems in defining and using performance shaping factors is found in the Human Factors Reliability Benchmark Exercise (Poucet, 1989). In this study 15 different teams were asked to perform a HRA for the same routine functional test and maintenance procedure, using the method of their choice. Nine of the teams chose to use the Success Likelihood Index Methodology (Embrey *et al.*, 1984), and therefore defined the performance shaping factors that were assumed to be important. This resulted in the following list:

(1) operator experience,
(2) training quality,
(3) time available,
(4) procedure quality,
(5) supervision quality,
(6) noise,
(7) accessibility,
(8) check off provisions,
(9) motivation,
(10) information quality,
(11) stress,
(12) communication,
(13) team structure,
(14) design,
(15) location,
(16) cognitive complexity,
(17) physical complexity, and
(18) perception of consequences.

None of the teams used all these performance shaping factors; the range varied from two (!) to 11. Even more interesting, no two teams used exactly the same set of performance shaping factors, even though they were analysing the same scenario using the same method. There were no clear definitions provided of the performance shaping factors, nor any consideration of whether they were dependent or independent. In some cases it seems as if two performance shaping factors were used interchangeably, e.g. "time available" and "stress". In other cases performance shaping factors were used together even though they clearly have overlapping semantics, e.g. "work conditions" and "team structure" or "motivation" and "stress". To make matters worse, the weighting and rating of the performance shaping factors was performed differently by different teams. The minimum requirement to a reliable method is that the underlying concepts are clearly defined and that unambiguous procedures for using them are available.

8.2.2 Innocence of PSF Calculations

The accepted approach to find the influence of performance shaping factors is to calculate, for example, the failure rate in the following fashion:

$$\log f = \sum_{k=1}^{n} PSF_k * W_k + C$$

where:

f = operator error rate
C = numerical constant
PSF_k = numerical value of PSF_k
W_k = the weight of PSF_k
N = the number of PSFs

According to this "model", the human error rate can be expressed strictly without making any assumptions about the cognitive functions of the person, who is considered as a black box. This approach implicitly assumes that performance shaping factors are linear and additive, i.e., that there is no interaction between them. These assumptions are clearly very unrealistic, as the examples above easily demonstrate. In order for performance shaping factors to make sense, it must be possible not just to describe how they interact but also to explain - or at least speculate - about how a performance shaping factor exerts its influence on performance. That, in turn, requires some kind of model of how performance is produced or caused.

Worse than that, many of the proposed performance shaping factors will have an effect on the performance which may easily dominate the more detailed contributions from estimates of individual event probabilities. This goes for universal performance shaping factors such as circadian rhythms to the task specific ones. Moray (1990) notes that:

> "Consider for example the massive body of data on the effect of circadian rhythms. The probability of error and accidents changes by more than an order of magnitude over a 24-h period. Consider the problem of motivation. If an operator is provided with an outstandingly well designed control room, but is obsessively worried about the possibility of being laid off, or whether his wife is being unfaithful, or some other emotionally upsetting problem, we know that the probability of accidents rises. There is physiological evidence that following a brief but intense stress the general physiological state of the body does not return to baseline for more than 24h, so that a near accident on the way to work means that the operator is in an undefined state of efficiency for many hours. Consider the problem of coordination between the members of a team. Will the fact that I know my work is to be checked make me more efficient (because I know it will be checked) or less efficient (because I can rely on someone else to check my work)? And how will this vary depending on the person with whom I am teamed? It is simply fantasy to think that the probability of human error is described by a single number unless the upper and lower bounds of uncertainty are made so wide as to render virtually meaningless the mean as an estimate."
>
> (Moray, 1990, p. 342)

The notion of a performance shaping factor is both intuitively correct and meaningful. It is beyond dispute that certain factors influence performance, and that these factors can be described as being either internal or external relative to the person. In certain cases it may also be warranted to suggest a well-defined dependency, e.g. that stress will have a distinct influence on the reliability of the performance in a task and that this influence may even be described in quantitative terms. The notion of performance shaping factors is nevertheless in many ways an artifact that has been derived from the decomposition principle in human reliability analysis and from the concept of the individual action. Just as the concept of the individual action may be questioned, so may the notion of performance shaping factors be subjected to a reconsideration. Probably one of the best ways of doing that is by considering another concept: the common mode failure.

8.3 Common Modes: Rule or Exception?

A common mode failure (or common cause failure) is defined as "a failure which has the potential to fail more than one safety function and to possibly cause an initiating event or other abnormal event simultaneously..." (Swain, 1989, p. xiv). Common mode failures thus denote cases where a single event (called a cause) can be the initiating occurrence for several other events which otherwise are independent. Common mode failures are one of the hard problems of system design (Hirschberg, 1990). Despite all attempts to design systems to avoid common mode failures, for instance by increasing diversity and redundancy, common mode failures always remain a possibility. This may be because they have been overlooked or been considered but rejected as being too improbable.

Common mode failures in technical systems are the exception rather than the rule. They do happen, but it is reasonable to consider them as exceptional cases and therefore on the whole exclude them from normal analysis. In human beings, on the other hand, common mode failures are the norm rather than the exception. In other words, if something goes wrong it will quite often be **because of** a common mode which has consequences for all subsequent actions, whether or not they otherwise are or were related. An example is a bias in interpretation (cognitive myopia, pet hypotheses), which will not be confined to a single step but carry through the whole sequence of events. Other examples are stress, information underspecification, information overload, fatigue, etc. In fact, most of the performance shaping factors may qualify as common performance modes at least by name.

The concept of the common mode failure is one which clearly differs in the decomposition principle and the cognitive viewpoint described above. The decomposition principle focuses on decomposition and (binary) event trees, and common modes are therefore used as a way of accounting for the combinations that are missed by the analysis. The cognitive viewpoint focuses on cognition and action as a whole, and common modes therefore become **Common Performance Modes**

(CPM), i.e., conditions which are common for the actions as a whole (e.g. a whole plan or procedure, a whole situation). It is conceivable that a Common Performance Mode may actually dominate the performance even in a quantitative analysis, i.e., that the contribution from an individual action is negligible compared with the contribution from the common mode. Whenever that is the case, it would then clearly make more sense to carry out the analysis on the level of the overall performance rather than on the level of individual actions.

The basic lesson is that human actions are not additive and independent, but rather conditional on some basic assumptions being true (no stress, no bias, maximum processing capacity, etc.). If the underlying conditions turn out to be wrong or are violated, then the result will most likely be wrong. This, however, cannot be accounted for by considering each action on its own; it can only be adequately accounted for by considering the actions as a whole sequence (a plan), and by considering how external conditions, or even worse conditions that become established or amplified because of that sequence, have an influence on the results achieved. This provides a true alternative to the decomposition principle, and will be the basis of the method described in Chapter 5.

9. Summary

This chapter has served as a broad introduction to the main themes of this book: performance, reliability, and unwanted consequences. The starting point is the unwanted consequences that invariably and inevitably happen in man-made systems, and the more so in complex ones. These unwanted consequences are generally due to the insufficient reliability of human performance - but in design, maintenance, and management as well as in operation. Since it is important to be able to analyse the reliability of human performance and to understand what the underlying causes may be, the endeavour is to provide a consistent description of human performance which can be used as the foundation for reliability assessments.

The first steps were to recount briefly the background for the emergence of ergonomics, and in particular note the steady increase in the number of cases where human action has been seen as the cause of accidents and incidents. It was pointed out that there was a possible coupling between the growing complexity of the systems and the rising number of unwanted consequences. One factor that might contribute to that is the predisposition for risk homeostasis, which means that technological improvements are used to increase the level of system utilisation rather than to increase safety and reliability.

Since accidents are an important motivation for doing research on human reliability, the next step was to present a general framework for the analysis of accidents - called the anatomy of an accident. Furthermore, a classification was proposed of the typical ways in which reactions to failures and accidents occur. In general the purpose is either to eliminate the cause of the accident or to prevent the unwanted consequences from taking place. It was argued that a thorough understanding of human performance and human cognition was required in order for such measures to be effective. The first step towards that is to scrutinise the concept of human reliability. According to the prevailing notion of the human as an information processing system, captured, for example, by the cognitive viewpoint, the cause of an inappropriate action can lie either in a knowledge mismatch or in the inaccurate execution of a plan.

The main purpose of human reliability analysis is rather pragmatically to enable specific **system changes** to be made in response to specific unwanted occurrences, as well as to be able to make better **predictions** of what will happen under given conditions, as an effort to improve the system design. As a precursor to that the difference between qualitative and quantitative analyses was discussed, and it was suggested that the difference is more formal than real. The quantitative analysis is at the root of the decomposition principle, which provides the foundation for most of the current analytical methods. The dominating influence of the decomposition principle was illustrated by going through an example: the President's Heart Attack.

One of the ubiquitous causes for unwanted consequences is the so-called "human error". It was argued that the term "human error" in itself is both misleading and too simplistic, and that a better term would be "human erroneous actions". The notion of erroneous actions points to the importance of the context in which the actions take place, rather than the action as an independent cause. This was illustrated by considering the so-called ant analogy. This analogy has been proposed to support the argument that the complexity of human performance is due to the complexity of the environment, rather than to the complexity of human cognition and of the human psychological system. It was argued that the ant analogy is misleading because it does not discuss the assumptions behind performance. An alternative view is that human cognition must be complex precisely because the environment is complex; this is consistent with the cybernetic Law of Requisite Variety.

Finally, the decomposition principle was looked at in terms of the artifacts that it produces. The two most important once are the notion of the individual action as an elementary component of analysis, and the concept of the Performance Shaping Factor. It was argued that both were inappropriate and inconvenient for a reliability analysis which takes the context into account. This argument was supported by considering how common causes or common modes differ between the performance of technical systems and human behaviour. For technical systems the common cause is the exception rather than the rule; for human behaviour the common mode is the rule rather than the exception. No analysis of the reliability of human performance will therefore be complete unless it pays due attention to the common performance modes, i.e., to the context in which the actions take place.

The remaining chapters will continue the several lines of thought introduced in Chapter 1. Thus Chapter 2 is concerned with the need for models as a basis for working with the Reliability of Cognition. It argues that current practice on the whole continues the procedure from the analysis of technical systems to the analysis of humans. This, however, makes certain assumptions about the nature of human cognition which unfortunately are wrong. The changes in the nature of work, from "doing" to "thinking", have made it necessary to have an adequate model of what human cognition is. Without such a model it will be impossible to analyse the Reliability of Cognition.

In Chapter 3, the topic is the nature of human reliability assessment. This includes the issues of obtaining appropriate data and of matching data to models. The principle approaches to data collection are discussed, based on a conceptual clarification of the nature of data. The practice of HRA is illustrated with the example of a Human Factors Benchmark study. This is rounded off by a short survey of the existing methods, which are seen in relation to an "ideal" method.

Chapters 4 and 5 develop the model and the method that are the basis of the contextual approach to human reliability analysis. Chapter 4 starts by discussing the common approaches to modelling and proposes a distinction between two types of models: procedural prototypes and contextual control models. The details of a specific contextual control model (COCOM) are developed. In relation to the Reliability of Cognition the main concept is the notion of control modes. The COCOM is described in detail, with special emphasis on how the model concepts can be made operational, to serve as the basis of a practical method.

Chapter 5 develops the specific method for human reliability analysis, called the Dependent Differentiation Method (DDM), by means of a detailed example. This method overcomes the limitations that follow from the use of the decomposition principle, and also offers an alternative to performance shaping factors in the form of Common Performance Modes. The DDM is based on a consistent method for task analysis which is presented in detail. Together the Goals-Means Task Analysis and the DDM provide an adequate qualitative basis for human reliability assessment. The problems of quantification are discussed - but not solved.

Chapter 6 rounds off the book by taking a high level view of human performance analysis. It considers the relations between attention and reliability in general and between the COCOM and the Reliability of Cognition in particular. It discusses some of the practical problems in performing an analysis, and suggests a scheme to evaluate human reliability analysis methods. It concludes by reiterating what must be the foundation for human reliability analysis: **that human performance cannot be understood by decomposing it into parts, but only by considering it as a whole embedded in a meaningful context.**

2.
The Need for Models of the Reliability of Cognition

1. Introduction

In Chapter 1 I have argued, firstly, that it is essential to achieve a better understanding of human performance and human cognition in order to reduce the number of situations where unwanted consequences may occur and, secondly, that human performance and human cognition are not necessarily simple phenomena. More specifically, there is a need to improve the scientific understanding of human cognition as a basis for developing a better grasp of the Reliability of Cognition - expressed in terms of a plausible theory and as better models. However, the question may well be put whether there truly is a demand a for something as grand as a THEORY OF COGNITION or a THEORY FOR THE RELIABILITY OF COGNITION. The endeavour should clearly not be a quest for scientific truths, but rather a search for something which can be turned into practical tools and methods. The problem has been described by Apostolakis (1990) as follows:

> "The quantification of human reliability is an active and frequently controversial research area. The increasing role that probabilistic safety assessment (PSA) is playing in the regulation of nuclear power plants, at least in the United States, has led to the need for the development of models that produce probabilities for human error rates. The main philosophy behind such efforts has been that decisions regarding the safety and continued operation of existing large

facilities cannot wait for the development of rigorous models and the proposed approaches are viewed as the 'engineering solution' to the problem. At the same time, researchers who try to understand human behavior and to develop models for the operators have a very negative view toward the use of such quantitative models, whose foundation they consider to be unacceptable."
(Apostolakis, 1990, p. 281)

As pointed out in Chapter 1, the purpose of studying the reliability of cognition is purely pragmatic, and the aim is to improve the basis for analysing actual and prospective system performance and to make specific system changes in order to reduce or prevent the occurrence of unwanted consequences. The result of this effort, whatever it may be, must be seen in relation to the established practice, commonly referred to as human reliability assessment, which at least is a workable "engineering solution" and which has the merit of being generally accepted as such. As Apostolakis points out, the engineering approach to human reliability analysis is, however, generally considered to be insufficient, not just by behavioural scientists but also by those who actually use the methods in practice, cf. the provocative paper by Dougherty (1990) and the many answers it generated in the same issue of the *Journal of Reliability Engineering and System Safety*. The reasons for this opinion are many and range from a concern about the conceptual basis for the approach to a worry about the lack of sufficient empirical data. Some of the reasons are covered in further detail below.

On the other hand, the hope that rigorous models will be developed may turn out to be in vain. The very need for rigorous models is based on the assumption that the current approach to PSA can successfully be carried over to human reliability assessment. I will argue throughout this book that the assumption is mistaken and that, consequently, there is no need to wait for rigorous models to be developed. In addition to that, the current scientific knowledge about human cognition is very far from being able to support the rigorous models of human reliability that PSA practitioners so desire. So even if we needed to develop rigorous models, we would be unable to do so for the foreseeable future. There is therefore every reason to try an alternative approach.

1.1 Definitions of Human Reliability

The definition of human reliability is an issue which can be either quite simple or very complex, depending on the point of departure. For those coming from the engineering world it is tempting to carry over the concept of reliability used for mechanical/technological systems. Here reliability refers to a measure of the probability of the malfunction of a system - or a measure of the time before a system (or an element thereof) will fail. (Defining reliability as the probability of malfunction of a system is not very useful, because it logically means that a system with high reliability is one which has a high probability of malfunction. It should really be the other way around.) Reliability is therefore something that clearly is **measurable**. For those coming from the world of behavioural science, reliability refers to a psychological quality or a personality trait. Reliability is the **quality** (or qualities) of a person in virtue of which he can be relied upon. The concept can be extended to cover artifacts as well as persons, for instance in speaking about trust in machines (Muir, 1988; Moray & Lee, 1990). Finally, for those coming from the worlds of statistics and measurement theory, reliability has a third and completely different meaning. It does not refer to an aspect of the object, but an aspect of the **method**. Reliability can be defined as the complex property of a series of observations or of the measuring process that makes it possible to obtain similar results if the measurement is repeated; a reliable measurement or method is therefore one which is free of random influence.

The objective for ergonomics and human reliability analysis has been defined as that of reducing the number of unwanted consequences in a system. It is consistent with this objective to say that reliability generally refers to the lack of unwanted, unanticipated and unexplainable variance in performance. A highly reliable system is therefore one that performs in the same way given the same initial conditions - or even across a range of different initial conditions - and given different working conditions. The latter is quite important because the detailed conditions under which the system may be called upon to function are not always known in advance. A typical definition of reliability in the technical literature is:

> "... the probability that an item will operate adequately for a specified period of time in its intended application."
> (Park, 1987, p. 149)

This definition of reliability can obviously be applied to artifacts such as technical systems or software systems because in both cases the specifications for adequate performance are known. But the definition is less useful when applied to human reliability - and for precisely the same reason. One of the differences between human beings and artifacts (machines, computers, etc.) is that while the latter have all been designed with an intended application in mind, the former have not. Furthermore, an artifact is designed to carry out a limited function or range of functions only and cannot normally be used for anything else. A human being, on the other hand, is assumed to be an all purpose system which can handle practically anything. There are no clear specifications for what a human being can do - although there may be clear physical and psychological limitations for what it is possible to achieve. Another important difference is that an artifact usually only responds to an external event (either directly or indirectly) whereas a human being may often act independently of what happens around him. An artifact usually does not emit actions spontaneously, but a human being does. This is evidently important for the issue of reliability.

The notion of an intended application, as used in the definition above, is on the whole inappropriate for human beings. This can be acknowledged by changing the definition of human reliability to be:

"... the probability that a person will perform adequately for a specified period of time."

This immediately makes more sense. It is normally known what to expect of a person in a particular situation and it is therefore possible to determine whether the performance has been adequate. The expected performance can either be based on practical experience from a range of similar situations or from the particular overall work design or be based on instructions or procedures that may have been provided in advance. The definition may therefore be modified even further:

"... the probability that a person will perform according to the requirements of the task for a specified period of time."

This agrees well with the definition proposed by Swain & Guttmann (1983), which defines human reliability as:

"the probability that a person (1) correctly performs some system-required activity in a required time period (if time is a limiting factor) and (2) performs no extraneous activity that can degrade the system."

The latter definition captures the engineering perspective on human reliability quite well; it does, however, demand that the system-required activities can be identified. This condition is fulfilled in, for example, procedure based operations, but will generally be more difficult to fulfil for unanticipated situations and disturbances.

In the case of human behaviour, reliability is most applicable to anticipated situations - whether they are normal or contingency operations. An important role of humans is, however, to handle unforeseen situations where pre defined procedures are usually either inadequate or completely lacking. In these cases the person may contribute significantly to the robustness and adaptability of the system. This contribution must be distinguished from reliability in the narrower sense as discussed above and is, in fact, only poorly described by it. In contingencies people often do things they have never done before and which they may never do again, and it is difficult to speak of the reliability of something that may happen only once.[1] The recourse is to focus on the reliability of the functions and processes that go on beneath the surface, i.e., the reliability of human cognition. The argument is that although manifest performance may show an enormous amount of variance, the underlying psychological and cognitive functions are more stable. The variations observed are therefore an effect of the external conditions (the context) as they affect human cognition. The consequence of this view is that the standard notions of human reliability must be supplemented by a cognitive modelling approach.

Just as important, the focus on cognition also means that a sensible definition of reliability has to take the current goal or target into consideration, i.e., reliability must be defined as following the instructions or achieving the desired outcome in a sense which implies an **understanding** of what the target is. In that way reliability becomes tied

---

[1] A case in point is the switching station black out in New York on September 17, 1991. Despite the lack of nearly all voice, data and radar systems, the air traffic controllers managed to keep the traffic flowing (or rather, flying) using many tricks (*Aviation Week & Space Technology*, October 7, 1991, p. 33).

to whether the systems works appropriately, not only on the level of individual components and sub-systems, but in the global sense as a joint system. This difference is important for analyses and models of reliability since it effectively distinguishes it from repeatability or reproducibility of single, observable actions.

2. The Decomposition Principle

The currently established practice, the "engineering solution", is based on the principle of (1) first decomposing the required performance into its constituent basic actions, then (2) assigning reliability estimates to the components, and finally (3) computing an aggregated result. The basic technique is to break the system down into its constituent elements (be they task steps or components) and to combine that with a detailed description and assessment of each element. I will refer to this as the decomposition principle.

The steps of impact assessment and quantification are obviously very sensitive to the preceding qualitative analyses as well as to the quality of the reference data. The detailed quantitative analyses are performed on the elements or structures that have been defined by the antecedent steps; if these steps leave something out the outcome of the analysis will necessarily be incomplete no matter how refined the quantification methods are.

2.1 Systematic Human Action Reliability Procedure (SHARP)

A typical example of the decomposition principle is the SHARP method (Hannaman & Spurgin, 1984) which contains the seven steps shown in Figure 2-1.

There are two clearly different ways of doing the first steps - from definition to breakdown - which usually are called task-driven and component-driven:

(1) The **task-driven approach** identifies all the task steps or actions which are required for the correct performance of the specified task. Each task step is then characterised in further detail with regard to, for example, objectives, requirements, Performance Shaping Factors, etc. Finally, the possible "error mechanisms" for each task

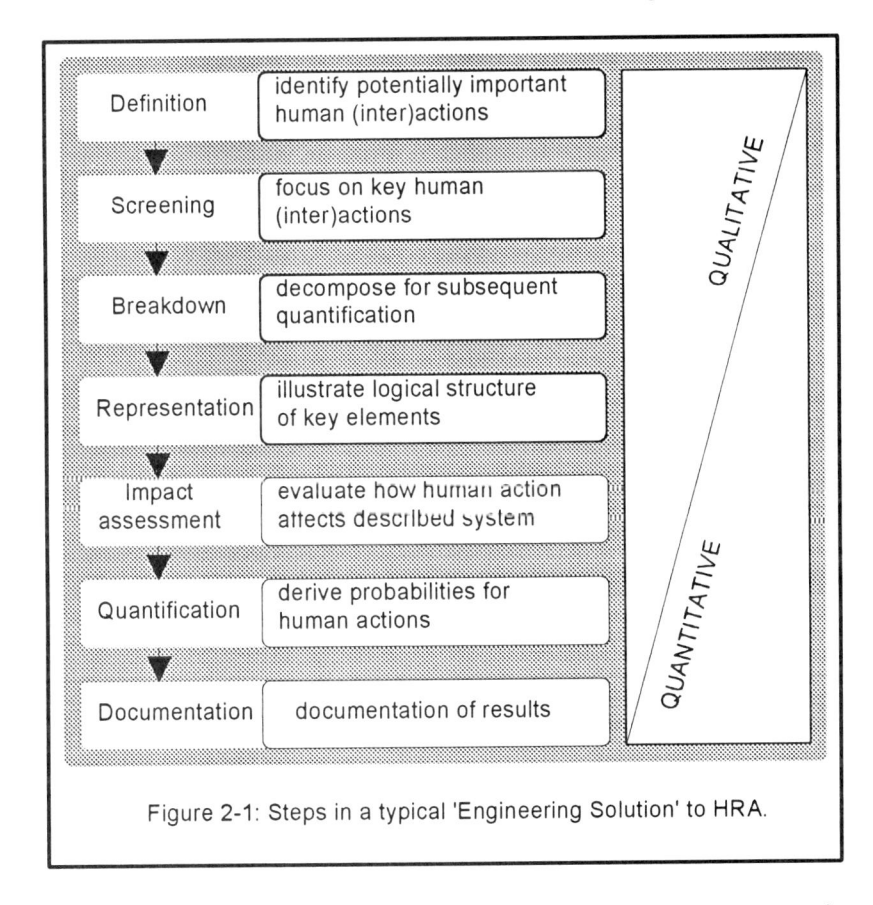

The Decomposition Principle

Definition	identify potentially important human (inter)actions
Screening	focus on key human (inter)actions
Breakdown	decompose for subsequent quantification
Representation	illustrate logical structure of key elements
Impact assessment	evaluate how human action affects described system
Quantification	derive probabilities for human actions
Documentation	documentation of results

The upper steps are QUALITATIVE; the lower steps are QUANTITATIVE.

Figure 2-1: Steps in a typical 'Engineering Solution' to HRA.

step are identified and the possible consequences for the execution of the task step are analysed.

(2) The **component-driven approach** begins by identifying the components of the system under investigation. It then uses these components as a basis for identifying the human actions that potentially may affect each component and bring it into one of its possible states. The starting point for analysing the procedures is thus the way in which the system is structured rather than the way in which the tasks are carried out.

The advantage of the task-driven approach is that it is not restricted to pre-defined components or component states. The disadvantage is that it is easy to include task steps and actions that do not have an impact on the specific situations being considered, i.e., the task-driven approach may suffer from the lack of a well-defined focus. The advantage of the component-driven approach is that it provides a clear focus for the analysis; the disadvantage is that it is unable to consider events that have their origin in human action or which otherwise not are included in the procedures. The choice between the two approaches therefore implies the common trade-off between breadth and depth of analysis.

3. Reliability Analysis and Event Estimation

It is usual to base the assessment and quantification of task elements on estimates of various kinds. The exceptions are the relatively small number of cases where specific frequency data are available. The estimates are typically derived from expert judgments, assumptions, simulations, or empirical data - in decreasing order of use. (This *de facto* order or frequency of use does not necessarily correspond to the order of importance of the various data types. The data types will be covered in more detail in Chapter 3.) Precise empirical data are very hard to come by and this is the primary reason why various types of estimates are applied. Although the use of estimates is widespread, this practice nevertheless presents some problems which are not easy to overcome:

(1) Probability estimates of psychological processes, in particular of the reliability of human cognition, need to be calibrated, either as true values (with ranges) or as parts of a sequence (cf. below). The basis for calibration is unfortunately extremely weak because these processes are inferred rather than observed.

The reality of cognitive processes is based on reasoning by analogy from subjective experience (introspection). I know what occurs in my mind when I try to solve a problem and it is not unreasonable to assume that something similar takes place in other peoples' mind. To identify the steps or parts of what happen in my mind is, however, not an easy task because consciousness is a stream rather than a state (e.g. Mandler, 1975). The very notion of a step is an abstraction: it cannot in itself be observed. The probability estimate is therefore of the mental "steps" that correspond to a defined overt action, rather than of the mental "steps" *per se*, using the assumption that the overt action was caused by the actions of the mind. Probability estimates of mental process or cognitive functions detached from a context should therefore be avoided if at all possible.

(2) Estimates can only be generated for known or anticipated events, and only for events which either have occurred or which are considered generally plausible within the established lore. Estimates of rare events are highly uncertain, because such events usually belong to neither of the two preceding categories. Even though evidence forces us to admit the possibility of their occurrence, estimates of their probability must remain pure guess-work.

(3) Even if the contributions from rare events are excluded, estimates only exist for a limited number of all the possible events. The decomposition principle encourages a careless creation of a multitude of single events, the more detailed the better. Even if the possibility of producing an exhaustive set of events is not seriously considered, the decomposition principle almost automatically creates a very large number of events. The traditional methods make it an arduous task to generate estimates for a large number of events - and in addition the results must also be categorised. It is therefore a pragmatic solution to consider a suitably defined subset. In this case the very nature of the decomposition principle seems to work against efforts to generalise across situations, because it produces an unnecessary increase in the burden of the analyst.

These problems are among the reasons why "purists" are somewhat dissatisfied with the "engineering solution" that Apostolakis referred to. For these reasons alone it will be necessary, and useful, to have a theory of the Reliability of Cognition which can provide a common basis for the estimates, hence improve their consistency. If we improve our understanding of human cognition and of the Reliability of Cognition, this will be of value for all the many ways in which estimates need to be considered, i.e., with regard to their generation, comparison, as well as calibration. In addition, an appropriate theory of the Reliability of Cognition will provide an alternative to the focus on components and elementary events, i.e., an alternative to the decomposition principle. As the discussion in Chapter 1 showed, such an alternative is badly needed.

3.1 Granularity of Decomposition

Conventional analyses of reliability and risk are, it seems, necessarily based on the notion of components - either as components in a physical system or as the component parts of a task, i.e., single events, as illustrated above. This view is so common that it is almost taken for granted; consequently, it is rarely disputed or considered analytically. I will, however, argue that although the use of a decomposition principle is widespread and seemingly unavoidable it is not really necessary. Neither is it adequate as a basis for the estimation of the Reliability of Cognition, as the following example will show. Assume that we have a step-by-step description of human cognition, say in the form of a model or theory of a specific aspect of cognition. (It might also be one scientist's pet theory against another scientist's pet theory.) For illustration, we can consider a simple way of describing the covert or "inner" events between alert and response as shown in Figure 2-2, or alternatively as shown in Figure 2-3.

The two descriptions or response paths correspond to the same overt performance or manifestation, for instance what is assumed to go on from the time a person becomes aware of an event (an alarm, a signal) to when a response is made. The main difference between the two descriptions is therefore the number of steps that are involved. Assume that we assign an estimate of the reliability of X for each step, say $X = 0.995$. (This number is completely fictional and is only used as an example. It does not suggest that the reliability of human performance generally is to be found in this range.) We can interpret this to mean that the probability that the person will correctly perform the step and achieve the intended goal is 0.995; or, conversely, that the probability of failing is 0.005. If we assume that the steps are independent, the overall probability of failure is $(4 * 0.005 = 0.02)$ and $(7 * 0.005 = 0.035)$ respectively. If we assume that the steps are dependent, the overall probability of not failing

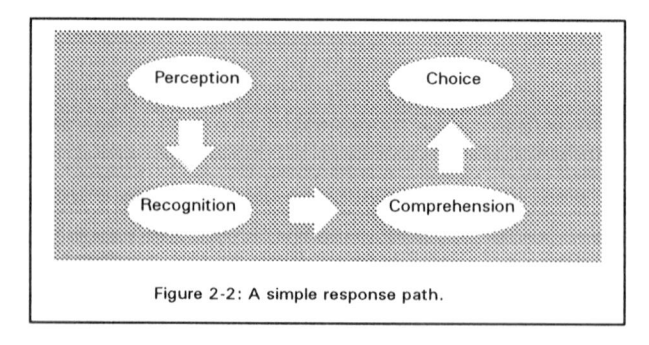

Figure 2-2: A simple response path.

Figure 2-3: A more complex response path.

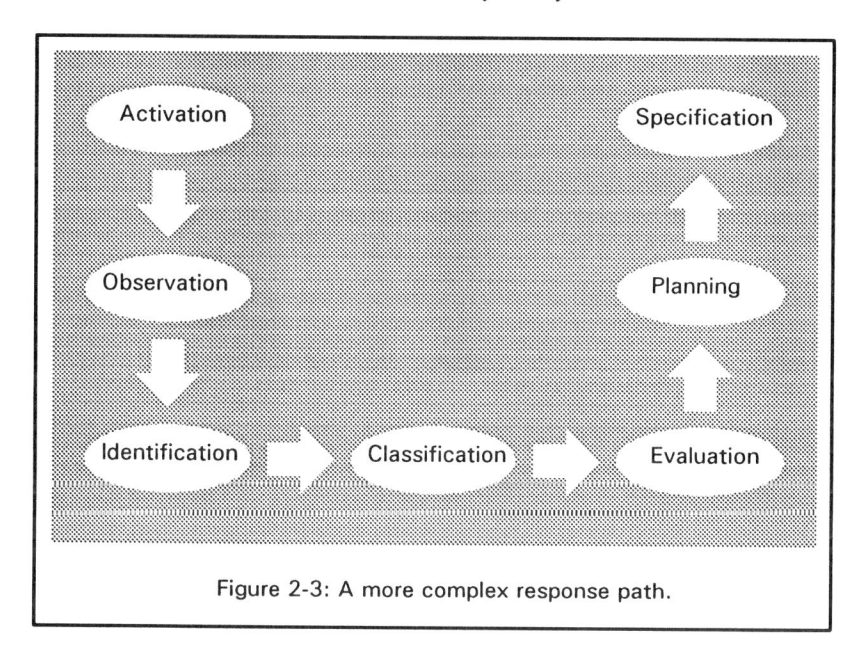

Activation → Observation → Identification → Classification → Evaluation → Planning → Specification

for the whole sequence is ($X^4 = 0.98015$) and ($X^7 = 0.96552$) respectively. Independent of the method of calculation there will obviously be a difference in the outcome for the two cases.

This difference of outcome is simply due to the different number of steps, which again is due to the chosen type (or theory) of performance description. Usually the value of X_i depends on the type of $Step_i$, i.e., the **type** of cognitive function. But the value of X should depend as much on the **number** of steps, i.e., the granularity or level of detail, as on the nature of the steps. Strict conformity with the decomposition principle would therefore require that the probability estimates for each step were a function of how many steps the cognitive function was assumed to be composed of. This would ensure that the result was not influenced by the numerical number of steps, which may be quite arbitrary. In general, however, the values that are obtained by this type of analysis depend only on the nature of each step, which is considered on its own according to its representativeness. This can obviously produce misleading results; it would clearly be more reasonable to consider the sequence as a whole, as the overall cognitive task, rather than decomposed into a more or less arbitrary number of component steps.

It is for these reasons that there is a need to establish a theory of the Reliability of Cognition. The theory, and the models that can be attached to it, must enable a description and understanding of human cognition and the Reliability of Cognition for meaningful sequences of actions, rather than for the decomposed task. This also means that if numerical estimates are going to be used they must take the whole context into account - not only the actual working conditions but possibly also past events and even anticipated future events. The basic assumption of a theory for the Reliability of Cognition is that **the choice and performance of an action is determined by the context**; the analysis must therefore include everything which reasonably can be assumed to be a part of the context in the sense that it can influence how the action is carried out. To do that requires a coherent theory and, if possible, a way of manipulating it symbolically, for instance in the form of a simulation. This is accordingly the rationale for the study of the Reliability of Cognition and at the same time a first definition of the goal.

4. Repeatability and Similarity

The problems that are met in human reliability analysis are certainly among the main reasons for taking an interest in the Reliability of Cognition. As described above, there is a clearly perceived need to improve the estimates of human failure rates. This is usually defined as a need for rigorous models that can be used to produce human error rates, i.e., quantitative expressions of the probability that a person, either a process control operator or, more generally, the user of an information processing system, will make a particular - and undesirable - action. The undesired action is normally classified as being either the non-occurrence of an action (an omission) or the occurrence of a substitute, but incorrect action (a commission). Although this binary classification is convenient and easy to use, it is unfortunately also too simple. The whole area of correct and incorrect actions, and of human erroneous actions, requires additional analysis and clarification which will be outlined later in this chapter. The need for models to generate probabilities tacitly accepts the underlying assumption, *viz.* that the quantification of error probabilities is the right way to approach the safety assessment of a process. This assumption is in turn based on the transfer of the dominant approach from the risk analysis of technical systems to the realm of human action as well.

Quite apart from the several debatable aspects of the decomposition principle, the transfer of a method from the technical to the behavioural sciences does require some justification. In this case it is relevant to consider two of the assumptions that lie underneath the quantitative approach: that of repeatability and that of similarity.

4.1 Reliability and the Cumulation of Effects

The reliability of a mechanical component can be assessed by repeating the component's function again and again until it breaks down or fails. Indeed, the same procedure can be performed on any number of identical components, thus establishing the reliability with a high degree of confidence (e.g. as MTBF: Mean Time Before Failure). This procedure is used for testing both the physical and the functional properties of a component. It assumes both that the external conditions are the same each time the event occurs and that there is a sufficiently large number of identical components. Repeatability assumes that only a single independent variable changes: the number of times the event has occurred. (The change in the performance of a component is, of course, caused neither by the number of repetitions nor by time itself. The change is due to the accumulation of changes in the component, e.g. wear and tear, metal fatigue, etc. Each occurrence leads to a small change in the internal state of the system, but this happens in a cumulative manner until a threshold is reached. The number of repetitions is therefore a convenient measure of the changes that takes place.)

If the same procedure is to be used for human actions it can only be done for a very limited subset, typically elementary motor actions (pressing a button) or basic perceptual actions (discrimination, cued reaction, etc.). In order for repeatability to be a valid approach it must be possible to ensure that everything else is equal - generally expressed by the phrase *ceteris paribus* or "other things being equal". Even for elementary psychophysical functions the *ceteris paribus* condition cannot be guaranteed, since the state of the organism sooner or later will change. One reason is the phenomenon which technically is known as satiation, defined as "a state of relative insensitivity to stimulation that follows exposure to a succession of closely related stimuli" (English & English, 1958, p. 473). The change that occurs when, for instance, a button is to be pressed repeatedly in response to a signal will make even this simple kind of performance unreliable. Thus, from one point of view the change invalidates the application of the procedure: *ceteris paribus* is no longer true. From another point of view it simply marks the point where performance becomes unreliable, which might be the result that was wanted. The problem with the procedure is, however, that the variations between situations may be very large, hence making it impossible to obtain a good numerical result. For example, the onset of satiation is not a simple function of the number of repetitions. Even if the same person is used in the same laboratory setting a uniform result cannot be guaranteed.

For more complex, hence also more meaningful and interesting functions, the repeatability approach is not a viable option under any circumstances. There are many reasons for that, but perhaps the most important is the effects of learning. It is a fundamental feature of human nature that the organisation of activities and tasks changes as we perform the repeatedly, because we deliberately try to find a better way of doing things, because we discover new relations or dependencies, or simply because of automation or skill development (e.g., Anderson, 1980).

4.2 Reliability and Situation Equivalence

The notion of reliability also hinges on the notion of similarity. One meaning of similarity refers to the situations from where data are collected. As pointed out above, the idea of repeatability requires that there is a significant number of identical components; otherwise the results from the analysis cannot be applied except as a historical record of what went on. The question is not so much whether human beings are identical (they never really are) but rather whether they can be expected to respond or perform in identical ways. In some cases the answer is a qualified "yes". An example could be reactions during panic, e.g. to a fire or an earthquake, where there is a limited set of stereotypical reactions. The "yes" is qualified because, even in panic, not everyone will respond in the same way. Furthermore, situations of this kind are not the ones that typically are the concern of human reliability analysis. In most other cases the answer is an unqualified "no", including practically all of the cases where we want to analyse human reliability. Even if the same person is put in the same situation the response may be different from one time to the other; anyone who has tried to observe people in, for instance, training simulators will know that this is true.

Another meaning of similarity refers to situations where the results are applied. The expression of the reliability of a component's function only has meaning if the situation where the performance occurs is similar or identical to the situation where the reliability measure was obtained. This the second side to the *ceteris paribus* condition; firstly, other things must be equal for the situations where the measurements are made; and then other things must be equal - or at least sufficiently similar - for the situations where the measure is applied. The notion of similarity therefore poses a particular problem for human reliability. It is generally acknowledged that work situations of interest are never identical. In particular, the events with serious unwanted consequences are always unique in some sense; if that was not the case, then the system must have been badly designed from the start. But if the similarity is very small, it is difficult to apply even the concept of reliability to human performance. The methods that work so well in the technical field are therefore confronted with two major problems: firstly, empirical reliability measures can, at best, only be established for very simple situations; secondly, these simple situations are not representative of realistic working conditions. It is not easy to find a way out of this; but a possibility is to go back one step and consider the assumptions for using this approach.

4.3 The Person as a System Component

Repeatability and similarity were valid assumptions for technical systems, but to use them in the case of human behaviour corresponds to considering the person as a system component. A mechanical component can be studied in isolation until it begins to malfunction. The same procedure can be repeated for other identical or similar components, and the environmental conditions can be strictly controlled. In that way a solid empirical basis can be established which allows the calculation of the component's reliability. A component can also be studied in realistic environments and the influence of environmental conditions can be assessed - within the limitations of the *ceteris paribus* condition. This procedure is sometimes difficult because the number of conceivable operating conditions can be very large. Yet even though there is no possibility of examining them all, a representative set can usually be defined. Furthermore, the envelope for normal operation can be derived from the requirement specifications; conditions under which the component is expected to be unreliable can therefore be avoided.

The same procedure cannot be applied to human beings, no matter how attractive it might seem. A human being is not a mechanical system or a component of a technical system, and neither can nor should be studied in isolation. Humans differ from machines in very many ways, even from the point of view of reliability analyses; this will be described in further detail in Chapter 3.

The difference between the two views can be emphasised by making a distinction between an engineering and a cognitive approach to the study of human reliability. The engineering approach is based on the principle of quantitative decomposition where the person is treated as a component in a complex system. The cognitive approach is based on the explicit use of models or theories of the cognitive functions which constitute the substratum for human behaviour. In the cognitive approach the assessment of human reliability is therefore based on categories that are meaningful because they have a common conceptual basis rather than on elements which, for one reason or another, have similar surface forms or manifestations.

4.4 Man as a Fallible Machine

Even if the human is not reduced to a component in a system or a cog in a wheel, there is still a strong tendency to perceive the human in mechanical terms. The notion of the human being as a fallible machine is inherent in the decomposition principle. The fallible machine is defined indirectly by posing the question: "What kind of information-handling device could operate correctly for most of the time, but also produce the occasional wrong response characteristic of human behaviour?" (Reason, 1990, p. 125). In other words, the fallible machine is a mechanism which sometimes produces correct and sometimes incorrect results.

The notion of the fallible machine is a step forward from even earlier attempts of explaining human erroneous actions which invoked the idea of specific error producing mechanisms (e.g. Rasmussen, 1986, 1988; Reason, 1988). In hindsight, the notion of an error producing mechanisms is obviously a mistake. If all psychological phenomena were explained by inventing appropriate mental mechanisms, the result would soon be an unorganised conglomeration of very diverse micro-theories which would be difficult to combine or integrate. The fallible machine is, however, still a very simplistic way of describing human cognition; it relies on the mechanical metaphor which assumes a host of unknown, but very convenient, functions. The fallible machine is an example of a theory which attempts to explain human cognition from a small set of elementary functions and/or components (cognitive structures) and therefore remains in good agreement with the decomposition principle. While there undoubtedly are some fundamental characteristics of human cognition to which we all can agree (cf. Simon, 1969), it is not yet possible to build a theory based on them - partly because they express limitations rather than capabilities, and partly because they are closely bound to particular experiments and therefore not pertinent for human performance in normal working conditions (e.g. Neisser, 1982). Cognitive psychology is still groping for a paradigm, and attempts at deductive theorising are very likely to fail at the present stage of knowledge. In particular, any assumptions about mechanisms as explanations for human performance may be premature - if not principally misguided!

Quite another point is that by viewing man as a fallible machine, the variations in human performance are seen as something that mainly has a negative rather than a positive effect. If the notion of a fallible machine is turned upside down, the human will be seen as a creative machine rather than a fallible machine. This is partly compatible with the view of "errors" as unsuccessful experiments in an unkind environment (Rasmussen, 1986, p. 150) and Mach's adage that "knowledge and error flow from the same mental sources, only success can tell one from the other" (Mach, 1905, p. 84). The occasional deviations from normal performance are not always "errors" and they do not necessarily lead to unwanted consequences. They may well match and partly absorb the unpredictability of the environment and therefore in time lead to new solutions. This view emphasises the advantages rather than the disadvantages of performance variations. Since it is impossible to ignore the fact that performance will vary, it is in my opinion better to take the optimistic view, i.e., that the performance variations are (at least) as advantageous as they are destructive.

5. Human Reliability and the Analysis of Erroneous Actions

The study of human reliability can go a long way just by using empirical observations and empirically established relations. A method for human reliability analysis may, in principle, be nothing more that a systematic classification of error modes followed by the assignment of probabilities (or error rates) from a database (e.g. Rosness *et al.*, 1992). Even this approach is not independent of theory, since the classification - and indeed the very observations - must be based on a coherent set of concepts about human performance. Yet in the long run the study of human reliability and the improvement of human reliability analysis will have to face the problem of developing a theory for the underlying phenomena - a theory of human reliability and the Reliability of Cognition.

The study of human cognition has one major obstacle: that human cognition is not directly observable. This does not mean that we cannot find out about human cognition; we do have access to our privileged knowledge (Morick, 1971) and subjects can usually produce both reliable and accurate reports of what they think and feel. Yet even such reports do not describe directly the cognitive functions that are of interest, and in particular do not describe the cognitive functions that a given theory assumes are the foundation for the observed performance. Even in controlled experimental conditions, the study of cognition is impeded by the lack of information about what actually goes on in the subject's mind:

> "...(I)n a traditional perception experiment, the experimenter has not only unlimited access when observing and describing the objective conditions of the experimental situation, but also unlimited control of that situation, by restricting the subject's opportunities for observation and description. ... In problem solving research, on the other hand, the conditions are almost completely reversed. Apart from the description of the task presented to the subject, and various measurements of the subject's overt performance, the investigator's control of the situation is severely limited, i.e., his control over the subject's choice of strategies and action sequences. But even more important and methodologically tricky, the investigator's specification of the problem must take account of how the *subject* conceives the situation and task in question."
> (Prætorius & Duncan, 1988, pp. 307 - 308.)

The situation is even more difficult when control is further relinquished, as in the study of human performance under naturalistic conditions or in the study of imprecisely defined phenomena such as the Reliability of Cognition. Nowhere is this more obvious than in the study of erroneous actions.

5.1 The Duality of "Human Error"

Actions carried out by a person can, logically, fail to achieve their goal in two different ways. The actions can go as planned, but the plan can be inadequate. Or the plan can be satisfactory, but the performance can still be deficient. Such incorrectly performed actions are often referred to as human errors, although this term is misleading in its connotations. It would be more correct to say that the actions are incorrect *vis-a-vis* the current goal because they produce unwanted effects and thereby make the goal more difficult or even impossible to achieve.

There is an important difference between the terms "human error" and "erroneous actions". "Human error" has become the established term to describe situations or events where undesirable consequences occur and where the cause can be attributed, in part or in whole, to some aspect of human action. (In PSA/HRA the search is usually confined to looking for the so-called **root cause**. However, in the field of human action the root cause is a myth, since it is always possible to go one step further.) The main problem with "human error" is that it may denote both a **cause** of an event and a special **class of actions**. Both uses of "human error" are important, but it is even more important that they are not confused (e.g. Senders & Moray, 1991). In order to avoid the problems in using a single term to denote two different things, I proposed the term **erroneous action** to characterise a certain type of action without implying anything about the cause (Hollnagel, 1991a). An erroneous action is an action which fails to produce the expected result and which therefore leads to an unwanted consequence. The expectations about the result, hence the criterion for a non-erroneous action, are normally those of the person(s) who carry out the actions. The expectations may, however, also be of persons who do not directly take part in the actions (supervisors, management, authorities), or expectations which reside in the system in some way (safety rules, production goals, procedures, etc.). The cause of the failure can logically lie with the person, the equipment (the system), the organisation, and/or the conditions under which the action was executed - as well as with acts of nature, sometimes referred to as Acts of God.

(Ultimately, there can only be two categories of causes: Acts of God and acts of man. Equipment failures are usually seen as unpredictable and unavoidable, hence as being caused by an Act of God. Yet even component failures such as, for example, metal fatigue, can be foreseen, and failing to do so can be ascribed to, for example, inadequate design, inadequate maintenance or inadequate communication within the organisation. In all cases where the possibility of the failure could have been imagined, the cause rightly belongs to the human category.)

5.2 The Systematic Study of Erroneous Actions

The study of the Reliability of Cognition must include the systematic study of erroneous actions. The traditional approach to erroneous actions has been based on casual and systematic observation or introspection, cf. Reason (1990). While this approach is convenient, it fails to make a clear distinction between **overt** human behaviour and the **covert** (cognitive) functions that are assumed to be the cause of behaviour. Behaviour in other people can be **observed**, classified, and analysed but the causes can only be **inferred** from the results of the observations. The systematic study of erroneous actions must clearly be based on what can be observed and verified. The consequence of this is not that one should remain with the observable and refrain from speculating about the unobservable, but rather that great care must be taken when both the observable and the unobservable are described by the same concepts and the same language. In particular, the implications must clearly be considered of basing operational classifications on common sense descriptions of the causes of erroneous actions. The fact that we have a privileged access to such descriptions and to many of the underlying factors must not mislead us to accept without question their validity as scientific concepts.

5.3 The Problem of Privileged Knowledge

Psychology differs from the rest of the empirical sciences because it has a unique opportunity to confuse manifestation and cause. This happens in the investigation of phenomena and events that are part of our conscious experience: in other words, in the study of **cognition**. The reason for this confusion is that we at the same time can observe a phenomenon and have first hand (or privileged) evidence of how it is consciously experienced. In order to avoid this confusion it is necessary carefully to maintain a distinction between the two sources of knowledge. Privileged (introspective) knowledge is not inadmissible as a source of data, but it must be treated appropriately. The many discussions about introspection, most recently in relation to the problem of knowledge elicitation, have been very worthwhile - not because they have solved the problem but because they have helped to clarify it (Nisbett & Wilson, 1977; Prætorius & Duncan, 1988). Introspection is a source of data which requires its own methodology. Observations are another source of data which requires other methods. The essential point is not to mix data at the source level, but only after they have been further refined through categorisation, classification and analysis: the combination should not happen at the level of raw data, but at the level of the formalised performance description (Hollnagel *et al.*, 1981).

As an example of the difference, consider the situation where a person fails to carry out an action. This will usually be classified as forgetting. Forgetting, however, indicates a cause; if a strict classification of the manifestation is used, the failure to carry out the action can only be seen as an omission. It is not possible to observe directly what the cause of the omission may have been. A second example is found in the use of terms such as information processing, diagnosis and decision making, action, and communication as error causes (e.g. Beare *et al.*, 1991). Only the last two are observable in any reasonable sense; the former are inferred causes. Yet another example of a misleading term is cognitive error which first of all refers to a supposed "mechanism" and furthermore denotes an internal cause which is highly hypothetical.

The distinction between different data sources and the effort to avoid confusing them is most important when the topic of investigation is human action and the reliability of such action. The several attempts that have been made so far of formulating a theory for action have all failed to maintain this distinction. This has led to considerable confusion in the terminology and therefore also in the concepts that have been made the basis for the theories. The emphasis on manifestations and causes is an attempt to reinforce the distinction between different sources of data by proposing a "pure" classification for one of them - the source of observable (empirical) data. Actions can be observed, action patterns can be identified and causes can be inferred. In addition, the result of the actions can be classified as unwanted consequences (errors). **But erroneous actions by themselves cannot be observed**; it is impossible to know that something is an erroneous action when we notice it, unless we are able to interpret it immediately in a context and observe or predict the consequences.

5.4 Phenotypes and Genotypes

There are two fundamentally different ways to consider erroneous actions. One is with regard to their manifestation or **phenotype**, i.e., how they appear in overt action, how they can be observed, hence the empirical basis for their classification. The other is with regard to their cause or **genotype**, i.e., the functional characteristics of the human cognitive system that are assumed to be a contributing cause of the erroneous actions. The genotype addresses the core issues of the Reliability of Cognition. In fact, imperfect reliability can be seen as one of the main, high-level genotypes.

It follows from what was said in the previous section - that erroneous actions by themselves cannot be observed - that phenotypes are the result of an analysis. It is only when the person acting is also the observer, or when the observer is intimately familiar with the situation, that the phenotype in the sense as the classification of the erroneous action, such as an omission, can emerge immediately. In these cases the genotype may also emerge immediately, and hence seem as if it was "observed" when the event happened.

It is interesting to note that one of the early definitions of human error describes it as a failure to perform a prescribed act (or perform a prohibited act) which could result in unwanted consequences (Hagen, 1976). This is clearly a definition of the manifestations of human error which does not imply anything about the causes. In contrast, later definitions of human error often mix manifestation and cause, much to the detriment of further research and understanding. Most authors, however, wisely refrain from giving a clear definition (Rasmussen, 1986, p. 149; Reason & Mycielska, 1982, p. 11).

5.4.1 Slips and Lapses

The need to maintain a distinction between phenotype and genotype is, quite ironically, emphasised by the definitions of the commonly accepted error types, summarised as "... (a) *slips and lapses*, in which actions deviate from current intention due to execution failures and/or storage failures ... and (b) *mistakes*, in which actions may run according to plan, but where the plan is inadequate to achieve its desired outcome ..." (Reason, 1990, p. 53). According to these definitions it is only possible to recognise a slip or a mistake if it is also known what the person's intentions are, which plan(s) he was trying to carry out, etc. It is not humanly possible to observe whether an action deviates from current intentions or whether an action runs according to plans unless one can also observe what the intentions or the plans were. The common definitions of slips and mistakes fails completely to maintain the distinction between manifestation and cause. The ensuing classification is therefore not operationally useful. Furthermore, unequivocally identifying a person's intentions is clearly beyond the ability of most people (and certainly beyond the ability of cognitive artifacts), hence not admissible as the basis for a systematic study of the Reliability of Cognition.

5.5 A Logical Classification of Phenotypes

It is possible to propose a logical classification of phenotypes, i.e., a classification which comes purely from considering the possible ways in which actions can be permuted and applying common sense terms to describe them. (This should not be confused with a theory of erroneous actions, i.e., an account of the genotypes, which clearly must follow different principles.) It does not take much analysis to see that very few categories are sufficient to categorise **logically** all possible forms of incorrectly executed actions. If we use as an example the sequence of actions called [... $Step_{i-1}$ $Step_i$ $Step_{i+1}$ ...], the basic categories can be defined as follows (cf. Figure 2-4):

(1) **Correct action**. In this case an action is correctly placed in the sequence of actions. In any sequence such as [... $Step_i$ $Step_{i+1}$ ...] the second action would be "correct" in relation to the first action. The condition can be extended to also include the time of occurrence of the action.

(2) **Jump forward**. In this case an action jumps forward in the sequence. Thus [... $Step_i$ $Step_{i+3}$ ...] is an example of jumping forward, because $Step_{i+3}$ is executed instead of $Step_{i+1}$.

(3) **Jump backward**. This category is the opposite of jumping forward, i.e., an action reverts to an already executed part of the plan. An example would be [... $Step_i$ $Step_{i-2}$ ...].

(4) **Intrusion**. In this case an action occurs which is not part of the current plan. It therefore constitutes an intrusion. The action may belong to another plan in the set of plans allowable for the application, or may be an unknown (hence also unrecognisable) action.

It is easy to see that the four types defined above make it possible to account for every possible combination of a sequence of actions. In that sense they constitute a complete set of phenotypes. Furthermore, they have the advantage that if we know what the correct sequence of actions should have been or can derive it from the goal descriptions, then it is easy to identify any possible deviation and classify it accordingly.

Yet even this simple classification carries with it a number of problems of a methodological nature such as the definition of what a correct action is when time is taken into account, the number of steps that

The Need for Models 72

Figure 2-4: The logical phenotypes.

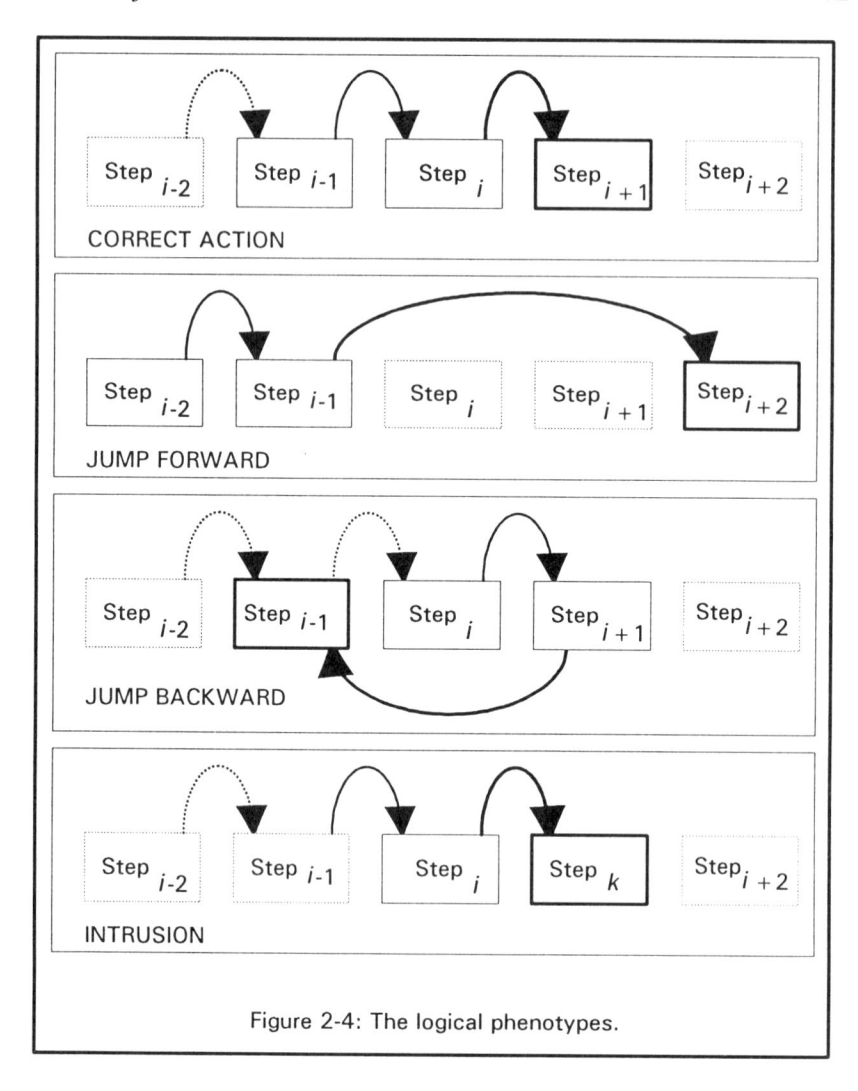

are skipped in either a forwards or backwards jump (leading to more specialised categories such as omission, repetition, and reversal, and what precisely an intrusion is. A more extensive discussion of these details can be found in Hollnagel (1993).

Although the classification has been restricted to consider single and simple sequences of action, these categories are not always intuitively meaningful and even potentially ambiguous. The advantage of the logical classification is nevertheless that it is easy to make. It only requires an identification of the current action and access to a (pre-defined) set of action sequences or plans (i.e., a plan library). Each action can therefore, in principle, be classified unambiguously as it occurs. This ease of rating reveals the underlying conundrum of the classification of erroneous actions. It is clearly desirable to obtain a precise classification not only in terms of the phenotype, but also pointing to the genotype, i.e., the reasons or causes for the erroneous action. Yet this requires a complex interpretation which must consider not only the action as it occurs, but also the context, the background knowledge and - in the case of a human observer - the complete experience. This experience is the only reason why we are able to see actions in their global context and make a fast yet meaningful classification.

5.6 Simple and Complex Phenotypes

One way of achieving a more precise classification in practice is to use a principle based on the idea of different levels of detection. Borrowing a well-known terminology (e.g. Shannon & Weaver, 1969), the phenotypes can be defined in relation to the level of elaboration or inference which is needed to determine them. Thus:

(1) **Zero order (0-order)** detection means that the phenotype can be decided by using only the current action compared with the expected action and/or the prescribed sequence of actions (plan or procedure). There is thus no reference to previous actions or developments (local context).

(2) **First order (1-order)** detection means that the phenotype can be decided by using the outcome of several 0-order detections. In other words, various combinations of 0-order detections which refer to the same sequence of actions or the same plan, can be combined to define a set of more complex phenotypes. This effectively means that the previous actions or developments are also considered.

(3) **Second-order (2-order)** detections can be defined in a similar fashion, i.e., as making use of two or more 1-order detections. In principle the process can be repeated, each step building on the results of the previous. The known proposals for phenotypes require only a limited level or nesting of detections and the elaboration can therefore be stopped after a few iterations.

With this as a basis, it is possible to propose a more extensive set of phenotypes which range from the simple to the complex. The simple phenotypes are:

(1) **Intrusion**, which is defined as the injection of a new action in a sequence of actions. The intruding action may be totally unrelated to the ongoing sequence, and may have serious negative effects. It may, however, also be neutral, hence not have any effects. A useful distinction can be made between intrusions and insertions, where the latter describes actions that do not belong to the action sequence but which do not disrupt it. This is the case when a single, unrelated action occurs in the middle of an action sequence (spurious insertion).

(2) **Replacement** occurs if an action in a sequence is substituted by an equivalent action. The replacement must not be fatal to attaining the goal - although this can only be determined when the following action has been carried out. Replacement may be seen as a special type of intrusion combined with omission, where the intruding action is benign rather than malign. It is, however, reasonable to consider replacement as a separate category.

(3) **Omission** occurs if an action has been omitted from the sequence. The sequence is incomplete and the goal cannot be reached unless the omitted action is reinserted or reproduced, provided that this is still possible. A special type of omission is a failure to complete the sequence. Experience has shown that this may be a frequent contribution to erroneous actions, particularly when more than one plan is carried out simultaneously. A symmetrical case would be the omission of the first action in a sequence. This is assumed not to be nearly as serious as omitting the last action since it only means that the action sequence is not started (or that the first action is missed; but if the actions can continue then, clearly, the first action was not crucial for attaining the goal, i.e., it was optional rather than required).

(4) **Repetition** means that an action has been carried out twice, i.e., it has been repeated. Depending on the situation, this may or may not be crucial for whether the goal can be attained.

(5) **Reversal** means that two actions have been reversed in their sequence. This may have serious consequences for attaining the goal

(e.g., reversing a left-right turn while driving a car). Reversal can actually be seen as an omission of an action followed by the intrusion of that same action.

With that as a basis, a taxonomy of the phenotypes of erroneous actions can be proposed which includes simple as well as complex phenotypes (cf. Figure 2-5). The taxonomy is based on the ability to distinguish between four main cases of incorrect performance or error modes: that an action is in the wrong place in a sequence, that an action is carried out at the wrong time, that an action is of the wrong type, and that an action is not included in the current plans. Reading from top to bottom of Figure 2-5, the following remarks can be made:

(1) **Action in wrong place.** Here the action belongs to the current sequence, but is placed incorrectly.

   (a) One simple phenotype is **repetition**. (This can only occur when the actions are contiguous. Otherwise the erroneous action will be considered as a case of forwards or backwards jumping.) The corresponding complex phenotype is **restart**, i.e., that an already completed segment of the action sequence is repeated, possibly from the very beginning, starting with a jump backwards.

   (b) A second simple phenotype is reversal, i.e., the next two actions in the expected sequence are reversed. The corresponding complex phenotype is jumping, i.e., segments longer than one action are reversed or more than one action is skipped (hence jumping ahead in the sequence).

   (c) A third simple phenotype is **omission**, i.e., that the expected action is missing. There are two complex phenotypes that correspond to this. If more than one action is omitted we effectively have a case of **jumping forwards**. If the rest of the action sequence is omitted, we have a case of **undershoot**, i.e., that an action sequence is not completed as planned. A special case of that is the omission of the last action of an action sequence.

(2) **Action at wrong time.** The simple phenotype omission could also occur if the action was not carried out when it was required. In fact,

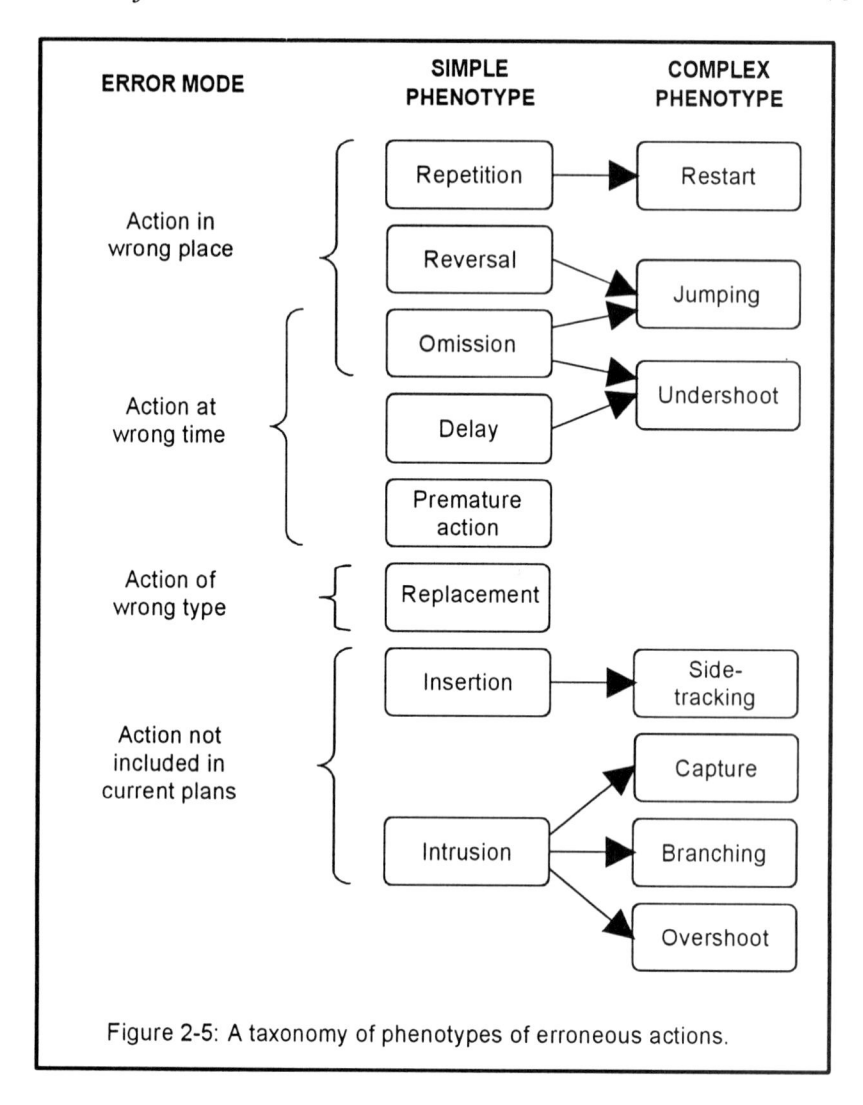

ERROR MODE	SIMPLE PHENOTYPE	COMPLEX PHENOTYPE
Action in wrong place	Repetition	Restart
	Reversal	Jumping
Action at wrong time	Omission	Undershoot
	Delay	
	Premature action	
Action of wrong type	Replacement	
Action not included in current plans	Insertion	Side-tracking
		Capture
	Intrusion	Branching
		Overshoot

Figure 2-5: A taxonomy of phenotypes of erroneous actions.

the classification of actions that occur at the wrong time can in principle be combined with any of the other classifications (wrong place, wrong type, wrong action sequence). There is, however, good reason to consider this classification in its own right. There are two additional cases:

(a) One simple phenotype is **delay**, i.e., that an action does not occur when it is required. A true delay means that the sequence of the following actions is maintained (although it may become subject to temporal contraction, cf. Decortis & De Keyser, 1988). Otherwise the absence of an action may be followed by any of the other phenotypes. If the delayed actions conclude the action sequence the corresponding complex phenotype will be **undershoot**, because the remaining actions in effect are omitted - at least within the time span of the observation.

(b) A second simple phenotype is **premature action**, i.e., an action occurs when no action was expected. The action may be either correct or of any of the known simple phenotypes. (Premature action may, in particular, occur during compression of actions to compensate for an earlier delay.)

(3) **Action of wrong type.** This category means that the action was incorrect although not so wrong that it disrupted the current plan:

(a) The simple phenotype is **replacement**, i.e., that the action is a proper substitute for the expected action. The two actions are functionally equivalent in the sense that the replacement does not invalidate the conditions that must remain for the continuation of the action sequence. The exact criteria for this will depend on the domain and the task. There is no corresponding complex phenotype, but replacement may be of a single action or a segment of actions. (Determining that an action is a replacement may require either additional input from or awaiting an evaluation of the consequences.)

(4) **Action not included in current plan.** Here the action does not belong to the current action sequence:

(a) One simple phenotype is **insertion**, which occurs if the action does not belong to the action sequence and if it does not disrupt it. This is the case when a single, unrelated action occurs in the middle of an action sequence (spurious insertion). The corresponding complex phenotype is **sidetracking**, where a segment or sequence of unrelated actions is carried out before the current action sequence is resumed. Note that this is not the same as a subroutine, which is not an erroneous action. It also differs from branching (and capture) because the original plan will eventually be resumed.

(b) A second simple phenotype is **intrusion**, which occurs if the action does not belong to the action sequence, and if it disrupts it. There are three complex phenotypes that correspond to this. The first is **capture**, which means that the action sequence is effectively disrupted and replaced by a rival (and probably more dominant) action sequence. The second is **branching**, which means than an action segment, which is common to two or more action sequences, is continued along the wrong action sequence (i.e., effectively taking the wrong branch at a decision point). The third complex phenotype is **overshoot**, which means that an action sequence is continued beyond its end point, i.e., after the goal criteria have been satisfied.

(The distinction between insertion and intrusion cannot be made when the event occurs but only after longer action segments have been analysed and the consequences ascertained. To be more precise, if the action sequence is successfully resumed, then the event was an insertion or side-tracking; otherwise it was an intrusion or one of the corresponding complex phenotypes.)

The topics of human action and performance are both at the core of a theory of the Reliability of Cognition, and it is therefore necessary to address the thorny issue of human erroneous actions. The phenotypes are essential for the systematic observation and classification of empirical data, and may in themselves serve as the building blocks of a human reliability analysis method - or as a modification of some of the established methods. Although the phenotypes in some sense are theory independent, or at least independent of specific theories of cognitive functions, they still refer to a set of concepts for human behaviour which conform to generally accepted theories of actions. In this way the phenotypes constitute an important criterion for a model of the Reliability of Cognition. The model must include a description of the genotypes (i.e., the causal mechanisms) that can be used to explain how the phenotypes can occur.

5.6.1   Errors of Omission and Commission

The concepts of genotypes and phenotypes are important as a basis for developing an operationally meaningful classification of erroneous actions.

Such a classification is not only needed as a complement to a model of cognition, but is also essential as the basis for a practical taxonomy of erroneous actions and for human reliability assessment methods.

In the field of HRA an often used separation is between "errors of omission" and "errors of commission". "Errors of omission" are defined as the failure to carry out some of the actions necessary to achieve a desired goal, while "errors of commission" are defined as carrying out an unrelated action which prevents the achievement of the goal. In relation to the terminology introduced here, both of these categories describe rough manifestations or phenotypes; yet they are mostly used as if they described causes or genotypes. "Errors of omission", as a general term, corresponds to several phenotypes of which an omission is only one (cf. Figure 2-5). Similarly, "errors of commission" also correspond to several phenotypes - for instance, repetition, insertion, and intrusion. An "error of commission" will, furthermore, often involve an "error of omission" as well; by doing an unrelated action the person may fail to do the necessary action. Altogether, the concepts of "error of omission" and "error of commission" fail to maintain a distinction between manifestation and cause, and must therefore be regarded as inadequate as a basis for HRA.

6. Human Reliability and the Reliability of Cognition

If we accept the definition that a reliable person is one who performs according to the requirements of the task, then human reliability is almost synonymous to the lack of unwanted, unanticipated and unexplainable variance in the performance. In other words, reliability denotes the degree to which it is warranted to assume that human performance - under given circumstances - will be stable and correct. A highly reliable system is one that performs according to the requirements of the task across a range of different initial conditions and different working conditions. This is quite important because we do not always know in detail all the conditions under which the system may be called upon to function. In this perspective cognitive reliability denotes the effect of cognition on the reliability of human performance as a whole; alternatively cognitive reliability may be seen as the Reliability of Cognition (as a process) *per se*. But it is quite consistent with the view adopted here - the pragmatic *versus* the scientific attitude - that the Reliability of Cognition should not be considered for cognition in isolation.

The use of the piecemeal, quantification approach may be questioned even for technical systems. It may be appropriate for systems which are non-interactive, but it does not follow that this is the case for interactive systems as well. The whole decomposition principle is based on the clockwork analogy which has dominated scientific thinking at least since the beginning of the industrial revolution (Dawkins, 1988). It would be incorrect to say that this approach has not been successful - it obviously has. But it may be sensible to question whether this approach should also be allowed to dominate in the behavioural sciences. It has put a distinct mark on industrial/organisational psychology in the guise of Taylorism (Taylor, 1911) - which is far from dead - and has heavily influenced other parts of the behavioural sciences as well, for instance in the notion of the mind as an information processing system which most recently has been epitomised in the Unified Theory of Cognition (Newell, 1990). The relevance of this underlying assumption - the clockwork analogy, the decomposition principle, or whatever it may be called - should therefore thoroughly be analysed and considered before it is brought to bear on something as complex and non-mechanical as human performance.

6.1 The Changing Nature of Tasks

Taylorism - or the Principles of Scientific Management - was based on the notion that tasks should be specified and designed in minute detail and that workers should receive specific instructions about how the tasks should be carried out. The steps in Scientific Management can be summarised as follows:

(1) The task should be analysed scientifically, if possible in quantitative terms, to determine how each task step should be done in the most efficient way, and how the task steps should be distributed among the people involved.

(2) The people (workers) should be selected to achieve the best possible match between task requirements and capabilities. Almost as a curiosity it was stressed that although workers should be physically and mentally capable of the work, care should be taken that they were not overqualified for the work!

(3) People should be trained carefully to ensure that they would perform the work exactly as specified by the prior analysis. In other

words, care should be taken to ensure that performance was reliable *vis-a-vis* the defined norm!

(4) Lastly, compliance with the instructions should be ensured by means of economic incentives or rewards.

With the exception of the last step, the principles of Scientific Management are consonant with the decomposition principle and the clockwork analogy. The person is seen as a relatively simple mechanism and the tasks to be carried out can be described and designed in the same way. It follows that the reliability of human performance can be analysed and assessed along the same lines. This conclusion is warranted to the extent that the work complies with the assumptions of Scientific Management. But the nature of work, and in particular the nature of tasks, has changed significantly since the beginning of the century. The development can be summarised in several ways (Welford, 1969), but will here be described by the following four phases:

(1) **Mechanisation** increased the efficiency of individual work and also led to a specialisation of the activities of the individual worker. Mechanisation introduced the use of power-driven tools on a large scale and was at the root of the industrial revolution. It was the paramount feature for more than a century, and brought the concept of tasks and task elements to the fore, as epitomised by the principles of Scientific Management described above.

(2) **Automation** gathered the elemental processes into clusters (sub-systems) that could be controlled by an automatic system rather than by a human worker. Simple regulators existed even before the invention of electricity and electromechanics (e.g. Watt's Governor), but it was not until the appearance of mechanised logic and electronic components that automation gained speed. Today's automation is dominated by information technology and microelectronics. Automation generated a concern for the proper specification and representation of process information, and for the development of control systems, particularly when it was reinforced by centralisation.

(3) **Centralisation** created large centralised systems that required high levels of automation. Centralisation is no longer confined by physical limitations, but may use information technology to link

components and sub-systems regardless of their geographical distribution; as an example, consider the use of satellites for remote sensing and communication. The complexity of the systems, however, has effectively defied attempts to achieve complete automation and spurred an interest in risk, system reliability, and decision making.

(4) **Computerisation** enabled a significant expansion of automation as well as of centralisation. In addition to increasing the speed and complexity of computation *per se*, computerisation enabled the use of very large databases in both off-line and on-line conditions. It also increased the possibilities for actively designing the interface between human and machine, and providing sophisticated support functions. Unfortunately, this created an equally large number of possibilities for making simple tasks unnecessarily complex and for creating task requirements that are clearly beyond human capacity (e.g. inherently unstable aircraft). The reliance on information technology, and in particular software, has quite rightly created a concern for human reliability, functional specification, and verification.

(Human Computer Interaction (HCI) has only appeared as a separate discipline during the last part of the computerisation phase. It does not address the fundamental issues of joint systems and the use of intelligence. But it demonstrates that the interaction between humans and artifacts has grown in importance.)

Through all these steps there was a corresponding change in the nature of human work, going from a dependence on manual skills to a dependence on knowledge intensive functions (i.e., tasks of a cognitive nature), cf. Figure 2-6. Human work can appropriately be characterised by a scale going from "doing" to "thinking": some tasks will require much "doing" and little "thinking", while others will require much "thinking" and little "doing". Examples of the former are found in manual skills, procedure following, etc.; examples of the latter, the knowledge intensive tasks, are found in diagnosis, planning, and problem solving.

The development of modern information technology has significantly changed the proportions of "doing" and "thinking" in human work; the trend is that the amount of "thinking" is increased and the amount of "doing" reduced. Yet at the same time the scale has moved so

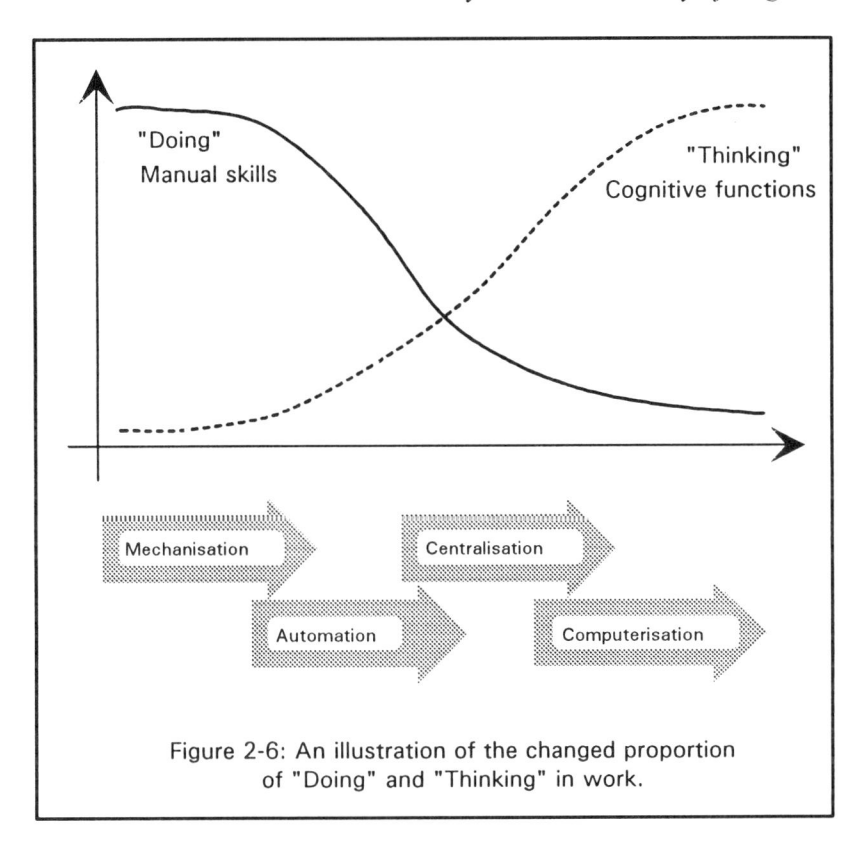

Figure 2-6: An illustration of the changed proportion of "Doing" and "Thinking" in work.

that what previously required "thinking" now only requires "doing"; a good example is calculations - simple as well as complex. As a consequence specific "doing" tasks have disappeared altogether, usually because they have been completely automated or been taken over by robots (newspaper typesetting, welding, spray painting, etc.).

Conversely, new "thinking" tasks have appeared which either did not exist before (e.g. programming) or which only were done to a very limited extent (e.g. long-term forecasting). Although there are many specific tasks which can become completely automated, there is probably an upper limit on how far automation can be taken for a system as a whole without jeopardising reliable performance. There seems to be a constant trickle of major accidents - some of which turn into real disasters - which in the main are due to an imbalance between the level of computerisation and the nature of the task. Current examples include cockpit automation, ships, anaesthesiology, and radiology. The inevitable

The Need for Models 84

dependence on human intelligence and human action for the safe and efficient use of complex systems underscores the importance of furthering the study of the Reliability of Cognition.

6.2 Task Change and Function Amplification

This can be illustrated by considering how each of the four above-mentioned stages has had effects on the nature of human work, on the interaction between man and machine, and hence also on the reliability of human performance in such interactive systems. A common denominator is the way in which each of the above-mentioned stages of development has led to a specific form for amplification.

(1) Mechanisation led to the amplification of force, reach, speed, of accuracy in movements and manipulation, etc. Although there had always been machines which could amplify human action (exertion of force and manipulation) as well as perception (e.g. the microscope and telescope), it was the industrial revolution and the mechanisation that followed which brought the amplification into practically every field. An outcome of this type of amplification was that the unwanted consequences of inadvertent actions also grew. This led to an appreciation of the new risks, for instance in insurance, and thereby also to a concern for how inappropriate human action could endanger the functions of a system.

(2) Automation led to an amplification in precision, reliability, efficiency (more steady maintenance of system states), and - without really intending to - complexity: automation reduced the need to follow all details of a process, hence increased both the number of processes that could be handled as well as their complexity. Automation started in the field of industrial production but has now spread to administration and control and communication in general. In most cases automation has been used to increase efficiency rather than safety, an unhappy development which has been expressed by the theory of risk homeostasis. One consequence of automation was that the speed of processes was no longer restricted to what the unaided human could control. Neither was the complexity, expressed in terms of coupling and dependencies between processes and sub-processes. Automation was quite often introduced on the premises of the technology rather than the premises of the person: a

common way of expressing this is by noting that automation more often was applied to what **could** be automated than to what **should** be automated. Thereby the systems actually became very dependent on how the person performed when they went beyond the envelope of automation - which could happen either because automation was limited or because it failed. The reliability of the overall system therefore had to consider the reliability of human performance as a controller and monitor.

(3) Centralisation led to an amplification of range, scope, and efficiency - but in abstract rather than physical terms. Centralised control made it possible to use resources more efficiently, to maintain higher productivity by coupling events at a large number of processing sites, and to improve safety. Centralisation went beyond automation by strengthening hierarchical command and control. This leads to a certain inertia in responding to changed conditions, and centralisation is therefore only feasible if the environment is sufficiently stable - either because of its inherent features or because it is effectively controlled. Centralisation reinforced the dependence on human reliability that was created by automation; and the importance of "thinking" functions grew as the person was further and further removed from the actual process.

(4) Computerisation has undoubtedly been the greatest boost to human work since mechanisation. Computerisation had effects on both mechanisation (which was brought to new heights), automation, and centralisation where it has increased the gains already made. Computerisation has also led to a substantial improvement in the use of human intelligence. Although computerisation has the potential for intelligence amplification *per se*, the effect so far has rather been that the availability and stability of knowledge and knowledge processing increased, e.g. through the use of expert systems of various types. The thrust has mainly been on automating functions which hitherto had been unattainable (essentially those depending on knowledge processing and cognition). Due to computerisation whole fields of specialisation or skill, such as typesetting, technical drawing, etc. have become obsolete. One particular effect of computerisation has been the introduction of artificial cognition - although still on a modest scale.

The Need for Models

One consequence of the development of sophisticated information technology systems, and the gradual introduction of intelligent artifacts, is that the issue of system reliability can no longer be viewed simply as an issue of the Reliability of Cognition. The traditional view is expressed in Figure 2-7. Here the reliability of the joint system depends on the reliability of the (mechanical) system, the design and functioning of the man-machine interaction and the interface, and the reliability of the person's performance and cognition. Each of these have been dealt with, with varying degrees of success, by established disciplines.

In an advanced information technology system the Reliability of Cognition applies not only to the person, but also to the intelligent artifacts in the system. In other words, we have to contend with the reliability of both natural and artificial cognition. Although each can be treated separately, one must also consider the interaction between them, i.e., the way in which human reliability depends on the man-machine interaction, which in turn contains a component of artificial cognition (Figure 2-8). In practise the reliability of artificial cognition is only treated indirectly, e.g. in terms of knowledge base verification and validation (Ayel & Laurent, 1991; Hollnagel, 1989b), and in software reliability. The

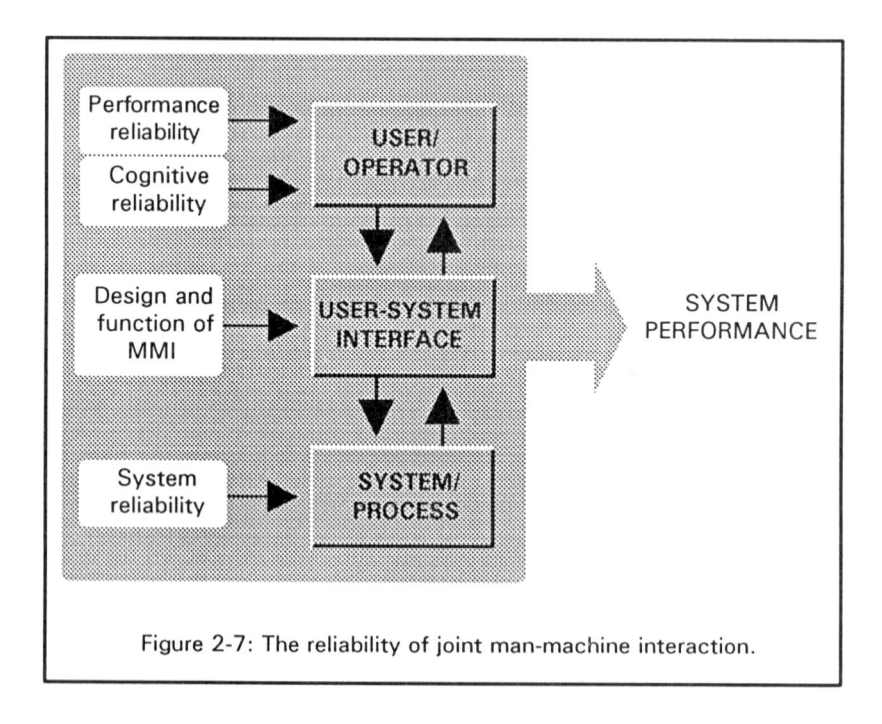

Figure 2-7: The reliability of joint man-machine interaction.

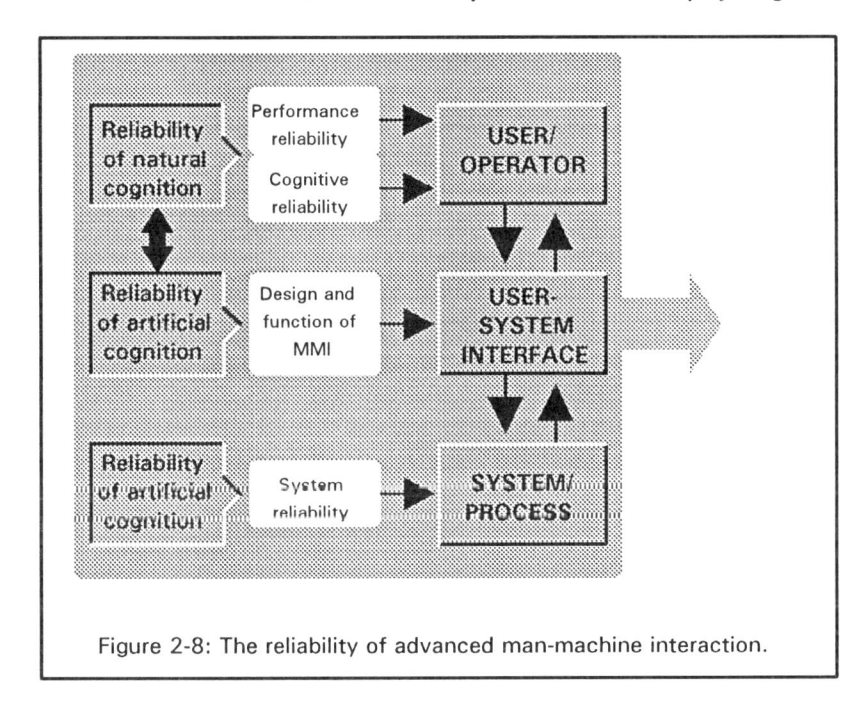

Figure 2-8: The reliability of advanced man-machine interaction.

inner or authentic issues of artificial cognition are not usually given serious consideration.

As a consequence of that it becomes even more important to pay attention to the context in which the work takes place. This context is not always stable and benign, but can be dynamic and relatively unpredictable. This means that the conditions for human cognition may depend on how an alien and unfamiliar "mind" behaves. In advanced systems the person is not only trying to cope with a dynamic process, but also to some extent with an artificial adversary who controls that process in ways which are not easily foreseeable. The precautions that are needed here should also carry over to less advanced systems; it is more reasonable to expect that a solution which can cope with a complex system can be used for a simple system as well, rather than the other way around.

6.3 The Social System Analogy

In order better to grasp the consequences of the decomposition principle it is useful to consider it in another context. Consider, for instance, the behaviour of groups, such as a crew in a control room. Groups - and beyond that societies - are made up of people. But no one assumes that beyond that societies - are made up of people. But no one assumes that the behaviour of a group or any other social unit can be explained by referring to the behaviour of individuals *qua* individuals. Even if individual behaviour was well accounted for, a simple aggregation would not be able to explain the behaviour of groups. Although it somehow makes sense to argue that group behaviour in principle must be explainable in terms of individual behaviour, it is not practically feasible.

The argument would go something like this: the behaviour of an individual in a group is determined by the interaction between the individuals in the group, both in terms of what the group does to the individual and what the individual imagines the group could do. Therefore, in principle, if we could account for all the ways in which actions and reactions could be perceived, understood and imagined, as well as the ways in which this determines individual behaviour, then we should be able to account for the behaviour of the group (cf. Figure 2-9).

As it is, there are a number of dominant and generally true phenomena that characterise group performance and which, furthermore, can be used to understand group performance (e.g. Porter *et al.*, 1975). While in principle these may be reduced to individual psychological functions, it is in practice sufficient to remain on the level of group functions. Nothing is gained by decomposing the functions of the group, because the principles of decomposition are insufficiently known. (Strangely enough, no one would think of trying to explain the behaviour of an individual by reasoning from the behaviour of the group - based on decomposition alone. The behaviour of an individual in a group may, of course, be dominated by group phenomena, but that is a different matter.) In other words, applying a decomposition principle without sufficient rationale is unsound. But that goes for the application of the decomposition principle to individual behaviour as well. Just as we cannot explain the group solely in terms of its components (the individuals) neither can we explain the individual in terms of its components (which are assumed to be the cognitive functions). Instead, the phenomena must be described on the level at which they occur. There is a significant difference between social and individual phenomena, hence also between the theories and models needed to account for them. Similarly, tasks cannot be decomposed into sub-tasks and actions which can then be treated individually. A task is a context, and must be treated as such.

Basically, human reliability analyses, and probably also probabilistic safety assessments, cannot be accomplished by decomposing a task into

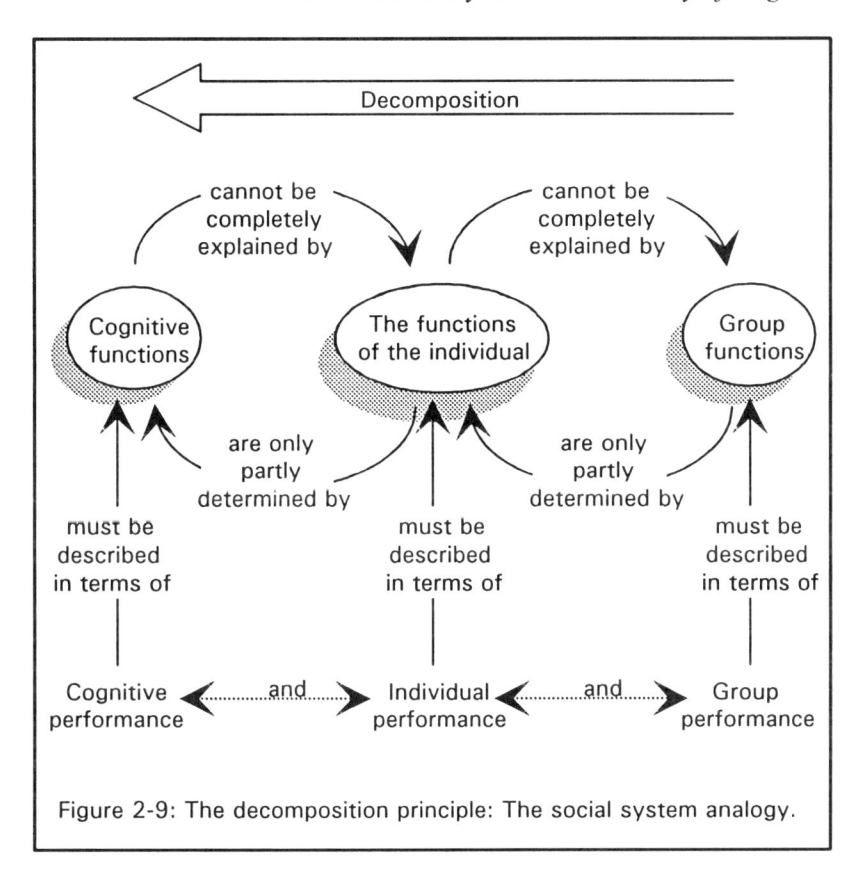

Figure 2-9: The decomposition principle: The social system analogy.

part of a context. Similarly, human reliability analyses should not be done as if human action was probabilistic. Human action may in some sense have a stochastic component (cf. the notion of residual erroneous actions), but it does not mean that human action as a whole is a stochastic phenomenon. If human action was probabilistic it would follow that given sufficient evidence or sufficient empirical material it would be possible to reduce the uncertainty to an arbitrary small level. However, if human action contains a stochastic component, the uncertainty of the description is not due to lack of knowledge but rather to the inherent nature of human behaviour. Therefore, the uncertainty of the description cannot be reduced. Furthermore, meaningful probability values cannot be given for single actions.

This has the interesting consequence that the notion of performance shaping factors becomes superfluous. Performance shaping factors are

This has the interesting consequence that the notion of performance shaping factors becomes superfluous. Performance shaping factors are introduced precisely because the task is decomposed into lower level actions and the context thereby is removed. The analysis must therefore be enhanced by performance shaping factors which, in a sense, brings back (although in a simplified form) what was taken away by the analysis. It follows that if the analysis is made in a context, focusing on the task as a whole, then performance shaping factors by definition become superfluous. Because there is no decomposition nothing has been removed, and it is consequently neither necessary nor possible to put it back in again.

7. Summary

The field of human reliability analysis and probabilistic safety assessment is dominated by a practical "engineering solution" which has been widely criticised - by HRA practitioners as well as by academics - as having inadequate theoretical and conceptual foundations. The "engineering solution" essentially carries over the practice from probabilistic safety assessment to human reliability analysis and applies the fundamental decomposition principle to humans and machines alike. Humans do, however, differ from machines in many essential ways, even if considered only from the point of performance reliability. Any definition of human reliability must clearly reflect that.

Human reliability analysis relies very much on the estimation of events, both which events are likely to occur and how probable they are. Although all analytical methods must begin by identifying and selecting a set of potentially important events, the current approaches to human reliability analysis soon turns into a quest for quantification. Many of the efforts in the practical methods are therefore aimed at improving the basis for event estimation, even to the extent of providing models to generate the probabilities.

When event estimation is extended from aspects of the physical system to aspects of the psychological and cognitive functions, several new problems emerge. Firstly, psychological functions and processes are inferred rather than observed; secondly, the level on which a function is described (the number and the nature of constituent steps) may produce a serious but unintended change in the results. This is partly due to the fact that the fundamentally engineering methods have been indiscriminately applied to cognitive processes. The study of reliability in technical systems is, however, based on two main assumptions which are not met in non-technical systems: repeatability and similarity. Repeatability means that the function of a component can be tested again and again under identical conditions. Similarity means that identical test conditions can be established, and also that the test conditions and the real conditions do not differ significantly. Neither assumption can be maintained for human action, except for very reduced or extreme conditions.

The decomposition principle implicitly makes some assumptions about the nature of human cognition and about what may cause a person to perform inappropriately. When unwanted consequences appear it is often because the people working in the system have done something which - in retrospect - turned out to be wrong. This has commonly been referred to as "human error", and during the 1980s there was a surge of interest for this phenomenon. It is, however, debatable whether the simplified concept of "human error" that most theories propound is very useful. "Human error" can be understood as being both a manifestation (a phenotype) and a cause (a genotype), and this duality is often confusing. The former should more properly be called a human erroneous action, and treated as a manifestation. The latter should be treated as the (assumed) cause for the manifestation and should be an integral part of an understanding of human cognition - rather than as more or less isolated speculations about specific error causing mechanisms. Observation and introspection are often confused as data sources, leading to a conceptually imprecise terminology.

In order to overcome this a systematic classification of the phenotypes of erroneous action is proposed, starting from a consideration of the ways in which actions can be permuted and described by common sense terms. The classification is based on what can be observed with a minimum of assumptions. It starts with four basic categories (correct action, jump forward, jump backward, and intrusion) and extends these to provide a number of definitions of both simple and complex phenotypes together with a suggestion for their application. It is illustrated how the commonly used concepts of slips, mistakes and lapses, as well as errors of omission and commission, all fail to maintain the basic distinction between manifestation and cause. This naturally makes them difficult to apply consistently, hence inadequate as a description taxonomy.

The decomposition principle implies that man can be seen as a machine - although possibly a fallible one. This view can be traced back to the principles of the so-called Scientific Management that flourished in the beginning of the 20th Century. The nature of work has, however, changed significantly since then. This was described by characterising four phases called mechanisation, automation, centralisation, and computerisation. Each of these were briefly characterised and discussed in terms of how they have changed and amplified human functions and in terms of their impact on the issues of human reliability. It was pointed out that the change in the nature of tasks can be described in terms of an altered balance between "doing" and "thinking" tasks, with a growing preponderance of the latter. This means that human performance becomes more dependent on cognition, and that human reliability assessment must take this into account in terms of analyses as well as event estimations. The situation is further complicated by the introduction of artificial cognition in the guise of expert systems and other types of artificial intelligence support. The natural and artificial cognition are closely coupled, which makes it even more important that human reliability analyses consider the whole task and the context rather than the task components. In order to do that it is necessary to have a proper model of the Reliability of Cognition.

3.
The Nature of Human Reliability Assessment

1. Engineering Quantification

Human reliability analysis is usually considered to be an engineering discipline or technique, and by virtue of that assumed to be intrinsically mathematical. Or perhaps it is more correct to say that many practitioners of human reliability analysis strive to be recognised as exponents of an engineering, quantitative discipline. One may, however, seriously question whether engineering is intrinsically mathematical. This is probably due to a confusion between engineering as a scientific discipline and the engineering methods; the latter are mathematical (quantitative) whereas the former is not. It may also be due to a common confusion between the meanings of the terms "mathematical" and "quantitative" and to a lack of appreciation of the real distinction between the meanings of qualitative and quantitative approaches.

Quite apart from that, while the quest for quantification may be sound within the realms of probabilistic safety assessment where analyses are confined to systems which do not include human-machine interaction as part of their functionality, it is not necessarily sensible in the case of systems which do depend on human action to achieve their aim. This restriction is quite serious because it both excludes the category of interactive or joint systems from consideration and means that the analysis is restricted to systems that are in operation; other parts of the life-cycle such as design and maintenance are ruled out since these, by necessity, include and depend on the performance of humans. The quest for quantification is particularly unhealthy in the case of human reliability analysis which, by definition, must include human action or even have it as the main focus. As Kantowitz & Fujita (1990) clearly point out, the current stock of psychological theories and models of human action (and with them the ubiquitous notion of a "human error") are not able to support a proper mathematical or quantitative approach. For that reason alone the ambition of engineering quantification must be abandoned.

This state of affairs may be lamentable for several reasons - one of them being that life would be so much easier for all of us if engineering quantification was possible. But it need not lead to despair either for the practitioners of human reliability analysis nor for the theorists. Furthermore, it should not automatically lead researchers to ponder better ways of developing rigorous engineering or mathematical models or new methods to generate quantitative data. The impotence of psychology and human factors to achieve the goal of a complete engineering quantification ought rather to make us wonder whether the right question has been asked and whether the right problem is being attacked. That, in turn, means that it is necessary to take a closer look at the underlying assumptions.

These assumptions were briefly introduced in Chapter 1, in the description of the artifacts of decomposition, and also described in Chapter 2. To simplify the argument, although thereby possibly overstating the case, the assumptions underlying the use of quantification can be reduced to the following:

(1) **The Atomistic Assumption**: Human performance can be adequately described by considering the individual elements or parts of the performance. A description of the total performance can be produced by aggregating the descriptions of the individual performance elements.

(2) **The Mechanistic Assumption**: The human mind can be adequately described as an information processing system, i.e., the human mind, as well as man in general, can be described as a machine.

The atomistic assumption is present in practically all models and theories of human performance that are applied today in the field of Man-Machine Systems (MMS), and it dominates current approaches to human reliability analysis. A prototypical representation is shown in Figure 3-1; it can easily be replaced by any of the common representations of operator action trees or event trees. The basic notion is that the whole is equal to the sum of the parts, and that one can therefore conveniently replace a study of the whole by a study of the parts taken one by one.

1.1 The Atomistic Assumption

There are two flaws in this argument. Firstly, the whole is usually more than the sum of the parts simply because the interaction between the parts is important. This interaction is, however, lost in the commonly used representations and is consequently also missed by the analyses. A diagram such as Figure 3-1 only shows the limited interactions which can be expressed by the arrows that connect the various boxes. The semantics of the arrows is usually not clearly defined; there is, for instance, an important difference between control and the transfer of information. Neither are the more complex types of interaction represented explicitly; for example, feedback, belief propagation, etc. The whole is, furthermore, an exceedingly complex system of relations of which we can only capture a finite, and usually rather limited, subset for analysis (corresponding to a homomorphism of the system). Even if the interaction between the parts played a negligible role, we would have great difficulties in constructing an adequate representation of the whole.

The atomistic assumption is usually attributed to the Greek philosopher Democritus, but in relation to the division of work a more recent source is found in the theory of Scientific Management (Taylor, 1911). As described in Chapter 2, the tenets of Scientific Management claim that a task can meaningfully be decomposed into segments, and that each segment can be described by its attributes. It follows from this that the task itself can be described by aggregating the descriptions of the segments. This notion of task decomposition corresponded well to the engineering view on a machine's functioning as composed of the functioning of the elements, which in turn is based on the idea of the human as a deterministic machine.[1] This view was reinforced by the apparent success when the decomposition principle was applied to the analysis of mechanical systems and processes. The atomistic assumption is therefore closely connected to the decomposition principle (cf. Chapter

---

[1] It is also worthwhile to point to the similarity with educational theory or didactics, according to which a topic can be taught by breaking it down into elements; however, in this view the relations between the elements do play a role.

The Nature of Human Reliability Assessment

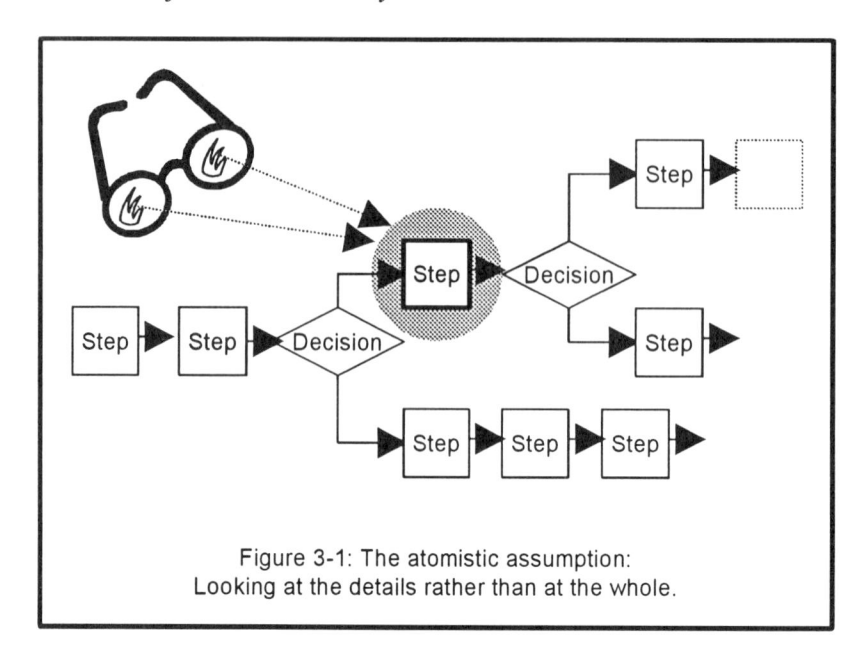

Figure 3-1: The atomistic assumption:
Looking at the details rather than at the whole.

1). The atomistic assumption can also be detected in the Stimulus-Organism-Response (S-O-R) metaphor of humans which is found in many different guises (discussed later in the chapter).

1.2 The Mechanistic Assumption

The mechanistic assumption is very clearly expressed by the information processing paradigm, according to which a human mind can be adequately described as if it was an information processing system. A classical version is found in Newell & Simon (1963) who conclude a description of the General Problem Solver by declaring that the information processing techniques:

> "finally reveal with great clarity that the free behavior of a reasonably intelligent human can be understood as the product of a complex but finite and determinate set of laws."
> (Newell & Simon, 1963, p. 293)

The mechanistic assumption has been with cognitive psychology ever since enthusiastic psychologists embraced information theory (e.g. Attneave, 1959; Miller, 1956), and has been reinforced by the formalistic

Engineering Quantification

approach of classical Artificial Intelligence (e.g. Newell, 1980) reaching its highest point so far in the so-called Unified Theory of Cognition (Newell, 1990). The mechanistic assumption is in itself an expression of the atomistic approach, because it assumes that all human performance can be decomposed into a number of elementary steps. This notion has be caricatured by Neisser (1976), as shown in Figure 3-2; the information processing view is also prevalent in the notion of a procedural prototype, discussed in Chapter 4. It unfortunately amplifies the atomistic assumption by providing a seemingly plausible theory of the human mind which furnishes the missing parts that make engineering quantification temptingly easy to apply.

The view of the human as an information processing mechanism may seem less simplistic than the black box approach that is part and parcel of the S-O-R metaphor, but is in reality not so. The information

Figure 3-2: The mechanistic assumption (the human information processor).

processing metaphor is certainly more elaborate in the sense that it can be used to depict information processing descriptions of subjectively complicated phenomena such as vehicle control, diagnosis, and decision making. But it can easily be shown that the power of this representation depends more on the interpretation and contribution of the observer/reader than on the representation in itself. Put differently, if the representation was interpreted literally by someone without a detailed knowledge of the field, the lack of details would become obvious; the semantics of the information processing paradigm are supplied by the user and are not carried by the representation itself. We interpret it as being more complex because it alludes to more complex phenomena in our experience, but the complexity is only apparent. (This problem is discussed again in Chapter 4.)

The criticism of the IPS view should not be taken to mean *a priori* that an accurate account of human cognition is impossible. If that was the case there would be no point in writing this book. The criticism means that it is not sufficient to base such an account on the assumption of elementary cognitive components. Detailed theories of particular phenomena can be expressed in terms of information processing, but the problem is precisely that they are just that - detailed theories of particular phenomena. Elementary components, whether of cognition or information processing, are not enough to explain what happens on the level of cognition at work. Information processing, furthermore, is insufficient to describe cognition (e.g. motivation, intention, personal knowledge). **It may sometimes be convenient to describe part of what happens in the mind in terms of information processing, but that does not mean that we actually process information!** (That phrase, by itself, is in dire need of an explanation.) The accounts we produce do not automatically become accurate if we use a specific terminology. It is more important to pay attention to **what** is being described than to **how** it is described. The only acceptable indicator of accuracy is compliance with the phenomenon as it is known and as it can be observed - rather than compliance with a specific theory.

2. Differences Between Humans and Machines

In the discussion of the reliability of man-machine interaction, it is important to be aware of the differences between humans and machines; a machine can here be any technological artifact. There are obviously many differences, depending on the point of view. In the context of human reliability analysis, the following are certainly important. Several of these will be discussed throughout the book:

(1) **Principle of function**: We may for pragmatic reasons assume that both humans and machines function according to the principle of causality although the assumption is only indisputable for machines. For machines we can further assume that the principle of function is strictly deterministic, whereas for humans it is mostly indeterministic - perhaps as a consequence of our lack of knowledge. For humans the principle of function can be inferred, but the outcome is strongly dependent upon the conceptual basis.

(2) **Design**: The design of humans is implicit and very hard to formulate. This is true on any level going from the DNA to the complete, living organism. Still, it is generally assumed that a principle of design can be found, and an impressive amount of basic research is dedicated to that - particularly in the biological sciences. In stark contrast to this the design of machines is explicit and usually very well formulated. The design is fundamental to the principle of function; if the design is known, so is the principle of function.

(3) **Availability of models**: Relevant models of humans are few and far apart (cf. the extended discussion in Chapter 4); furthermore, they are mainly descriptive. Models of machines are many and varied, and can be normative as well as descriptive. Models used for design are obviously normative; models used for analysis can be normative as well as descriptive.

(4) **Range of applications**: A human is clearly an all-purpose system who is able to adapt to a wide range of tasks and working conditions. A machine has few applications with a limited range, but usually does them well. If a new task needs to be solved it will normally require either substantial modifications to the existing design or the design of a completely new machine.

(5) **Failure modes**: It is possible to provide a systematic description of the failure modes of humans, although there is no single classification that is generally accepted. Human failure modes are, however, highly unpredictable both with regard to type and occurrence. The very notion of failure modes for humans may,

indeed, be a misconception derived from the mechanistic assumption. As argued in Chapter 2, even the notion of the human as a fallible machine is misleading. In contrast to that failure modes for machines are both describable and to a large extent predictable, for instance based on calculations or models of gradual degradation.

(6) **Capacity**: The capacity of humans will vary over time, but may both increase and decrease. The direction of the variations is unpredictable and the range may be considerable - and certainly beyond what most system designers seem to imagine. The capacity of a machine usually is constant with a fixed upper limit, but may diminish slowly over time - or abruptly disappear altogether.

(7) **Reliability**: The reliability of humans is generally low and is only partly measurable. The reliability of machines is usually high, because they have been designed for that, and is furthermore measurable.

It may seem surprising to claim that the reliability of humans is low rather than high, since many people will have the opposite opinion. The seemingly high reliability of humans is due to the fact that we are extremely good at recovering from erroneous actions and of finding a way out of tricky situations. However, if we seriously consider the number of erroneous actions, large and small, that we make during a day, including the multitude of unreported near-misses, I think it is warranted to state that human reliability generally is low. Consider, for instance, all the artifacts (memory aids, calculators, personal assistants, etc.) that we need to get us through the daily drudgery.

Taken separately each of these differences, summarised in Figure 3-3, serves as a reminder that it is no straightforward matter to transfer methods from the engineering world to the study of human reliability. Taken together they make the rather strong point that is not possible to treat the human operator in the same way as a machine or an engineering process; the differences are many and fundamental for the methods that could be applied. A mechanical component can be studied under strictly controlled conditions, alone or as part of an assembly, until it begins to malfunction. The procedure can be repeated with any number of small variations. The important parameters can be defined and manipulated so that a solid empirical basis can be established which allows the calculation of the component's reliability. Even if it is impossible to examine all

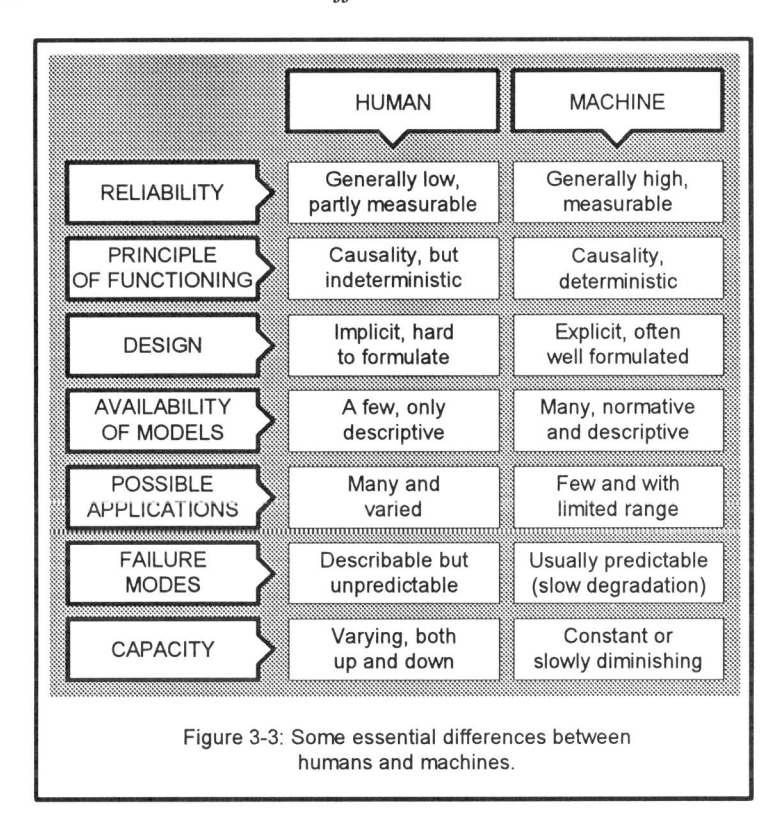

	HUMAN	MACHINE
RELIABILITY	Generally low, partly measurable	Generally high, measurable
PRINCIPLE OF FUNCTIONING	Causality, but indeterministic	Causality, deterministic
DESIGN	Implicit, hard to formulate	Explicit, often well formulated
AVAILABILITY OF MODELS	A few, only descriptive	Many, normative and descriptive
POSSIBLE APPLICATIONS	Many and varied	Few and with limited range
FAILURE MODES	Describable but unpredictable	Usually predictable (slow degradation)
CAPACITY	Varying, both up and down	Constant or slowly diminishing

Figure 3-3: Some essential differences between humans and machines.

combinations and all environment conditions, a representative set can usually be defined. (In this context, the term "mechanical" is used to refer to components that are physical objects. As the discussion in Chapter 1 pointed out, there may be similar problems for the reliability of non-mechanical components such as software.)

The same procedure cannot be applied to human beings. We only have vague ideas about their principle of function, we know nothing of their design, we have few models available, we cannot delimit their range of applications, we are uncertain about the failure modes, there are no fixed upper and lower bounds for their capacity, many of the essential functions are inferred rather than observed, and the reliability is only partly measurable. A human being is definitely **not** a mechanical system and neither can nor should be studied in isolation. It must be a consequence of the differences described above that the human cannot be modelled in the same way as a component or an aggregation of components; neither the atomistic assumption nor the mechanical assumption are correct. Human performance is a result of a complex interaction with the context which lies beyond the range of simple deterministic modelling. The solution is therefore not to search for better mathematical or quantitative models, but rather to recognise the fundamental differences between humans and machines, and develop a basis for human reliability analysis which takes these differences into account.

The difference between the two views can be emphasised by making a distinction between a decomposition approach (read: engineering) and a holistic approach (read: cognitive) to the study of human reliability. The essential feature of a holistic approach is that the dynamics of a living whole cannot be explained as resulting from independent elements. The decomposition approach is basically quantitative and the human operator is considered as a component in a complex system. The holistic approach is based on the explicit use of theories or models of the cognitive functions which constitute the substratum for human behaviour. The assessment of human reliability is therefore based on causally meaningful categories rather than on elements which only appear similar in the way they manifest themselves.

3. Identifiable Models *versus* Curve Fitting

The importance of the problems outlined above is illustrated by the discussion of quantitative models of cognitive psychological theories by Kantowitz & Fujita (1990). They point out that the problems arise because, on the one hand, "engineers would like to treat the human like any other system component" (Kantowitz & Fujita, 1990, p. 325) while, on the other hand, existing models and theories of human cognition and action are not nearly well enough developed to support quantification. They illustrate the problem by making a distinction between curve-fitting (estimation) and model identification, as illustrated in Figure 3-4. Curve-fitting consists in fitting some mathematical model to a set of data. This may be a laborious exercise, but can usually always be done. The possible success in achieving a fit does not, however, say anything about whether the model is a proper description of the process behind the data. Model identification, on the other hand, consists in finding a one-to-one correspondence (an isomorphism) between the structure of the model (the model parameters) and the data set. In this case the model must clearly define the relationship between a parameter and the corresponding model

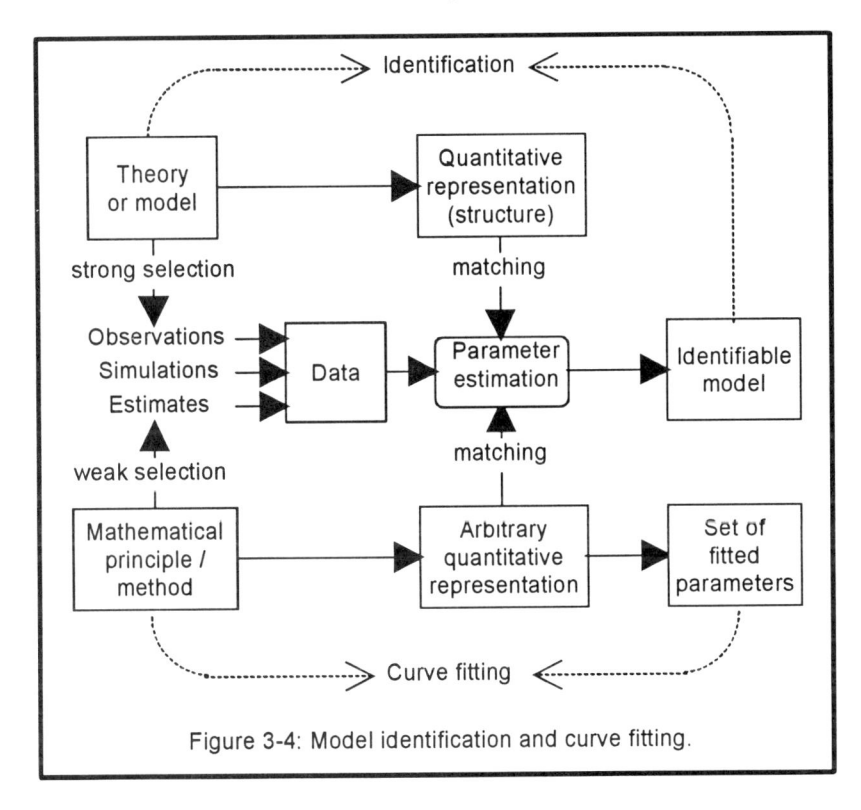

Figure 3-4: Model identification and curve fitting.

function (for instance, a type of mental process). Here the inability to obtain a model identification essentially falsifies the model, whereas a successful model identification increases the belief in the model (without, of course, in any way proving that the model is correct).

The distinction between the two methods is illustrated by means of a study using a full-scale simulator of a nuclear power plant to investigate the Human Cognitive Reliability (HCR) model (Hannaman & Worledge, 1988). The HCR model was developed to enable PSA analysts to quantify the reliability of operator crew responses for accident sequences in nuclear power plants. The HCR is basically a normalised time-reliability curve; a time-reliability curve (or time-reliability correlation) expresses the probability of a response (or of a non-response) as a function of elapsed time after an event. Thus the longer the time, the smaller the probability of a non-response. The shape of the curve is determined by the categories of cognitive processing that are associated with the tasks being performed. The HCR model actually consists of a set of curves, corresponding to the categories of skill-based, rule-based, and knowledge-based performance (Rasmussen, 1986).

The outcome of the study described by Kantowitz and Fujita was that although it was possible to obtain a satisfactory fit between a three-parameter Weibull function and the latency data, a model identification of the HCR model could not be made. The conclusion was that "attempts such as the Human Cognitive Reliability model are premature. Quantification must await better theory." (*Ibid*, p. 324). Later studies have only confirmed this conclusion (Spurgin & Moieni, 1991).

This outcome of the experiment is used by Kantowitz and Fujita to discuss the way in which human reliability analysis should develop. The discussion is very frank, and reveals the basic assumptions (although the authors do not go on to question them). Kantowitz and Fujita maintain that "HRA, like any respectable engineering technique, is intrinsically mathematical." (p. 318); they also state that "the equations that explain complex human behavior are too complex for analytical solutions." (p. 326); this leads to the "final" problem of how engineers should incorporate psychology into pragmatic engineering models.

The fundamental assumptions put forward by Kantowitz and Fujita are that, to most engineers, human reliability analysis is intrinsically mathematical, i.e., quantitative, and further that psychological models and theories should be expressed in terms of sets of equations. It is hardly surprising that these assumptions create problems for engineers or others who want to do human reliability analysis. In the preceding sections I have tried to show that the assumptions are incorrect. But more important than that is, as I will argue throughout this book, that the assumptions are not really necessary. Firstly, although a human reliability analysis should preferably provide results in a quantitative form - since that is a clear requirement from the other disciplines with which they must merge - this does not mean that human reliability analysis in itself must be a mathematical or engineering discipline. It is probably more important that human reliability analysis provides a systematic and consistent way of attaining pertinent human performance data which allow the production or construction of quantitative expressions. Mathematics itself is, in fact, not intrinsically quantitative. Secondly, models or theories of cognition - and of psychological phenomena in general - need not be expressed as a set of equations. It is entirely possible to obtain the same rigour and precision in a different way. The model described in Chapter 4 of this book is an example of that.

4. The Need for Better Models

In Chapter 1 the Law of Requisite Variety was briefly mentioned. The Law of Requisite Variety was formulated in cybernetics in the 1940s and 1950s (Ashby, 1956) and is concerned with the problem of regulation or control. In everyday terms the Law of Requisite Variety expresses the principle that the variety of a controller should match the variety of the system to be controlled, where the latter usually is described in terms of a process plus a source of disturbance (cf. Figure 3-5). The Law of Requisite Variety states that the variety of the outcomes of a system can be decreased only by increasing the variety in the controller of that system. Consequently, if the controller has less variety than the system, then effective control will not be possible.

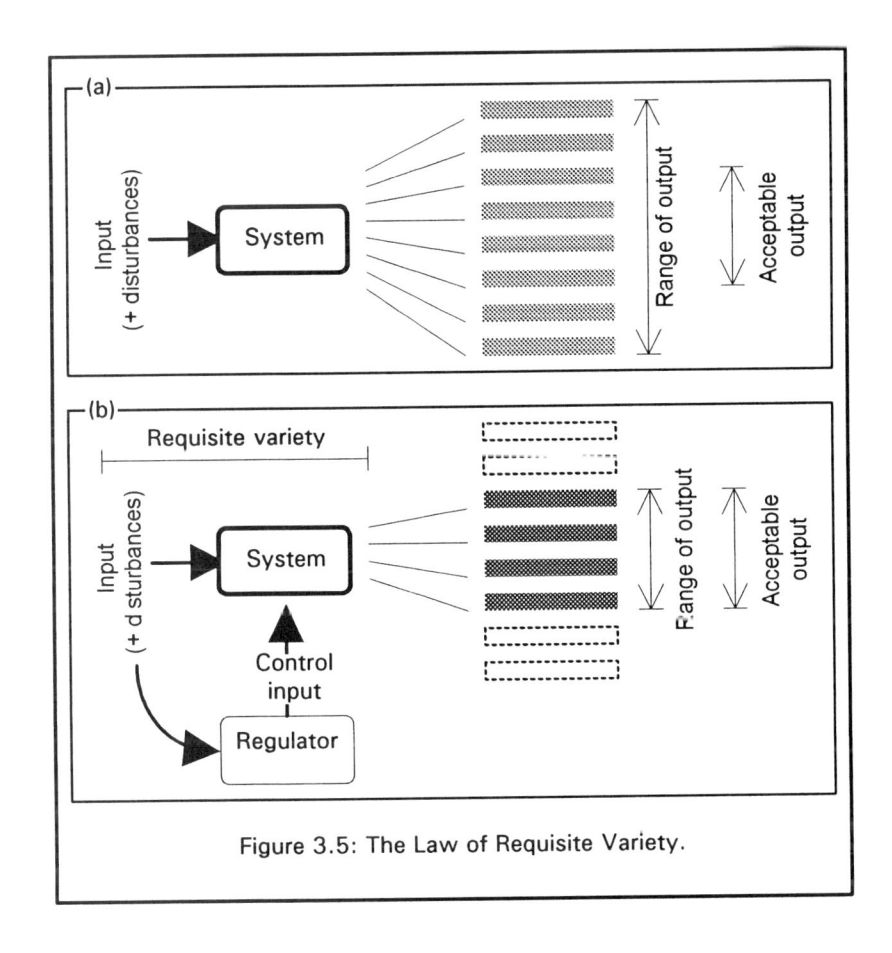

Figure 3.5: The Law of Requisite Variety.

The importance of the Law of Requisite Variety for questions of modelling can be seen if the modelling is regarded as a form of control. In this case the model is the basis for generating control input to the system, in order to keep the variety (of the output) within given limits. The success of the model can therefore be judged from whether there is sufficient agreement between the predicted performance and the actual performance. Conversely, we can only hope to reduce the occurrence of unwanted consequences if we have adequate models. So there is indeed a need for better models. The question is what we need better models of? In the modelling of cognition the variety to be controlled (and modelled) is the variety of human performance as it can be ascertained from experience and empirical studies. What is needed, therefore, seems to be enough variety to match the observed variety of human performance. If the main problem is unwanted consequences, then the question is what the main causes of these unwanted consequences may be: this should be the focus of the efforts to develop better models.

4.1 Parameter Uncertainty and Model Imprecision

The purpose of making a probabilistic safety assessment is to answer the following three questions:

(1) firstly, what can possibly go wrong in the system?

(2) secondly, if something can go wrong, how often is it likely to happen? and

(3) thirdly, if something goes wrong, what are the likely consequences?

The probabilistic safety assessment refers to three factors: the **scenario** which provides the context for the analysis, the **probability** that specific events may occur, and the **consequences** that may derive from that. Each of these factors can vary and none of them can be described with absolute certainty. The ideal situation for a probabilistic safety assessment is a completely described scenario into which an event with a known probability can be injected (e.g. the failure of a component), and where the consequences then can be precisely calculated or predicted. But this ideal situation can only be approximated. The goal of probabilistic safety assessment is therefore to secure a better appreciation of the uncertainty which is found in either of the three main factors.

It is sometimes claimed that the uncertainty is due to the lack of good models. This does not refer to the known deficiencies of the models, e.g. that probabilistic safety assessment neglects the influence of common modes as well as the effect of organisational factors or that probabilistic safety assessment generally makes the pessimistic assumption that the operator is a negative influence, and therefore cannot take into account the situations where the operator actually contributes positively to the success - over and above following procedures. The lack of good models means that even if only limited applications are considered they may still be incomplete or imprecise and the parameter values may be uncertain.

The unspoken assumption in probabilistic safety assessment is that the systems we try to model are inherently deterministic and that the modelling approach therefore must also be deterministic (cf. the differences between humans and machines discussed above, Section 2). It follows from this assumption that the more details one can get, the better. If the system is deterministic, then it is in principle possible to get all the information necessary to describe it. Thus, if enough information is obtained, the uncertainty can be reduced to an arbitrary low level; in other words, if we improve our models (of the physical phenomena) we can obtain the desired level of precision. This amounts to an almost Laplacian faith in the adequacy of information: if the functions of the physical system can be described precisely and if, furthermore, exact values for the parameters can be provided, then the complete behaviour of the system can be described and predicted! (In addition there is a fundamental difference between uncertainty and imprecision (Dubois & Prade, 1989), although this distinction is usually neither recognised nor applied in PSA/PRA studies.)

This assumption is, unfortunately, fundamentally wrong also for physical systems, as clearly demonstrated by Chaos theory (e.g. Stewart, 1989). Even if a precise model of the system's function is available, the question of obtaining and representing sufficiently precise values of the parameters is a problem. The assumption is similarly wrong for humans, i.e., for psychological systems. Here we may even assume that obtaining the precise values would not be enough. The chaotic (read: unpredictable) performance of humans is not due to the same phenomena that are treated in Chaos theory. It is rather due to the fact that too little is known about the functioning of the human mind; attempts to describe the human mind with analogies taken from the technical world, from Sigmund Freud and onwards, have not yet produced any satisfactory results. Even if we knew all about cognition, we would still have to consider, for instance, motivation or learning. The practice of probabilistic safety assessment struggles with the problem of handling uncertainty: estimating uncertainty, assessing uncertainty, and controlling uncertainty. But that problem does not in itself make any assumptions about how it should be solved; accordingly one need not assume that the solution must have a deterministic basis.

An alternative is a holistic view. According to this the control of uncertainty can best be achieved by considering the problem as it presents itself - without making any assumptions about how it eventually should be explained. The first priority therefore becomes one of **description** rather than of **modelling** and the objective is therefore to develop adequate ways of describing the situation as a whole in a context - on the level at which it appears rather than as an aggregation of little parts.

The problem is the variations and imperfections of human performance and the corresponding phenomenon is the reliability of human performance. Furthermore, it is reasonable to assume that the occurrences of inadequate action are not due to a specific, exotic fault producing mechanism such as GEMS (Reason, 1990), but rather are the effect of a combination of the usual "mechanisms" of action in combination with the specific circumstances, in a sense the "sneak paths" of actual over potential performance. The focus must therefore be on the description of human action - and this may eventually result in a theory and a model that can be used in the further work.

Human action is, however, a rather wide concept, and can be taken to imply almost everything about human nature. It is therefore sensible to be a bit modest and consequently be more specific in outlining the target. I will here take that position that the target is human action in work environments rather than in experimental or controlled conditions. This involves humans acting as single persons as well as humans acting in groups (as members of a group). I will further maintain that there is no significant difference between human action in or out of groups. This follows the same line of reasoning as used to discard human error or incorrect actions as a separate domain (Hollnagel, 1993). Whether a person acts alone or together with other people, the underlying principles of cognition are the same. (But as discussed in Chapter 2, this does not imply that individual cognition should be used to describe collective behaviour.) To be more precise, human action in work environments will always be assumed to be **interaction** - either with an artifact or with one or more other persons. Human action in isolation is therefore excluded from the treatise given here; that goes for painting, writing, contemplating nature, etc. The interaction must furthermore be in situations where unwanted consequences can occur.

(Clearly, the problems of definition are considerable. Even in the case of the lonely painter there can be unwanted consequences, *viz.* that the result is not what the painter had in mind. This can be due to the lack of skills and experience, to an improperly formulated goal, to imprecise plans for work, faulty materials, etc. In other words, the situation can in many ways be described in exactly the same terms as those applied for the working situations I do consider. I will, however, not make any claims to that level of generality, and will therefore allow myself to exclude situations that are not clearly work related, as defined in the text.)

5. The Art of Human Reliability Analysis

Much of the effort in human reliability analysis, as it is currently practiced, is dedicated to reducing the uncertainty about human performance by producing or obtaining more precise estimates of the numbers used, i.e., refine the probabilities for the defined events.

The common approach is not to start from a specific model but rather from a specific or generic description of events, e.g. either particular examples of performance elements (slips, mistakes, decisions, observations, etc.) or characteristic modes of functioning such as skill-based, rule-based, and knowledge-based (Wakefield, 1988). The definition and selection of events can be based on a loose model of performance or on a reference description of some sort (e.g. SLIM-MAUDE, OATS, etc., cf. below). Assuming that the identified "units" are relevant and valid, the quest is to obtain reasonable numerical data for them - because numerical data are what the human reliability analysis procedures need.

This is done in a number of ways, the most prominent being expert judgment or expert evaluation. There are many ways in which expert judgment can be elicited, and many ways in which one can take care to get judgments that are as sound as possible. This does not, however, reduce the basic flaw of the approach: firstly, that expert judgments are an uncertain and imprecise source of information, secondly that the whole idea of looking at discrete events is based on the decomposition principle, hence rests on the fallacious atomistic and mechanical assumptions. Numbers may be easier to fit into current methods, but numbers at best play only an intermediate role. The real purpose of human reliability analysis is not to determine the probabilities of unwanted events as such, but to use that to point to aspects of the system which are undesirable. **"HRA (and PRA) is not about estimation, but about design"** (Moray, 1990, p. 343, emphasis added). The reduction of unwanted occurrences can only come about through redesign, and numbers alone are not very useful for that. Numbers therefore only serve the intermediate step of helping to identify the points where a redesign is needed. Since this obviously can be done by other means, numbers are not strictly necessary.

5.1 The Problem of Quantification

In order for something to be quantifiable, it must be based on a causal and consistent model, i.e., it must be identifiable in the sense that:

> "... there is a one-to-one correspondence between the structures of the model and the permitted probability measures within the data. If there is a many-to-one correspondence between structures and probability measures, then the model is not identifiable."
> (Restle & Greeno, 1970, p. 331).

It is relatively easy to make a model which can be quantified, if all that is needed is that the parameters can be assigned numerical values. But it is important that the parameters are meaningful; otherwise the model cannot be identified from the data and the analytical exercise simply becomes a question of curve-fitting as discussed by Kantowitz and Fujita (*op. cit.*), cf. Figure 3-4. If model identification cannot be completely achieved then it must at least be possible make the identification for a subset of the parameters and to show a strong empirical correlation between the empirical data and the remaining parameters of the model. As an absolute minimum it must be possible to establish an empirical correlation between the data and the salient features of the description (the theory). This solution, however, leads to a paradoxical situation: in order to know where to look for data it is necessary to have a model (at least in a loose sense as a set of guidelines or a set of operational concepts). The model and the operational concepts must both, in turn, be based on a qualitative description or a conceptual basis.

The paradox of quantification, as discussed in Chapter 1 and illustrated in Figure 1-8, is thus that quantification cannot be achieved without first having made a qualitative analysis or having recourse to the results of one. The data are selected on the basis of a model, no matter how vague it is, and a certain dependence is therefore already defined. There is obviously no value in finding "random" empirical correlations, since these must be expected to happen every now and then - depending on the chosen level of statistical significance. Even purely empirical investigations require a model description, which is then corroborated through the data. Attempts at "objective" data gathering are scientifically naive.

One part of the quantification consists in comparing estimates of uncertain events. Apart from the problem of getting the assessments or judgments in the first place (cf. below), the problem is whether it is sensible to perform the comparisons. This can clearly only be the case if the contexts are related, i.e., if the numbers can be given the same semantic interpretation. If that condition is violated all that is left is simply to compare numbers as numerical quantities. Such an exercise is a numerical formalism which at best is useless and at worst is misleading: it is arithmetic machination rather than probabilistic safety assessment. If the comparison of values is to have any meaning it must be based on an interpretation or a common context, and is thus basically a qualitative issue (although the qualitative and quantitative aspects are not separate): the use of numerical values is just a convenient way of expressing a qualitative statement. A numerical value must have an unambiguous semantic interpretation in order to be valid.

6. Obstacles to the Study of Human Reliability

The analysis of the Reliability of Cognition and of human performance in general requires data. Data are necessary both as a basis for developing models and theories of human reliability, to allow inferences to be made and hypotheses to be formed, to support methods for human reliability analysis, and for the quantification and calibration of reliability assessments. The only inherently valid source of data is systematic experience and observation of human performance under natural conditions rather than in controlled experiments, i.e., as observation of performance as it actually occurs. There are, however, a number of reasons why this is difficult to do.

6.1 Observation

It may be difficult to observe the right type of events. This is often used as an argument against making naturalistic observations at all. The interest in human reliability analysis has often been focused on performance during accident conditions, i.e., in situations where the effect of unwanted consequences on the system could be considerable. In any given system or application accidents are fortunately quite infrequent - but unfortunately also very unpredictable. It therefore stands to reason that the probability of observing a "significant" event is so small that naturalistic observation cannot be considered as a practical solution. It is consequently inefficient to collect data through observations of work.

If the purpose of data collection is restricted to include only the observation of rare events, then this line of reasoning is valid. But it is still based on the tacit assumption that human reliability can be seen as analogous to the reliability of technological systems. It is thereby assumed that the person's performance under adverse conditions is so different from performance under normal conditions that they cannot be compared. This means that the interest is restricted to how the person functions under adverse conditions, i.e., how he reacts and responds to an accident. While it is warranted to be concerned about how people respond to adverse conditions and accidents, this very approach promotes a segregated, mechanistic view of human cognition.

In contrast, I shall argue that the appreciation of the Reliability of Cognition must be based on an understanding of cognition under normal circumstances, hence also on the study of performance under normal conditions. This is consonant with the notion of the human as a fallible machine, although it does not accept the premise of that view. The efforts should therefore be concentrated on systematic naturalistic observation; even if this does not give precise data about responses to accidents, which partly may be because the existence of such data is an illusion, naturalistic observation will still provide important data about the reliability of human performance - in a sense the base rate. Since, in fact, many accidents can be traced back to either inadequate performance under normal conditions or near misses under disturbed conditions, these data have a value in themselves. There is, however, no denying that naturalistic observation is difficult to achieve, that it is often expensive and tedious, and that it requires the collaboration and consent of many people. Furthermore, the analysis of the data is far from straightforward (cf. below). Yet the alternative, to perform data collection under experimental conditions, should only be used with great prudence.

6.2 Registration of Data

The actual registration or recording of data may be difficult and may even interfere with the event. The advantage of naturalistic observation is that the situations, by virtue of being the real ones, contain all possibly relevant information. The disadvantage is that there is far too much to observe and/or to record. Furthermore, it may be difficult to register the data that one needs, even if these have been defined (cf. below). Actual working environments are generally not designed for easy data collection, and the possibility of making changes and modifications may be very limited, for instance, in terms of modifying existing hardware, software, physical environment, working procedures, etc.

An additional impediment is that most forms of observation are obtrusive in one way or another. Even very discrete methods, such as hidden video or audio recordings, will have some effect on the performance. (I assume, of course, that observations are never made without informing the subjects about it.) While it is true that people at times may easily forget that they are being observed, there is no way of determining to which extent this is the case and to which extent their performance is influenced by the observation. This effect is more pronounced if the data collection includes direct interviews (or post-event analysis), direct participation, debriefings, etc. In many cases such methods are needed to get the necessary details, but they will undoubtedly disturb the observations themselves, for instance by sensitizing the persons to certain aspects or conditions of the work which they otherwise might not have considered. Observation always implies a measure of control, however limited. The more the situation is controlled, as for instance in experiments, the more likely it is that the performance somehow is affected.

6.3 Specification of Data

A final obstacle to the study of human reliability is that it may be difficult to specify precisely in advance which data to look for. In order to make good observations and data recordings one needs a coherent set of concepts or a good theory (Hollnagel et al., 1981). The concepts and the theory define what to look for, how to observe it, and how to analyse it afterwards. The field of human reliability analysis is, if anything, characterised by a lack of good theories (Swain, 1990). The basis for defining relevant data categories is therefore very meagre. The situation is somewhat easier in the case of accidents because the search may be narrowed down to specific (assumed) causes but, as argued above, these cannot feasibly be made the target of naturalistic observation.

One often tried solution is to record everything possible in order not to miss something interesting. This approach is, however, self defeating. First of all, it is not possible to record everything and not even possible to define what "everything" is. Secondly, even if all available sources of measurement and information were used, or even if only all reasonable sources were used, the problem would only be moved from data collection to data analysis. There is a clear trade-off in all data collection between the efficiency of data collection and the efficiency of the following analysis, as described in the following section. In order for the analysis to be efficient, the data must be well filtered and selected either at the time of observation or at an intermediate stage. The less concentrated the data collection is, the more inefficient will the analysis be. Conversely, if the data collection is too restricted, the following analysis may be rendered impossible.

The dependency between the observable data and the underlying concepts is sometimes neglected by people who apparently believe that data can exist in an objective manner, i.e., as pure data which reflect the true state of the world. Any efficient and meaningful data analysis must, however, recognise the close links that exist between data and concepts. These are described in more detail below.

6.4 Data Collection and Data Analysis

The types of data may vary from one source to another, from routine event reports (such as Licensee Event Reports) and field studies to sessions from training simulators and protocols from controlled experiments. Similarly, the purposes of data collection and analysis may be quite different, from human reliability analysis over decision analysis and training analysis, to the evaluation of specific design features. This means that the way in which the raw data are analysed depends upon their type as well as the purpose of the activity. Fortunately, this does not lead to completely different types of analysis, but rather to several different modes of analysis which have a considerable overlap because they can be based on the same conceptual foundation. The different modes also clearly show the dependence between data collection and data analysis. It is important to acknowledge this common conceptual foundation because it provides a basis for designing a proper analytical method as well as a possibility for cross-checking results.

The observation and analysis of data on human performance may conveniently be described by a series of steps, derived from the work reported in Hollnagel *et al.* (1981):

(1) **Raw data**. The raw data constitute the basis from which an analysis is made. Raw data can be regarded as performance fragments, in the sense that they do not provide a coherent description of the performance, but rather the necessary building blocks or fragments for such a description. Raw data can be defined as the elementary level of data for a given set of conditions. The level of raw data may thus vary from system to system and from situation to situation.

(2) **Intermediate data format**. This format represents the first stage of processing of the raw data. In this stage the raw data are combined and ordered in some way, typically along a time line, to provide a coherent account of what actually occurred. It is thus a description of the actual performance but is given in the original terms, i.e., as a professional (domain expert) description rather than a theoretical (analysis expert) description. The terms and language used are the terms from the raw data level rather than a refined, theoretically oriented language. The step from raw data to the intermediate data format is relatively simple since it basically involves a rearrangement rather than an interpretation of the raw data.

(3) **Analysed event data**. In this stage the intermediate data format has been transformed into a description of the task or performance using formal terms and concepts. These concepts reflect the theoretical background of the analysis, which in current human reliability analysis typically is the assumptions of the decomposition approach. The description of the performance is still ordered along a time line which is specific to the situation in question. The transformation has, however, changed the description of the actual performance to a **formal description** of the performance during the observed event.

The step from the intermediate data format to the analysed event data may be quite elaborate, since it implies a theoretical analysis of the actual performance. This step is not purely data driven because much of the activity has not been observed. To make up for the lack of data the analyst makes inferences of what he assumes must have happened - based on his own knowledge and experience. The transformation is one from task terms to formal terms. The emphasis is also changed from providing a description of what happened to provide an explanation as well.

(4) **Conceptual description.** At this stage of the analysis, the description is no longer specific to a particular event but rather aimed at presenting the common features from a number of events. By combining a formal description of performances one may end up with a description of the **generic** or **prototypical performance**. The prototypical performance may still be described as a sequence of activities ordered along some time line, or as an operator event tree, but the underlying dimension is independent of the specific event. On the other hand, the description of the performance in a specific event may be seen as an example or a variation of the prototypical performance or the **procedural prototype** (Hollnagel, 1992a). Thus generic descriptions of human error mechanisms and the notion of the fallible machine are, in fact, descriptions of typical deviations from the prototypical performance. The step from the formal to the prototypical performance is typically quite elaborate and requires an analyst with considerable experience, in addition to various specialised translation aids.

(This also suggests a way in which the prototypical performance description may be put to the test, by determining whether the formal description of an actual performance can be subsumed under the prototypical performance, or by comparing it with the predictions of typical performance made from the prototypical performance.)

(5) **Competence description.** The final stage of the data analysis combines the conceptual description with the theoretical background. The description of **competence** is concerned with the basic concepts, such as mental models, decision strategies, performance criteria, preferences and heuristics, problem solving strategies, etc., which in a given situation are combined to produce the performance. The description of competence should be relatively context independent: it is the description of the

behavioural repertoire of a person independent of any particular situation though, of course, still restricted to a certain class of situations. As soon as a context is provided the description of competence can become a description of prototypical performance and, pending further information, a description of the typical performance. The competence description is thus essentially the basis for performance prediction during system design.

The step from the conceptual description to the competence description may be quite elaborate and require that the analyst has considerable knowledge of the relevant theoretical areas as well as a considerable experience in using that knowledge. The analyst must be able to provide a description in task independent terms of the generic strategies, models, and performance criteria which lie behind the observed performance.

The relations between the five steps defined above can be shown as in Figure 3-6. The right side of Figure 3-6 describes the steps in going from raw data to competence description. The basic trend is an aggregation of the various data types and a removal of the context, i.e., going from the specific to the generic. The left side of Figure 3-6 shows the complementary development from the level of competence to the level of performance fragments. The basic trend is here an increasing level of detail and context, i.e., going from the generic to the specific. Figure 3-6 also illustrates the dependence between observation and analysis, and why it is futile to engage in an effort to record everything possible. The performance fragments clearly do not exist independently of a description of the performance, which in turn depends on the conceptual description of the generalised performance. If the filtering or selection is not made at the time of data collection, it must be made at a later stage. The advantage of making it at the stage of data collection is that it is less likely to be hidden.

The problems of observation, registration, and specification all contribute to making data collection difficult and costly. Neither of these reasons should, however, deter anyone from trying. They emphasise that data collection should be planned with great care, and that the job is only halfway done when the data have been registered. Data should only be gathered if the purpose is well-defined and if the principles for classification and analysis have been established. A willy-nilly programme of experimental data collection, in the hope that something will turn up, is

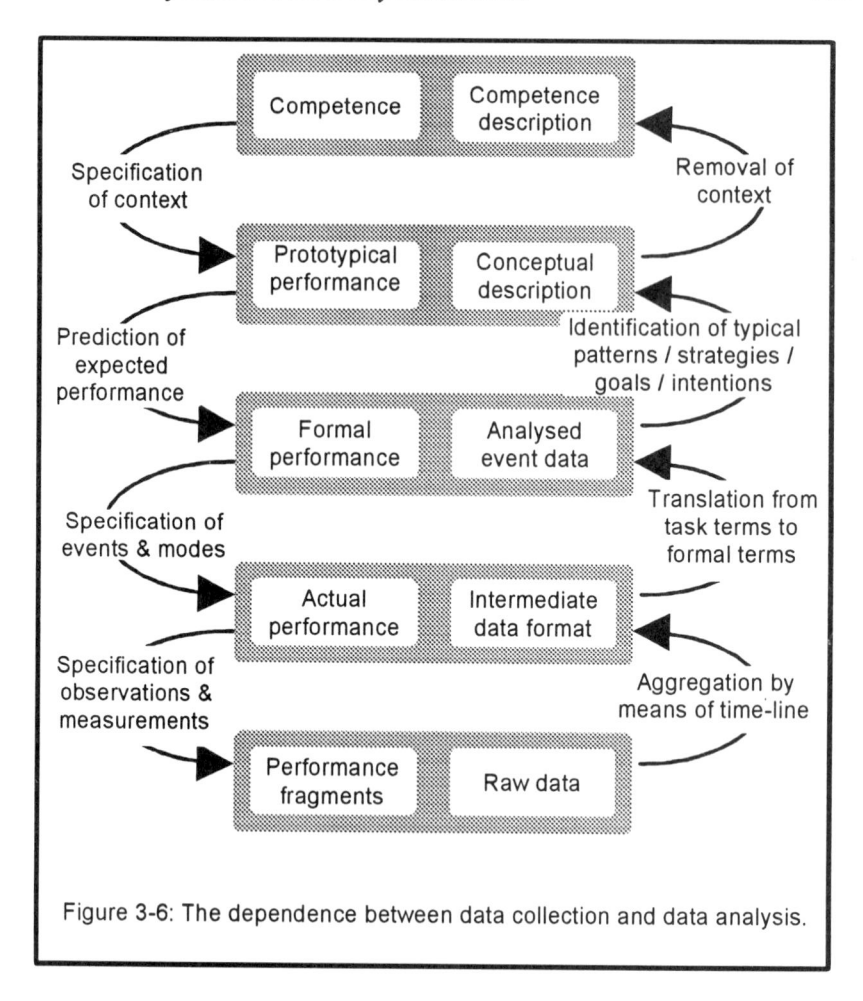

Figure 3-6: The dependence between data collection and data analysis.

futile and waste of resources. Quite apart from being scientifically unsophisticated because it is based on a deficient understanding of what an experiment is, it will at best give a false sense of assurance. At worst it may produce seriously misleading results. The need for meaningful data is so great that efforts should not be wasted if it is at all avoidable.

6.5 Experimentation: The Use of Micro-Worlds

Data are needed to develop models and theories of human reliability, and to sustain quantification of reliability assessments. In order to evaluate the specific predictions from a theory (or a model) it is necessary to established controlled performance conditions. (Strictly speaking, of course, the purpose of experiments should be to reject a so-called specific hypothesis. But given the relatively low level of theoretical development in the study of cognition, even a confirmation of a hypothesis may be counted as a success.) In order to get data about specific events, e.g. accidents, it is also necessary to create situations where the occurrence of such events can be controlled. The systematic study of human reliability and the Reliability of Cognition therefore makes it necessary to complement naturalistic observation with the use of situational artifacts, i.e., explicitly designed experimental working contexts which will serve a given purpose. In relation to theory development such contexts are often referred to as **micro-worlds**. But there is an important difference between micro-worlds and the experimental situations that are traditionally used in behavioural science. In the classical experiments the emphasis is on controlling all the independent variables, varying a few of them, and observing the outcome (the dependent variables). In micro-worlds the emphasis is more on providing a context which has some face validity with the real world (and thereby also some content validity), but which can still be manipulated.

It is in good agreement with the preceding discussion that the notion of using classical laboratory experiments should be abandoned as an approach to research on the Reliability of Cognition. The micro-world approach recognises the futility of trying to control all independent parameters, and instead hopes that establishing situations with sufficient similarity to real situations will do the job (Moray & Sabadosh, 1992). In relation to data collection the contexts are usually based on simulations with varying degrees of realism. There is little intention of doing experiments rigorously to test theoretical predictions; the aim is rather to gather data about performance in situations that in one way or another can be described as representative of the actual working context. This could be in the sense of offering rare situations, better observation and recording facilities, no-risk actions, etc. Such simulations are different from the use of micro-worlds in, for instance, problem-solving research (e.g. Dörner, 1989) because the former put the emphasis on the face validity of the situation while the latter are more concerned with replicability. However, for the purpose of this discussion both approaches will be included under the notion of micro-worlds.

While micro-worlds do provide a convenient way of getting specific data, they also establish a dilemma which sometimes is formulated as follows: in micro-worlds it is not possible to get **interesting conclusions**, whereas in the real world it is not possible to obtain **definite conclusions**. (This is, of course, an overstatement.) The conclusions that may be drawn from micro-worlds are not interesting because the micro-worlds remain impoverished and insufficiently valid. The conclusions that may be drawn from real worlds are not definite because of the problems mentioned above with regard to representativeness and data collection.

Since there is very little that can be done about the real world, the problem becomes one of whether it is possible to establish an "ideal" micro-world could provide both definite and interesting conclusions. This demand should serve as the basic requirement to the micro-world research and be used to identify the characteristics of real worlds that one would like to keep, as well as those that one would like to avoid. The fundamental characteristic of real situations is that they are so complex that a person, be it an operator or an analyst, cannot hope to achieve a complete understanding of them. Even extensive formalised descriptions, such as sets of differential equations or qualitative models, only provide part of the truth. The main goal for the person is therefore to identify those fundamental constraints which enable a reasonable degree of predictability and control. The person is, in other words, trying to identify the fundamental regularities or the requisite variety of the context (Ashby, 1956; Hollnagel & Cacciabue, 1991); if he succeeds it will enable him to plan and execute meaningful actions, rather than having to resort to uncertain expectations or even trial and error.

One criterion for an ideal micro-world is therefore that it must include the underlying regularity of the dynamics that characterise the real world. **The requisite variety of the micro-world must match the requisite variety of the real world.** It is of little value if the experimenter/investigator can change or control the micro-world as he pleases unless the dynamics develop in a reasonable or naturalistic way. One of the goals for the user is precisely to identify the dynamics of the situation, and this should remain valid in the micro-world.

A second criterion is that the micro-world should not seduce the person to establish a complete "model" or overview - nor should it let the person suspect or believe that such an undertaking is possible. It is a prominent characteristic of real worlds that they are too complex to allow a complete understanding. Humans realise that *a priori*, and therefore usually abstain from the attempt. Much of their effort is in fact dedicated to develop a simplified representation (a mental model of the world) which will enable them to cope with the given complexity - to make sense of the situation. If the micro-world does not evidence this characteristic, it will radically change the person's goals and strategies, hence be unsuitable as a vehicle for investigation of reliability - whether of cognition or of performance.

A third criterion is that the micro-world should display the same variation in information quality as that found in the real world. Many micro-worlds produce information (e.g., as process displays) which are always correct; this is particularly the case when they are used for theoretical investigations. However, the quality of available information is always a crucial factor. Although a person usually trusts the information that is available, he also always knows that it may be incorrect or unreliable. The quality of information can change in a number of ways: true/false, changing correctness, competing information, simultaneous sources and tasks, conflicts between sources, temporal changes (non-monotonic quality), etc. If the information given by the micro-world is always correct, the micro-world fails to provide face validity on this very important aspect.

If the reasonable assumption is made that in real life a person is trying to establish the constraints that will enable him to work in a sensible way (i.e., establish sufficient control of the environment), then the micro-world should both allow and induce the same kind of behaviour. The micro-world should preserve the essential traits of real systems, and not induce behavioural artifacts that are only due to the experimental and artificial nature of the situation, hence only true under those very conditions.

One very important point is that the micro-worlds should elicit the use of multiple strategies - rather than a single, correct strategy - as well as the use of multiple representations (models) and the characteristics of coping - including higher level control actions (also know as meta-cognition). Users are generally Cartesians rather than Newtonians: they try to find the causal explanations or mechanisms that will enable them to control how things develop rather than try to establish the set of equations (or logical expressions) which provide a complete description of the world. Performance is more often a case of muddling through than of rational planning and decision making (Lindblom, 1959; Hollnagel, 1992b). To serve its purpose well, a micro-world must induce the same conditions.

7. Assessment of Human Reliability

The assessment of human reliability depends very much on the types of data that are available as input to the specific method. The possible data types can be divided into the following three main categories, based on their main source: empirical or historical data, simulation data, and expert assessments.

7.1 Empirical Data

Empirical data are, in a sense, the *sine qua non* of human reliability analysis. Even if other types of data are used, they must eventually be corroborated with empirical data. The two primary sources of empirical or historical data are:

(1) **Incident Reporting Systems**, which may provide information on human erroneous actions leading to accidents or near-accidents. Incident reports may, for example, be used to identify deficiencies in established systems and procedures and to select measures which may reduce erroneous actions.

(2) **Human Error Rates** (HER), which are ideally determined from error frequencies as the ratio between the number of errors of a given type and the number of opportunities for that particular error. Both the numerator and the denominator are difficult to obtain, and in practice most available data on human error rates have not been compiled according to very strict criteria. Human error rates are furthermore only useful when the decomposition principle is applied to human reliability analysis. Data found in handbooks are often estimates based on simulator exercises, experimental studies, or expert judgments.

Although empirical data clearly are the best basis for assessing human reliability, they are costly and difficult to obtain. Several factors contribute to this problem:

(1) Erroneous actions that do not lead to unwanted consequences may go unnoticed unless special efforts are undertaken to secure this information.

(2) The scope for generalising the data is often limited because the situations may be widely different

(3) Incompatible classification systems for tasks, action types, or casual and contributing factors may interfere with the transfer or translation of data to new applications (Bagnara *et al.*, 1989).

(4) Exceedingly long periods of observation may be needed to obtain reliable data on low-frequency erroneous action types.

(5) The exact boundary between acceptable performance and erroneous actions may be difficult to define in the case of non-proceduralised tasks.

Quite often the situation will be either that appropriate empirical data are missing, or that the data available are not sufficient for the purpose. This difficulty is increased when the analysis gets more fine-grained and looks at the details of a task rather than the global aspects. The problems in getting sufficient empirical data have made it necessary to rely on suitable replacements. These have been found in process simulations and expert judgments.

7.2 Data from Simulators and Simulations

The use of simulations of human behaviour can be traced back to the analog simulation methods developed for the design of aircraft controls during the 1950s. These led to studies of control capabilities compared with requirements for accuracy and speed, to a large amount of work on mental load, and to significant work on presentation and control modes. This work was initially almost centred entirely on routine activities. The use of simulations has later been extended to other applications covering more complex activities, as well as the interaction between several people in an organisation. The most recent development is the use of qualitative simulations, which still is in its beginning in the behavioural sciences.

Simulations are used either to (re)produce the situations that are required to study specific types of performance and/or to provide a somewhat controlled environment. In the former case the primary purpose is to overcome the limiting aspects of empirical data collection, as described above; this is particularly useful in studying dangerous situations, e.g. in aviation. In the latter case the purpose is to overcome the imperfections of empirical field research, notably the lack of control over the experimental situation. By subjecting persons to the controlled conditions of a simulator it is hoped that enough quantitative data may be established to allow valid inductions to be made. An example is the Operator Reliability Experiments project (Beare *et al.*, 1991). The control is, however, limited to establishing the initial conditions and/or the triggering of specific events as the simulation develops. Even well-trained subjects will vary in their performance, thereby making the simulation diverge from the intended developments.

Simulations of a situation are mainly a simulation of the physical process, including a replication of the control system, the interface, etc. Simulations, may, however, be extended also to cover the humans who are involved in the process, i.e., the operators. The rationale is clear: if it is possible to develop a simulation of a system, this can be used to study how the system performs under different conditions. Furthermore, the simulation will enable unlimited controlled repetitions, thereby complying with the requirements of classical empiricism. A simulation thus serves as a very complicated model. Consequently, if a comparable type of model or simulation of the human operator can be developed, a very important step will have been taken towards resolving the quantification problem of human reliability.

7.2.1 Cognitive Modelling

Cognitive modelling is a particular type of simulation with roots going back to work on psychological modelling, natural language understanding, and Artificial Intelligence (e.g. Newell & Simon, 1972). An additional influence was the development of cognitive systems engineering in the early 1980s (Hollnagel & Woods, 1983; Norman, 1986). Cognitive modelling is central to the understanding of human reliability and the Reliability of Cognition because it provides the possibility for a causal description of performance - and in particular of erroneous actions. While the technique is generally applicable for all kinds of human action cognitive modelling is, at present, the only technique which allows systematic treatment of erroneous actions in dynamic, interactive environments.

The cognitive modelling approach to human reliability tries to look behind the observable action and to construct a model of the cognitive functions that can be used to explain overt behaviour, as done by Woods *et al.* (1988) or Hollnagel *et al.*, (1992). The reliability assessment is therefore contingent upon a reasonable model description of the cognitive functions. The data needed are determined by the requirements of the model as well as the task analysis of the domain. The strength of the cognitive approach is that it tries to account for the complex actions that occur during contingency operation, as well as for the elementary actions. On the level of automatic behaviour (a.k.a. skills) it can therefore, in principle, be calibrated by the results from studies based on the decomposition principle. On the levels of more complex behaviour, it can only be calibrated by experience and empirical data.

No existing cognitive simulation is able in practice to cover the full range of problems that are found in work studies - indeed only a few have been applied experimentally with even a reasonable degree of coverage (cf. Apostolakis, Kafka & Mancini, 1988). Paradoxically, the greatest problem with the cognitive modelling approach is not that it does work, but rather that it works too well. For any given event, normal or not, a reliability analysis can find literally hundreds of "causes" at the level of cognitive functions. This is because the backwards tracing in the event tree is done without any empirical reference or context; in post-accident analyses the situation is better controlled. It is of little help that the failure cause analyses can be supported by case studies which show that the problems are real, i.e., that the causes are all plausible. Given enough samples, any mixture of causes and conditions is likely to be found. Cognitive simulations must therefore be better focused to be of practical use. Gathering information about the actual system, the man-machine interface, and persons can be used to improve the cognitive simulation, the basis for analysis, as well as reduce the abundance of possible causes. This allows the more likely or problematic causes to be singled out, but only at the cost of additional information gathering. Another solution would be to develop better theories of cognition which could serve as a basis for the simulation.

7.3 Expert Judgment

A fundamental problem with all analytical quantitative methods is a severe lack of data that can support the assignment of human error probabilities to individual task elements. The normal procedure is to rely on expert judgment or estimates. This leads to a combination of the analytic approach with the methods for subjective probability estimation developed in normative and descriptive decision theory, and to more advanced procedures for aggregating and normalising the estimates. Expert judgment often uses a database of human error to obtain the base rates for the error probabilities. But the use of such databases raises questions of what should be done about error rates for erroneous action types that are not included, and whether there are sufficient weighting factors (or performance shaping factors) included to reflect properly the variations in error rate found in practice. The two questions are connected. Many erroneous action types related to non-trivial tasks are not only fairly rare but, worse, also have a large number of significant factors which affect their probability. This makes it very difficult to collect sufficient data to allow statistical rates to be given. Expert judgment is therefore required, but since adequate experience with these erroneous actions is missing, the expert judgment itself becomes a "best guess" based on the expert's implicit model of operator behaviour.

When expert opinion is needed because the underlying models are incomplete or imprecise, the experts are in fact asked to provide data that the models cannot provide! Expert evaluations are thus used as a way to get around the inadequacies of the models, but the manner in which this is done is completely beyond control.

There is a certain irony in the attempts to model expert judgments, which are subjective opinions about something which is only incompletely modelled. Expert judgment belongs to one type of phenomena (call them psychological or type-M) and are statements about another type of phenomena (call them physical or type-P). But although type-M and type-P are clearly different in nature, it appears that the formalised attempt of improving expert judgments treats type-M as if they were type-P, i.e., as belonging to the same category, hence subject to the same regularities. This is clearly not the case.

The irony of this is perhaps best seen in the attempts to train experts to follow rigid procedures (and further to select the experts on their ability to do so). This amplifies the myth that expert evaluations are an alternate source of detailed data and seems to neglect the fact that it would be much better to automate the procedure, if it could be described. The fact that most procedures cannot be turned into algorithms seems to indicate that they cannot be precisely described. It is only because the human mind can supply the missing details that the procedures can be used to "mechanise" the human performance. It is important to note that decision theories require point probabilities and point values (rather than, for instance, ranges or descriptions). This amplifies the quest for precision and numerical data, i.e., it is a need which is created not by the problem but rather by the methods by which the problem is solved. One may seriously question whether the (implicitly) chosen solution, the decomposition principle, is the correct one. One should rather consider what the underlying problem is and take the assumptions up to revision. From a pessimistic point of view, expert judgment is either a social game of "who knows best" or a futile exercise in making humans do what a machine can do much better.

The use of expert judgment makes the further assumption that there is a true value (either a point value or a range or an interval) which the expert tries to comply with. But this assumption can easily be challenged. The expert judgment should not be seen as an alternative approach to finding a true value. Expert judgments should rather be seen for what they are - opinions about something which are used because there are no other data to be found. All attempts of either aggregating expert judgments or training people to follow rigid procedures should consequently be given up. True values may exist for events that have occurred, i.e., for specific cases, but not for general classes of events. The notion of a true value is even more meaningless in the context of human performance, because the human mind is not subject to the restrictive conditions that apply to machines.

7.4 A Procedure for Using Expert Judgment Data

To provide a feeling for the magnitude of the problem, it is illustrative to describe a typical procedure used to obtain data via expert judgment.

(1) **Requirements**: First of all, a clear description must given of which data are being sought. There may be an existing data set which is incomplete or insufficient in some way, for instance because it is not known exactly under which conditions they were obtained, hence whether the currently considered situations are sufficiently similar for the data to be used. By describing why the (possible) available data are not applicable, one has also given an indication of what is being sought.

(2) **Means**: The next step is to identify the existing pools of expertise that can be used to provide the data. The expertise may reside with human experts alone or require the combination with other knowledge sources such as documents, recordings, etc. A decision must then be made about which source(s) of expertise to use,

depending, for instance, on availability, accessibility, cost, dependability, etc.

(3) **Tools**: The third step is the preparation of the questions or means by which the required expert judgments are to be elicited. This in itself is a phase which may draw on a number of methods regarding questionnaire construction, interview techniques, data conditioning, base rate identification, etc. Since the elicitation of expert judgments must produce results which conform to certain standards (e.g. the values, rates, uncertainty ranges, etc., being used), it requires considerable preparation and elaboration.

The expert should ideally give answers which can directly be used in the required format. That puts the onus of transformation on the expert, and in a sense hides it from the analysis; the advantage is, of course, that all judgments are similar in their format. If this is not possible, then all the "free form" judgments have to be transformed to a common format, but this time with full visibility. Whereas this may actually be preferable from a scientific and methodological point of view, it makes life more difficult for the analyst.

(4) **Collection**: The next step is actually to elicit the expert judgments from the experts. (Even if the source of expertise is documents, etc., it may still require an expert as an intermediary.) The actual elicitation is based on the prepared materials and techniques. It is important in this phase to make sufficient cross checks with regard to completeness, consistency, and coherence of the judgments. There are usually very few opportunities for repeating the exercise. The actual elicitation should take into account the differences that may exist, for example, in the ability to provide numerical results, use visual aids, comply with the sequencing of the interview, etc.

(5) **Aggregation**: The final step is the combination of the expert judgments thus obtained, and the feedback to the experts. The combination is necessary in order to obtain results which can be trusted as being representative of the situation (task, problem) rather than of the single expert. Here the concordance between the various experts can be determined and possibly quantified. The feedback serves the useful function of confronting the expert with the results, and eliciting their reaction to it. A further purpose may be to "calibrate" the experts with respect to each other - the danger,

of course, being that in this way a contrived agreement is established.

Altogether these steps indicate how difficult it is to obtain expert judgments even for very simple cases. Furthermore, all expert judgments must in the end be calibrated or corroborated by empirical data - either historical data, simulations, or follow-up studies. That usually requires additional effort and is therefore often conveniently pushed aside.

7.5 The Value of Data

The need for data is defined by the methods that are used for human reliability analysis. If the methods focus on individual events and the plethora of causes and conditions that may affect the outcome, then data will be needed about these elements. If the method focuses on the context and the dominant solution strategies, then data will be needed for that. It is therefore important at all times to keep the purpose in mind. If not, then the process of providing or generating data may become something that is done for its own purpose: art for art's sake. The method of eliciting expert data becomes the focus and how the data are going to be used may become a secondary concern.

Data are valuable if they can be used to instantiate a structure from a model. The data should certainly be both valid and reliable, i.e., they should not be picked willy-nilly from personal experience nor be based on individually held assumptions or beliefs. But the concern for reliability in the data expressed, for instance, as the concordance of expert opinions, the calibration of expert judgments, or the repeatability of simulator sessions, must not be allowed to dominate the efforts. If that happens the data will be of little value because they do not relate to the situations that are analysed. In extreme cases that the basis for choosing a human reliability analysis method may even be determined by the way in which data and estimates are generated rather than by the requirements of the analysis. This is obviously not very desirable.

8. Human Factors Reliability Benchmark Exercise

The many problems that make the assessment of human reliability a complicated affair are clearly demonstrated in the Human Factors Reliability Benchmark Exercise (HF-RBE). The HF-RBE was organised by the Joint Research Centre at Ispra, on behalf of the European Commission, to appraise the state of the art in human reliability modelling and assessment. It was the third in a series of similar benchmark exercises, the two previous ones having been directed at systems analysis (Amendola, 1985) and common cause failures (Poucet et al., 1987). The purposes of the HF-RBE were (Poucet, 1989, p 2):

(1) to compare the various procedures and approaches used to identify various human failure possibilities and mechanisms;

(2) to compare the modelling techniques and data used to quantify the probabilities of human failures;

(3) to assess the degree of consistency in the results and the advantages and limitations of the various techniques used to obtain the results; and

(4) to achieve agreement on the current state of the art of human reliability assessment and, if possible, identify consensus analysis procedures.

Fifteen teams took part in the study representing eleven countries (eight EC countries plus Sweden, Finland, and the US) as well as industry, utilities, licensing organisations and research institutes. The details of the HF-RBE study are described in Poucet (1989) and shall therefore not be repeated here. For the present discussion it is sufficient to note the following:

(1) The benchmark exercise addressed two basic issues, defined in two study cases. Both study cases were described in great detail, and ample documentation was provided. More importantly, all teams were given the same information, including video tapes of crews performing the tasks.

  (a) The first was a test and maintenance study case, where the purpose was to look at how operators performed following routine written procedures. The specific issues looked at were test induced failures, the probability that failures remained latent or undiscovered, and the possibility that a transient would occur because of errors performed in the test.

(b) The second study case was an operational transient. The focus was how operators would diagnose a transient and select their response strategy in a situation where time constraints played a role.

(2) For each study case the teams would first perform the analysis individually, using their own procedures, methods, and data. This served to provide each team with a global overview of the study cases. Following that, a set of common boundaries and assumptions were defined in order to eliminate the differences between the teams in terms of how the analyses were performed.

(3) Although a number of different methods were known to the teams, most of them used at least either THERP (Swain & Guttmann, 1983) or SLIM (Embrey *et al.*, 1984). This is important because it allows a comparison of the results between the teams.

The successful use of a human reliability analysis method can be due to two factors: the efficiency of the method or the experience of the analyst. As indicated above, the exercise was planned in order to provide the greatest degree of commonalty in the analyses, i.e., to test the power of the methods rather than the experience of the teams. The first study case was carried out in three phases:

> "(the) first phase gave an insight in the qualitative approaches used and some indications on the variability that can be expected in quantitative results.
> In the second phase, the scope was limited to selected steps from the third test procedure with the aim to get insight in the cause of the variability in quantitative results... In order to analyse the impact of the data assessment, a third phase was performed in which a problem from the second working phase was recalculated on the basis of a common adopted breakdown and modelling."
> (Poucet, 1989, p. 26).

In view of these precautions it would not be unreasonable to expect that the teams would produce estimates of the events that were very close. They were, after all, using the same methods on the same study cases, and a considerable effort was made to ensure that the analysis was based on the same assumptions. Despite these precautions, the teams varied considerably in the estimates they produced, between teams as well as within teams, typically with a factor 10 or more. (The differences were even larger for the first quantification.) The discussion of the results of the first study case concluded that the major contribution to the observed variability was the differences in modelling, which in turn were caused by the teams using different assumptions, different levels of decomposition, different recovery mechanisms in the models, and a different dependency structure between the interactions (of the tasks).

The second study case used a postulated accident sequence which again was documented in great detail and the procedure was again to go through three phases: qualitative analysis, breakdown + modelling, and quantification. The main methods used were THERP, SLIM, and HCR (Hannaman & Spurgin, 1984). All teams tried to produce probability estimates for not corresponding correctly in the described scenario. The general results were the same as for the first study case, but the spread in the results was larger than before; even if extreme values were excluded the results differed by more than two orders of magnitude!. (The extreme values were 0.5 and 0.005.) Even more interesting was that different single quantification methods apparently gave quite different results. Thus THERP consistently resulted in lower estimated values than the HCR method; in addition, HCR also showed the largest team variability. The third of the major methods, SLIM, gave results that always seemed to agree well with the results from the other methods, although this might be due to the way in which SLIM depends on calibration anchor points (Poucet, 1989, p. 116).

The HF-RBE is valuable because it has clearly demonstrated the strengths and weaknesses of current approaches to human reliability assessment. In particular, it has clearly pointed out how quantitative methods depend on the preceding qualitative analysis, and how both steps depend on the assumptions that are made by the analyst. None of the methods used in this exercise were robust enough to compensate for the differences between individual analysts and analysis teams. It emphasises the point of view there are no "true" values of human reliability estimates that an analysis can find, but that the main purposes of any analysis rather are to clarify the structure of the tasks, the context in which they are carried out, and the conditions that may influence the performance. The weak part of human reliability analysis is the quantification, and the reason is that the quantification is carried out on the wrong basis. As argued above, the quantification of human performance is approached as if it was the analysis of a mechanical artifact or a technological system. But humans are not machines, and should not be analysed as if they were. This is clearly expressed by the conclusions of the HF-RBE (the emphasis is mine):

> "...the problems linked with human reliability analysis are much greater than those in systems analysis. The typical approach used in systems analysis, i.e. to use decomposition, collect data on the component level and integrate those data again in a system model, does not work for analysing complex human interactions. **Man is not a machine and complex interaction cannot easily be decomposed and modelled deterministically into a structure of elementary actions without loosing subtle feedback, feedforward and other dependency mechanisms.** Human behaviour is extremely context dependent and only recently some important factors such as organisational framework and culture have been recognised. The incorporation of such dependencies into quantitative models, if ever possible and desirable, is not for tomorrow."

(Poucet, 1989, p. 118)

9. Short Survey of Human Reliability Methods

Several comprehensive and comparative surveys and evaluations of the most often used human reliability analysis methods have recently been performed (e.g. Swain, 1989; Spurgin & Moieni, 1991). In Swain's study fourteen different human reliability analysis methods were systematically evaluated by means of three main criteria (usefulness, acceptability, practicality) each of which were specified into a number of more detailed criteria as shown in Table 3-1.

Each method was rated by one or two out of a group of eight experts; the intention was to produce a systematic evaluation of the methods using a common set of criteria. These ambitions were somewhat counteracted by the fact that some of the experts were also the authors of some of the methods, and in several cases evaluated the methods they themselves had developed! The outcome was that only three methods

Table 3-1: Classification of Human Reliability Analysis Methods (Swain, 1989)

Usefulness	**Availability** of published information and data
	Quantitative outputs relevant to PRA
	Qualitative outputs useful to NPP operating safety/effectiveness
	Completeness (comprehensiveness) in scope
	Validity/accuracy
	Consistency (reliability)
	Flexibility
	Ease of use
	Traceability
Acceptability	**Evidence** of acceptability
	Responsiveness to qualified criticism
Practicality	**Monetary** cost and time required
	Utility support needs
	PRA team support needs
	Data and information requirements
	Overall considerations
	Summary evaluation of total resources

were rated as acceptable on all criteria. The methods were the Accident Sequence Evaluation Program HRA Procedure (Swain, 1987), the Systematic Human Action Reliability Procedure (Hannaman & Spurgin, 1984), and the Technique for Human Error Rate Prediction (Swain & Guttmann, 1983). Although the effort that went into this study was considerable, the value of the outcome must be judged on a relative rather than an absolute scale - on an individual basis according to needs and experience. Suffice it to say that the authors of these three methods in all cases were among the experts who evaluated them.

Although the criteria used in this study all seem justified they were hardly complete. When the purpose is to evaluate how well a method functions, additional considerations must be included. The most important ones for human reliability analysis methods are robustness and sensitivity.

(1) **Robustness** refers to the ability of the method to accept and withstand non-standard input and non-standard use.

A method will usually work in the hands of an expert, particularly if that expert is the person who has developed the method. But it should also be possible for a non-expert to use the method, i.e., a person who only knows the method from how it is described, for example, in the instructions or user manual should be able to achieve the same results as the expert. In this sense robustness indicates the extent to which the outcome depends on the method itself or on the person who uses it; clearly, the former alternative is more attractive than the latter. Robustness also indicates the method's ability to withstand non-standard input and to use data which are not ideal test-case examples. This is a question of how data are interpreted and where the responsibility for the interpretation lies. If methods mostly are used manually, the interpretation lies with the user. The rules and principles of interpretation must therefore be described explicitly by the method. If methods are used with the assistance of a computer, the rules and principles of interpretation will to a large extent be imbedded in the software. This makes robustness even more of an issue, since the ability to accept possibly misleading input should be considerable.

(2) **Sensitivity** refers to the minimum variation in input needed to change the outcome.

Sensitivity is clearly very important for a method that tries to predict the probability of a certain event or a certain effect. There are two sides to sensitivity. Oversensitivity means that the method is too sensitive or too reactive to changes in the input; this will lead to a distortion of the results, for instance as probabilities that are out of range. Reduced sensitivity means that the method does not respond to changes in the input that should have changed the output; the result is that the outcome may be misleading and give a false sense of safety. In both cases the sensitivity is very dependent on the possibility of calibrating the method, i.e., comparing it to actual results and adjusting the method accordingly. This emphasises the need to have adequate empirical data. The sensitivity should not be adjusted using expert estimates, nor using simulator results. In both cases the data sources are themselves in need of calibration.

The criteria of robustness and sensitivity are closely related to the reliability and validity of the method. Reliability and validity were included in the criteria proposed by Swain, but only on a secondary level. From a practical point of view robustness and sensitivity ought to be more important than, for example, usefulness or practicality. There is little gained by having a method which can be used in practice and which is cost-effective if the results cannot be trusted. To put it very bluntly, the primary purpose of a human reliability analysis method must be to provide information which can be used to improve the functioning of the system, rather than to provide sufficiently many numbers.

In another study the methods were grouped in three main categories called simulation models, expert judgment methods, and analytical methods (Mancini & Lederman, 1989). If this is combined with the distinction between methods based on engineering and cognitive models, we achieve the grouping shown in Table 3-2.

Other surveys have made distinctions between, for instance, time-oriented models and rating-oriented models, which somewhat cut across the one used here. Notice, by the way, that the name of a model or method does not necessarily correspond to the underlying principles; thus, the Human Cognitive Reliability method is based on the engineering rather than the cognitive model, despite its name.

It is clear from this overview that the majority of methods are derived from the traditional, decomposition principle while only a few are based on the cognitive approach. Without judging the merits of any individual methods, the general problems which most of the methods face can be summarised as follows:

(1) Human performance is too complex to subject itself to simple models like those used for component and system reliability; human action and performance cannot be decomposed in a mechanical fashion.

(2) Analytical and expert judgment methods are highly dependent on the judgment of the analyst/expert and inter-judge reliability may be quite low. Therefore, approaches that rely heavily on expert judgment as the basis for establishing Human Error Rates often present insurmountable problems with regard to data reliability.

(3) Methods based on event or fault trees are limited to a small number of well-described cases, because the practical problems and manual efforts needed to perform the analyses is of a considerable size. Simulation methods (whether quantitative or qualitative) are usually more versatile, and can be accomplished in a semi-autonomous fashion.

Table 3-2: A General Classification of HRA Methods.

	Methods based on Engineering Models
Simulation Models	MAPPS (Maintenance Personnel Performance Simulation Model) SAINT (Systems Analysis of Integrated Networks of Tasks) DYLAM (Dynamic Logical Analytical Methodology)
Expert Judgment Methods	Paired Comparison Absolute Probability Judgment Rating-oriented Methods SLIM (Success Likelihood Index Methodology) Influence Diagram Approach STAHR (Socio-Technical Approach to Assessing Human Reliability)
Analytical Methods	**Time Dependent Activities** ASEP (Accident Sequence Evaluation Program) THERP (Technique for Human Error Rate Prediction) SHARP (Systematic Human Action Reliability Procedure) HCR (Human Cognitive Reliability) OAET (Operator Action Event Trees) CM (Confusion Matrix) TALENT (Task Analysis-Linked Evaluation Technique) "Tree of Causes" Variation Trees **Time Independent Activities** Confusion Matrix Component-specific errors Configuration errors
	Methods based on Cognitive Models
AI Models	**CES** (Cognitive Environment Simulation)
Psychological Models	**SRA/GEMS** (System Response Analyser/Generic Error Modelling System). **SHERPA** (Systematic Human Error Reduction and Prediction Approach) **AEAM** (Action Error Analysis Method) **COSIMO** **SRG** (System Response Generator)

(4) A description of actions in terms of binary success or failure states does not account for the full range of human performance. A static description, in the form of a diagram, is usually ill-suited to represent the possibility of alternate choices, as well as the interactions between task steps.

(5) There is, even for engineering methods, a lack of a simplified model that accounts adequately for the various parameters that may affect human performance. This lack is equally severe for cognitive methods.

(6) There is presently a serious lack of data on human behaviour which can be used across situations and populations (cultures and countries).

The inherent weakness of all methods to date is the lack of appropriate empirical data; this has been emphasised by Swain (1989, 1990). This lack of data is due to a number of things, among them the absence of an established operational classification (a recognised error mode taxonomy) and the complexity of the phenomena coupled with the low frequency of relevant events (i.e., accidents). The ordinary solution is to employ expert judgments or estimates in lieu of empirical data. A more costly solution is to use simulators to provide the missing data. But regardless of which approach is used it is important for any human reliability analysis method to replace human judgment and simulation results with empirical data as far as possible, or as a minimum to calibrate derived data with actual data.

9.1 An "Ideal" Method for Human Reliability Analysis

The problem with most of the methods listed above can be illustrated by considering an "ideal" method for human reliability analysis. The "ideal" method is constructed by formulating a set of requirements based on the experience from different methods, analytical studies, and currently accepted cognitive modelling. One proposal for an ideal method is the following, adapted from Rosness *et al.* (1992):

(1) The method should start from the individual steps of an operating procedure, and the error modes which can arise. The steps can be found either through analytical methods or by means of scenario

simulation. The advantages of this choice are that it concentrates effort at the point where the effects of actions can be predicted with some certainty, and where the effort consequently can be concentrated on serious problems. It is also sufficient to find one significant cause, if this is serious enough to justify prevention measures. It is not necessary to identify all possible causes.

(2) If consequences are significant but prevention measures are difficult or expensive, the method should provide an in-depth search for possible causes.

(3) The method should take into account particular patterns of multiple actions, which can arise from knowledge intensive tasks such as judgment and interpretation.

(4) The method should take into account situations where seemingly simple actions may interact with latent failures in the system, thereby triggering serious accidents.

(5) The method should identify general action patterns and general problems, and accumulate these to indicate points of design weaknesses in the man-machine interface.

(6) The method should be equally sensitive to normal and contingency operations and should employ the same framework for analysis in both cases.

(7) The method should use currently accepted knowledge about human erroneous actions and cognitive modelling, but should focus on the erroneous actions which are both serious and relatively probable.

(8) The method should give explicit guidance on the prediction of error probabilities. It should also be flexible and permit modifications based on new experience and/or concepts.

On a first reading most of the requirements of an ideal method seem quite reasonable. The main structure is not too different from the typical "engineering solution" described in Chapter 2; that should not be a surprise, since both examples refer to established practice. A closer look will reveal that the reasonable requirements address the least specific aspects. It is obviously desirable that the method should be able to account for patterns of multiple actions and for the interaction between actions and latent conditions; also that it should point to general design

The Nature of Human Reliability Assessment

weaknesses, be equally good for normal and contingency operations, and be flexible and open to modifications. But even though these are generally accepted requirements, few of the current methods score high on these points.

Another requirement is that the method should focus on serious and relatively probable actions. This makes sense because the analysis cannot be exhaustive; it is therefore necessary to focus on those parts of the performance where unwanted consequences are most likely to occur. All existing methods recognise this requirement and comply with it, although in different ways. An important difference is, for instance, whether the focus is determined prior to the analysis or as a part of it.

A final set of requirements is that the method should begin from the individual steps in an operating procedure, that it should provide an in-depth search for possible causes, and that it should provide explicit guidance on the prediction of error probabilities. These requirements reflect the basic thinking of the decomposition principle and are therefore the most contentious according to the line of arguments promoted here. It is perhaps not altogether surprising that these requirements are also those with which most existing methods comply in one way or another. Some methods may put the emphasis on the search for causes, some on the prediction of probabilities. None of these features are, of course, wrong; any method to analyse human reliability must address the possible causes of unwanted consequences as well as produce an outcome that can be used in practice. But this need not lead to the traditional step-by-step approach which, in many ways, is counter-productive.

One serious disadvantage of focusing attention on individual actions and the individual person is that it becomes difficult to account for combinations of actions (multiple actions) and patterns of interaction. Interdependence between actions, and between error modes, can arise for several reasons:

(1) The actions can be connected in an incorrect procedure, or a correct procedure either chosen to satisfy an erroneous goal or applied under inappropriate conditions.

(2) Two parallel series of actions, possibly carried out by different persons, can come into conflict.

(3) Demands for attention between two conflicting tasks or activities can produce patterns of error.

(4) Erroneous diagnoses, or planning errors, can lead to multiple erroneous actions.

(5) Finally, multiple erroneous actions can be caused by time compression (Decortis & Cacciabue, 1988) where an omitted action is remembered and an attempt is made to recover (sometimes known as the "forget/remember" syndrome).

The "ideal" method combines the virtues and vices of most of the known methods in one. There is probably no method which corresponds completely to the "ideal" method, but it is easy to find similarities with every known method. The "ideal" method is, furthermore, not an ideal in the sense of being something to strive for, but it is useful as a way of highlighting the weaknesses of the established approach. These weaknesses can only be overcome by enlarging the scope from the individual actions or procedure steps to the working situation as a whole. But this requires that the decomposition principle is permanently put in the background.

10. Summary

Human reliability analysis is usually considered to be an engineering discipline whose practitoners strive to comply with the ideal of quantification that has been taken over from the engineering sciences. The basis for the use of quantification was seen as being derived from two main assumptions: the **atomistic assumption**, which means that the whole can be described by aggregating descriptions of the parts, and the **mechanistic assumption**, which means that the human mind - as well as man in general - can be described as a machine. The goal of quantification has proved very hard to achieve if the quantitative results are to have any meaning, and the question is raised whether that goal is indeed the right one. A brief comparison of humans and machines in relation to the set of characteristics that are pertinent for reliability analysis (such as design principle, model availability, capacity, and reliability) concluded that the differences are so fundamental that the straightforward transfer of methods from one area to the other is not advisable.

The nature of human reliability assessment is nicely illustrated by considering how data analyses can aim at either curve-fitting or model identification. In the former case the exercise is mostly a demonstration of mathematical skills; given enough patience it is usually always possible to find a mathematical model that will fit a set of data. In the latter case the purpose is to find an isomorphic relation between the data and the model. This is much harder because the set of parameters are defined and given meaning by the underlying model; there are thus fewer degrees of freedom for the analyst.

Practitioners of human reliability analysis often seek to develop better models. Before this is done it is, however, necessary to specify what the models are needed for, i.e., what they should be about. Using the Law of Requisite Variety it was argued that the model should focus on the uncertainty of the system's performance and try to capture that. This raised the question of whether the uncertainty is due to insufficient information about an otherwise deterministic system, or due to the inherent uncertainty of the system itself, in particular when the system is or includes the human mind. The epistemological choice can conveniently be sidestepped by concentrating on describing the system rather than explaining its inner mechanisms. The target was consequently defined to be the development of adequate ways of describing the performance of an interactive system as a whole in a context - in stark contrast to the atomistic and deterministic assumptions.

A further issue is that it is not always easy to get the data that are needed for the study of human reliability. There are three reasons why data gathering is not a straightforward matter: it may be difficult to observe the right type of events, it may be difficult actually to do the data collection, and it may be difficult to specify in advance which data to look for.

Data collection is not simply a matter of observing and recording everything that can be found in a situation. Data collection is only the first step in the path from events to concepts, and it must be properly tied to a set of conceptual descriptions. Data collection always represents a choice, hence a selection; it is important that the nature of this selection is carefully matched to the following analysis and to the overall purpose of the study. One way of accounting for that is by using a description of the various stages that data/information pass through: raw data, intermediate data format, analysed event data, conceptual description, and competence description. This description highlights the dependencies between the different levels and the way in which filtering and processing gradually makes the transformation from observation - that which is recorded of the event - to inference - that which is assumed to be the underlying causes.

A specific issue in the study of human reliability is the use of controlled experimental conditions, which here is referred to as micro-worlds. It is argued that micro-worlds must meet the following three conditions: a micro-world must include the underlying regularity of the dynamics that characterise the real world; a micro-world should not seduce the person to develop a complete understanding of the situation; and a micro-world should display the same variation in information quality as the real world. In practical problems people basically try to make enough sense out of the situation to be able to act. Although we introspectively have a reasonable feeling for how this is done, it is still far from easy to turn this into prescriptions for data collection and paradigms for investigation.

The practical assessment of human reliability very soon runs into the problem of getting the data needed for quantification. There are three main sources for such data: empirical data based on observations and reports from the actual working environments, data from simulators and simulations, and expert judgments of various types. The predominantly quantitative approaches create a substantial need for data for all of the identified "elementary" events, preferably in the form of human error rates. Since the diversity of events is overwhelming, neither empirical data nor simulator studies can fulfil the need. Only expert judgments are flexible enough to be applied to any conceivable situation. This, unfortunately, is of limited value unless the estimates can be calibrated and the whole problem therefore easily becomes very hard to manage. The problems were illustrated by going through a characteristic procedure for using expert judgment data.

Practical experience, supported by a benchmark study, shows that different HRA methods usually give different numerical results - and even that the same method may give different results when used by different analysts. It is sometimes argued that this is of secondary importance, since it is the insight gained by doing the HRA that is the most important. If different HRAs result in the same recommendations despite the dissimilar numerical outcomes (probabilities), then one may rightly ask why these outcomes are needed. Conversely, if different methods result in different recommendations, the question becomes which method is to be trusted.

The number of known methods for human reliability assessment is surprisingly large given that the whole field is relatively young. There have been several attempts at classifying and comparing the various methods, but none have provided sufficiently objective results. Among the many criteria that can be used to describe the candidate methods, two seem to be rather important: robustness and sensitivity. A method is robust if the outcome depends on the method rather than on the person who uses it; it is sensitive if there is a balanced relationship between variations in the input and the output.

The best known methods were listed and their common features were briefly characterised. The problems with most of the methods were illustrated by considering what the requirements to an "ideal" method for human reliability analysis might be. Although the requirements pointed to aspects of the method that were clearly significant, the requirements were also rather imprecise in the sense that they would be hard to turn into concrete specifications or procedures. This is probably quite characteristic: the methods each excel in their own way, but their individual advantages may be a question more of feasibility than of necessity. Most of the methods present convenient technical solutions to a problem, without putting much effort into considering whether the problem they solve is indeed the right one, hence what the nature of human reliability assessment really is.

4.
Fundamentals of the Model

1. Metaphors and Models of Cognition

In order to assess and analyse the Reliability of Cognition it is necessary to have a clear position on such issues as human cognition and the nature of work. This is usually expressed by saying that it is necessary to have an adequate model of the human - a cognitive model or a metaphor for human information processing. The need for a model has already been mentioned, e.g. in the discussion of model identification *versus* curve fitting and in the discussion of data collection (both in Chapter 3). In this chapter the discussion will focus on ways of describing human performance, in particular the ways in which human performance is explained and what the underlying assumptions are. This will be done by presenting three different metaphors. A **metaphor** is a way of conveying a specific meaning, i.e., to speak of human cognition as if it was like something else. A **model** is a deliberate or intentional ordering of components or concepts, for example, to facilitate the understanding of something. Strictly speaking a model is therefore not the same as a metaphor. Two of the three metaphors mentioned in the following are also the basis of typical cognitive models, while the third rather corresponds to a paradigm. Although the three metaphors described below are not exhaustive they do represent three very characteristic ways of considering humans and human action.

1.1 Stimulus-Organism-Response

The most pervasive model of humans is probably the Stimulus-Organism-Response (S-O-R) paradigm. Another way of stating this is by saying that

Fundamentals of the Model

the **response** is a function of the **stimulus** and the **organism** (or rather, the current conditions of the organism), i.e.,

R = f (S, O)

This is a classical psychological paradigm which, in one way or another, has been present in psychological theories from the very beginning. It is very similar to a black-box approach or to the view of the human as a Finite State Automaton (e.g. Arbib, 1964). The S-O-R paradigm is, however, considerably older than the notion of the Finite State Automaton. The S-O-R is no longer widely accepted as a proper paradigm in psychology, although it still rears its ugly head in different disguises, e.g. information processing psychology and decision making (cf. below). It is nevertheless extensively used in the technological fields, as the following quotation shows:

> "The common sequence of three psychological elements that is basic to all behavior, namely perception, information processing, and action represent what is conventionally referred to by psychologists (*sic!*) as the S-O-R paradigm: stimulus-input, organism-mediation, and output-response. These three behavioral elements are the essence of most human activities in the sense that a stimulus acts upon an organism to effect a response."
> (Park, 1987, p. 13).

This view may seem simple-minded when stated as bluntly as here. But a little thought will show that it is the same notion that is at the root of common engineering descriptions of operators, as, for example, in supervisory control theory. An example of that is the typical model from optimal control theory (e.g. Johannsen, 1990) or the Observer/Controller/Decision model as described by Stassen (1986), cf. Figure 4-1.

The model depicted in Figure 4-1 represents the relations between a supervised system and a human supervisor model. The supervised system, or the process, is depicted by the system dynamics and the display. The inputs to the supervised system are either external disturbances, control actions, or requests for information (observation actions). The outputs to the human supervisor are simply the observations. (Presumably there are other outputs from the process, but these need not concern us here.)

Metaphors and Models of Cognition

Figure 4.1: The observer/controller/decision model.

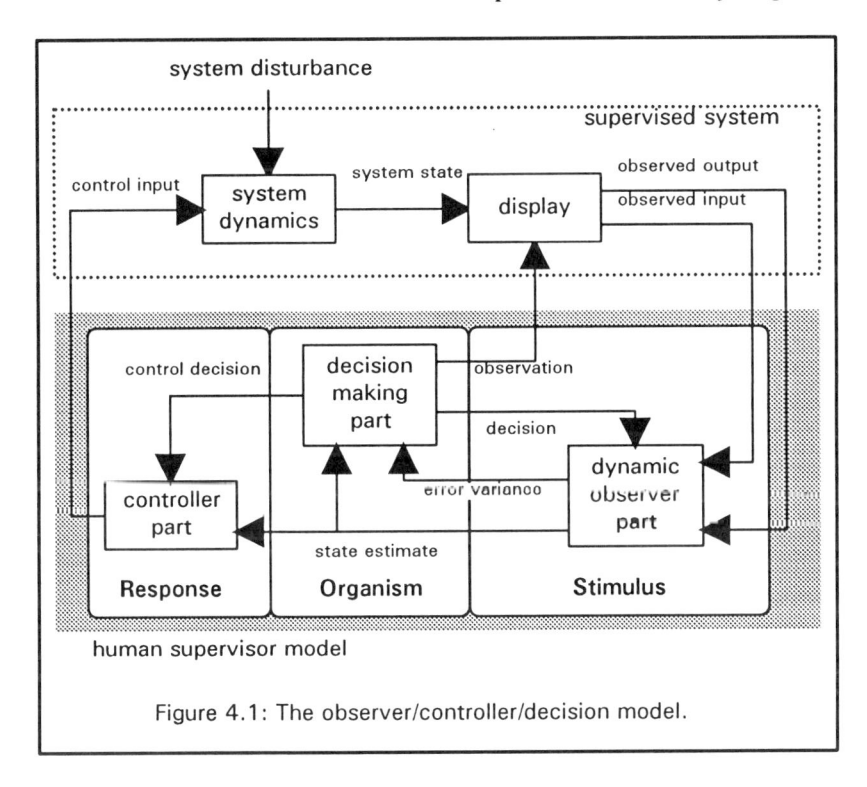

The human supervisor model in Figure 4-1 is clearly separated into three parts called stimulus, organism, and response. The stimulus part processes the observations - which may have been requested and the processing may therefore also be biased by the request. The observations are processed and the processing may give rise to a decision, which can lead to a control decision or a request for observation. The processing may also be a "simple" state estimation which generates a response, i.e., a control action of some sort. The contents of the S-O-R parts of the model will in this case presumably be described in quantitative terms, e.g. as sets of differential equations; they could in principle be describe as Finite State Automata. But despite the level of description of the individual parts, the fact remains that the model as a whole basically is an S-O-R model.

1.2 The Human as an Information Processing Mechanism

A more recent metaphor is found in the view of the human as an Information Processing System (IPS or information processing mechanism). Although apparently less simplistic that the S-O-R metaphor, it is in reality not so. This is most clearly seen in the tenets of the so-called computational psychology which is the main proponent of this idea. According to this view, mental processes are considered as rigorously specifiable procedures and mental states as defined by their causal relations with sensory input, motor behaviour, and other mental states (e.g. Haugeland, 1985) - or in other words an automaton, or even a Finite State Automaton. This corresponds to the **strong** view, that the human **is** an information processing system or a physical symbol system. There is also a weak view, according to which the human can be described **as if** it was an information processing system; there is a world of difference between these views. The strong view has been advocated by, for example, Newell & Simon (1976), while the weak view has been championed by Weizenbaum (1976) or Searle (1980). In this decade long debate everyone seems tacitly to have accepted that a human is **at least** an information processing system. The arguments have essentially been about whether a human is **more** than an information processing system, i.e., whether there is a need for intentionality (Searle, 1980) or "thoughts and behaviour" (Weizenbaum, 1976). I will maintain that although it may often be useful to describe cognition **as if** it was information processing, there is no need to assume that it really **is** that. Rather than discussing what the inner mechanisms of cognition might be we should be concerned with how cognition controls performance.

The classical information processing view has been caricatured by Neisser (1976) and shown in an extended form in Figure 4-2, but is still found in the common information processing models. The above mentioned Observer/Controller/Decision model is also an example of that. So is the well-known Step-Ladder Model (SLM) of decision making described by Rasmussen (1986). In an IPS model the internal mechanisms are typically described in far greater detail than in an S-O-R model - at some point focusing on the **O** almost to the exclusion of the **S** and the **R** (as in the notion of skill-based, rule-based, and knowledge-based performance). It is nevertheless not difficult to appreciate that they have two fundamental similarities: the sequential progression through the internal mechanism, and the dependency on a stimulus or event to start the processing.

One of the latest versions of this metaphor is the notion of human cognition as a fallible machine (but none the less a machine!). This idea is most clearly expressed by Reason (1990), who provides an indirect

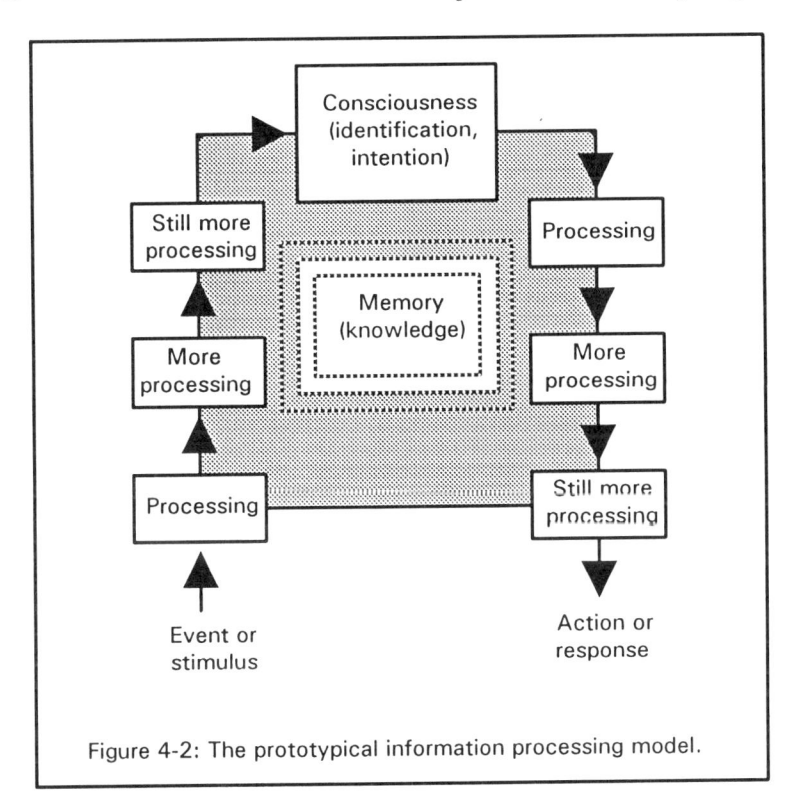

Figure 4-2: The prototypical information processing model.

definition of the fallible machine by posing the question: "What kind of information-handling device could operate correctly for most of the time, but also produce the occasional wrong response characteristic of human behaviour?" (Reason, 1990, p. 125). In other words, the fallible machine is an information handling device, a mechanism, which sometimes produces correct and sometimes incorrect results. The notion of the fallible machine is a step forward from the earlier idea of specific error producing mechanisms, but is still a very mechanistic way of describing human cognition, which assumes a host of unknown, but very convenient, functions.

1.3 The Cognitive Viewpoint

A third metaphor is the cognitive viewpoint which has already been mentioned in Chapter 1. The cognitive viewpoint is exemplified and explained by cognitive systems engineering (Hollnagel & Woods, 1983;

Fundamentals of the Model 150

Woods & Roth, 1990). The main difference from the S-O-R and the IPS metaphors is that cognition, which in all cases is considered to be the basic phenomenon to explain, is viewed as **active** rather than **re-active**, for instance as a set of self-sustained processes or functions that occur simultaneously. The consequence of the cognitive viewpoint is to focus on **overall performance** as it appears rather than on the **mechanisms** of performance. This is partly because behaviour is not simply a function of input and system (mental) states, and partly because the complexity of the inner "mechanisms" is too high to be captured adequately by a single theory.

One fundamental difference between the cognitive viewpoint and the S-O-R type paradigms is that human cognitive functioning, hence also human performance, is seen as cyclic rather than sequential. Cognition does **not** start with an external event or stimulus and end with an action or

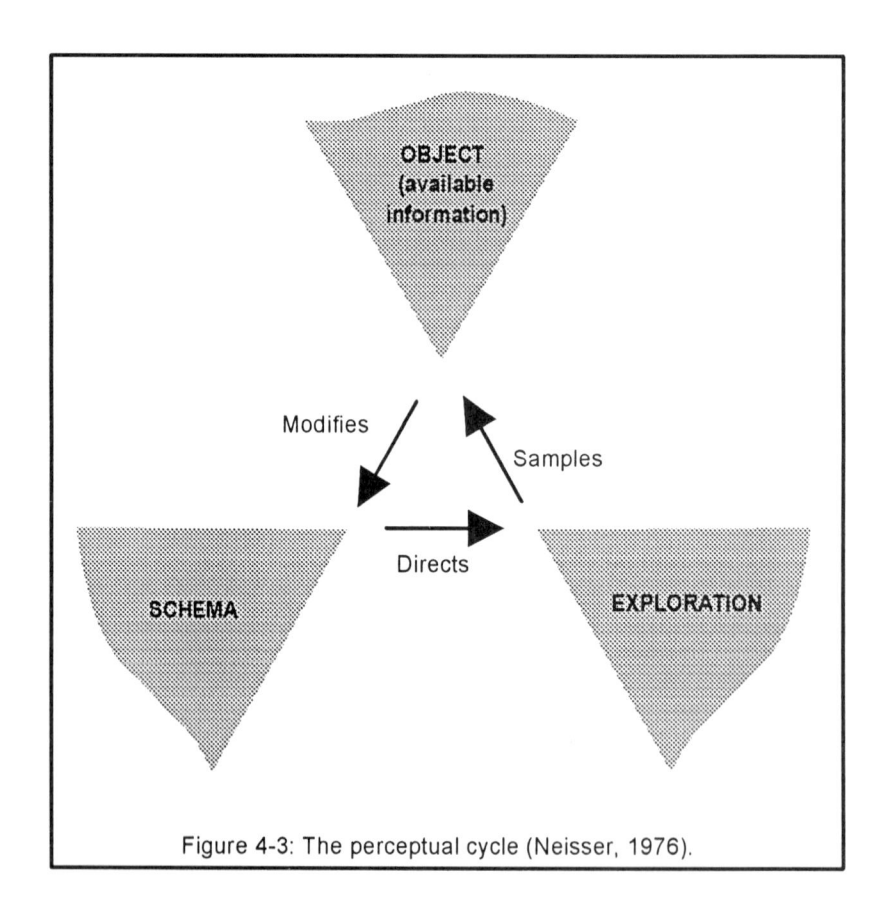

Figure 4-3: The perceptual cycle (Neisser, 1976).

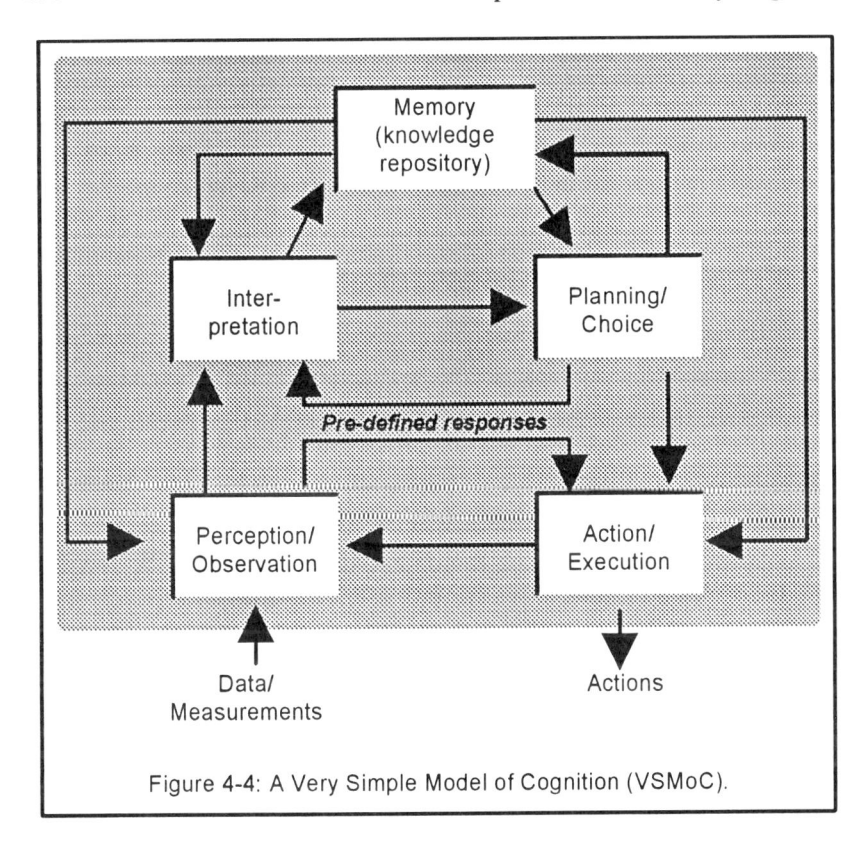

Figure 4-4: A Very Simple Model of Cognition (VSMoC).

response. In a very simple sense this can be shown as in Figure 4-3, which presents the classical exponent of this view, the perceptual cycle proposed by Neisser (1976). A more elaborate version is Figure 4-4 which shows the Very Simple Model of Cognition (VSMoC) described in Hollnagel & Cacciabue (1991). The fundamental difference between the cyclical and sequential views springs from the acknowledgment that human action is determined as much by the context (the task and the situation) as by the inherent traits and mechanisms of human cognition. In this view humans do not passively react to events; they actively look for information and act based on intentions as well as external developments. We may easily be mislead to think so, because we **observe** events and reactions and **interpret** them using our deeply rooted model of causality. Yet an observable action does not need to have an observable event as a cause; conversely, an observable event does not need to result in an observable action.

The difference between the two views is not just a matter of words. In order to show how deep it goes, the remaining part of this chapter will be used to compare two models which are typical of either view. This will finally lead to the formulation of a model of cognition which can be used to develop a method for analysis of the Reliability of Cognition.

2. Procedural Prototype Models of Cognition

One way of classifying models of cognition is to make a distinction between those which emphasise the sequential nature of cognition and those which view cognition as being determined by the context. The former can be called procedural prototype models and the latter contextual control models.

(1) A **procedural prototype model** implicitly expresses the view that there exists a characteristic or pre-defined sequence of (elementary) actions which represents a more natural way of doing things than others, or even that a certain sequence or ordering is to be preferred. According to this view the expected next action can, at any given time, be found by referring to the ordering of actions implied by the prototype. A procedural prototype model of cognition is therefore a **normative** description of how a given task should be carried out; an example is the description of decision making referred to in the following.

(2) A **contextual control model** implies that actions are determined by the context rather than by an inherent relation between them. It is therefore not possible *a priori* to describe procedural prototypes or "natural" relations between actions. The contextual control view instead assumes that the choice of the next action at any given point in time is determined by the current context (the situation cues) - although that choice need not be a deliberate or conscious decision by the person. A contextual control model therefore concentrates on how the **control** of the choice of next action takes place rather than deliberating on whether certain sequences are more proper or likely than others.

The two views do not imply that human action is either of one or the other type. The aim is rather to make clear that the actions we observe can be analysed and understood in different ways, of which procedural prototypes and contextual control simply are one set of alternatives. The two views differ in how they account for the underlying causes of actions, hence in their notion of what cognition is and which role it plays in human performance. This has further consequences for how the reliability of cognition is described and for the ways in which it can be analysed and assessed.

Both views recognise the fact that there are recurrent sequences or patterns of action. Any systematic observation of human performance will clearly show that we tend to do things in certain ways, that actions may follow each other in specific sequences. To some extent this is due to the inherent logic of what we do: trying to hit something it is obviously necessary first to take aim (Hollnagel, 1984). Thus, in some cases we cannot do action B before we have done action A, because certain preconditions have to be fulfilled: I cannot bake a cake unless the oven has been heated (cf. the discussion of the Goals-Means Task Analysis in Chapter 5). Another reason is the constraints that are in the environment. If object X is supported by object Y, I can usually not remove object Y without first doing something about object X, either removing X itself, providing an alternative kind of support or suspension, etc. (This has been the main focus in much of the classical research in robotics, e.g. Winston, 1984.) These constraints may be due to the laws of nature, or be the result of the way in which systems have been designed. In fact, system design often has the purpose of forcing the person to do things in a certain sequence, by introducing clever constraints in the control options. This goes from controlling a nuclear power plant to the notorious fool-proof camera.

The difference between the two views is therefore not in the phenomena they address but rather with the way it is done, i.e., the concepts and relations that are used for descriptions and explanations. It is an important part of the IPS view that there is a clear sequence or serial ordering to mental operations (Anderson, 1980, p. 13). The procedural prototype models appear to see the sequences as necessary and to take the recurrent patterns for granted; as a result the sequences or patterns are reified as templates or standards for action. Much effort is consequently spent on describing the ways in which actual performance may deviate from the ideal prototype. The contextual control models instead may look

Fundamentals of the Model 154

to the factors that produce the recurrent patterns. The investigations are consequently shifted towards the way in which actions are actively selected, as part of the person's coping with the current situation.

2.1 The Step-Ladder Model

A good example of a procedural prototype model of cognition is the description of decision making, commonly is known as the Step-Ladder Model (e.g. Rasmussen, 1974; 1986); this model is shown in a simplified version in Figure 4-5. In this figure the eight steps of decision making are shown as a straightforward sequence, going from activation to execution, only "interrupted" by the possible loop between interpretation and evaluation.

(The Step-Ladder Model is convenient to use as an example only because it is so well known. The choice does not in any way imply that it is in better or worse than other sequence models. It is useful simply because it very clearly demonstrates the main features of sequence models. For a discussion of the origins of this type of model see Hollnagel

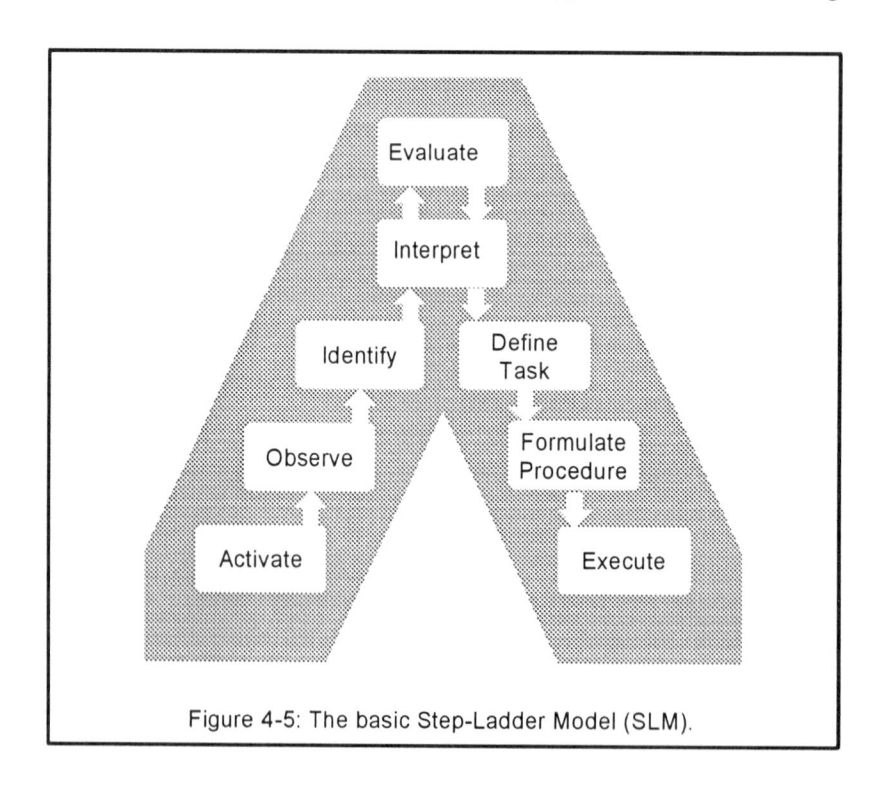

Figure 4-5: The basic Step-Ladder Model (SLM).

(1984) or Sanderson & Harwood (1988).)

The SLM clearly describes a procedural prototype; but while one might say that decision making ought to take place as a series of steps, it rarely does so in practice. This is acknowledged even in the classical version of the SLM (Rasmussen, 1986) which introduced a considerable number of by-passes or short-cuts between the steps to account for empirical findings. The caption from the original figure is interesting reading because it clearly states the underlying assumption:

> "Rational, causal reasoning connects the 'states of knowledge' in the basic sequence. Stereotyped processes can by-pass intermediate stages. Together with the work environment, the decision sequence forms a closed loop. Actions change the state of the environment, which is monitored by the decision maker in the next pass through the decision ladder."
> (Rasmussen, 1986, p. 7)

The SLM contains a basic sequence of steps which is carried out in turns, i.e., one time after another. Changes in the environment are, in terms of the SLM, described as changes in the input to decision making, but do not directly have any effect on how the steps are ordered; their assumed "natural ordering" is maintained from one cycle to the next. The by-passes are variations of or deviations from the prototypical sequence; they may skip one or more steps but the underlying step-by-step progression is immutable.

The shortcomings of the SLM become clear when one considers more elaborate versions which contain a large number of by-passes; an example is shown in Figure 4-6. The introduction of by-passes brings the problem of control into view by prompting the question of what causes a by-pass to be made. In the simple version of the model the control of actions was an integral part of the model: when a step had been completed, the person would go on to the next step, until the decision had been made and the action carried out. Because the control was not explicitly described, it did not really have to be considered (Hollnagel, 1984). But in the extended version the control issue looms large and is basically unsolved. Even the more complicated versions of the SLM are incomplete because they still make the assumption that the person somehow starts a pass with the first step and continues until it has been completed, with or without by-passes. Yet practical experience clearly shows that people often interrupt their decisions, leave them unfinished,

Fundamentals of the Model 156

Figure 4-6: The extended Step-Ladder Model (SLM).

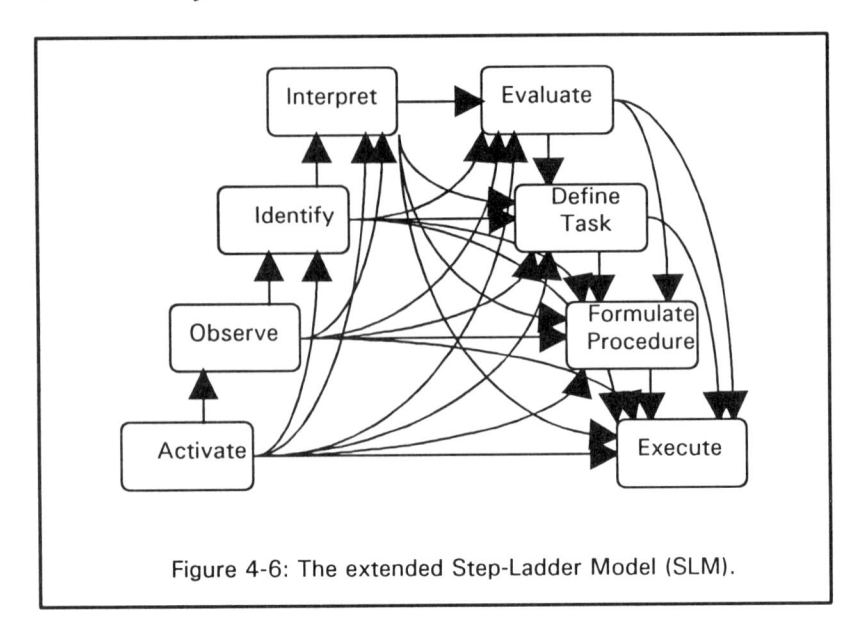

or imbed one decision in another. All this makes the control problem more difficult to solve. The straightforward solution of assuming that the control is carried out as a higher level decision by recursive use of the SLM only aggravates the problem, as shown by Lind (1991), for example. The procedural prototype model describes decision making as if the person attempted to make rational progress through the various phases. But since this rarely happens we are faced with a model that only poorly matches the variety of the target system (the person). The procedural prototype model is therefore inadequate.

2.2 Predominance of Procedural Prototype Models

Although all procedural prototype models suffer from the same shortcomings as the SLM they seem to dominate current thinking and current models of cognition (as well as models for most other psychological phenomena). There may be several reasons for that.

Firstly, any behaviour or performance looks sequential once it has occurred. Most performance descriptions are organised with time as the basic dimension, hence order the events by the sequence in which they occurred. Events are naturally sequential because they are executed one by one. But this is not sufficient to assume that there is an underlying causal mechanism or organising principle which imposes this ordering.

Secondly, there may be functional dependencies or relations between different actions. These relations correspond to what in task analysis are called pre-conditions, i.e., conditions that must be fulfilled before an action can be carried out. (This is developed further in Chapter 5). If we continue to use decision making as an example, one such relation is that before the decision maker selects an alternative (or before two or more alternatives are compared), the alternatives must have been formulated and identified. There is thus an implied sequentiality between "alternative selection" and "comparison" because the former is a pre-condition for the latter.

If the principle of functional dependency (pre-conditions) is combined with the empirical or practical sequentiality of performance, it is very tempting to assume that whatever governs performance - and therefore also the nature of models describing performance - must include sequentiality. Put differently, the fact that past performance without fail can be described in one-dimensional time misleads one to describe potential or future performance in the same way. Hence, procedural prototype models arise. But as the preceding arguments have suggested, this notion of procedural prototype models may not be strictly necessary; as a matter of fact, it can even be suggested that the idea of sequentiality is misleading both for theory and analysis. Anyone who has ever tried to analyse performance protocols knows that unbroken lines of actions which correspond to, for example, decision sequences are few and far apart. The procedural prototype model of decision making is therefore a model which ill fits the observed variety, and the principle of sequential ordering may therefore be considered an artifact of the method of description - amplified by the ingrained notions of determinism and causality which dominate Western European thinking and philosophy.

2.3 Loose Ordering (TOTE)

Although procedural prototype models of cognition prevail, they need not do so. The classical Test-Operate-Test-Exit (TOTE), as shown in Figure 4-7, is a good example of a model description which includes sequentiality but only in an loose sense (Miller, Galanter & Pribram, 1960). In this model the T(est) must precede the E(xit), but this is due to chosen formalism rather than because of inherent characteristics of the T(est) or

Fundamentals of the Model

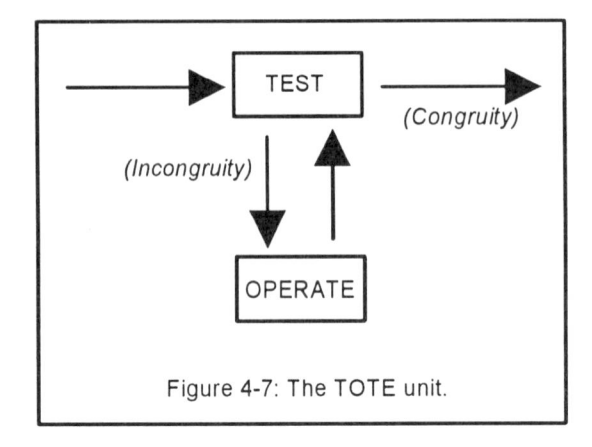

```
        ┌─────────────────────────────────┐
        │            ┌──────┐              │
        │    ──────▶ │ TEST │ ──────▶      │
        │            └──────┘   (Congruity)│
        │      (Incongruity) │  ▲          │
        │                    ▼  │          │
        │            ┌─────────┐           │
        │            │ OPERATE │           │
        │            └─────────┘           │
        │                                  │
        │       Figure 4-7: The TOTE unit. │
        └─────────────────────────────────┘
```

the E(xit) operations.[1] Similarly, the sequentiality of the TOTE is due to the conventional ordering of tasks and pre-conditions rather than because of characteristics of the T(est) or the O(perate) functions *per se*; the emphasis is on control rather than on patterns of action. The notion of pre-conditions define a functional dependency which in turn imposes a certain sequence, in the loose sense, between events. But this is a far cry from the detailed sequentiality that is imposed by a decision model such as the SLM and is not sufficient to account for that. In fact, the TOTE is so abstract that it does not refer to any specific context but rather describes universal features of performance - or rather, of control. It can therefore easily be reconciled with the notion of a contextual model of control where the ordering of events cannot be derived from characteristics of individual (elementary) actions.

The value of using a loose ordering principle, such as the TOTE, can be seen by applying it to the widely used concepts of skill-based, rule-based, and knowledge-based behaviour (cf. Rasmussen, 1986). The assumption of the three levels is intimately connected with the SLM, hence with the inherent sequentiality of that: each level represents a subset of the complete SLM sequence.

The definition of skill-based behaviour is that it is smooth, automated, and highly integrated performance that does not require attention or conscious control (Rasmussen, 1986, p. 100). However, even the most routine performance requires attention in some measure - either regularly or at intervals. Take, for instance, walking. We can walk and

---

[1] That the T-E sequence is not strictly necessary can be seen from the cases where the person abruptly exits from the task.

talk at the same time, but if we think of something difficult or try to remember something, we involuntarily stop in the walking. The stop is not only because walking requires steady awareness of the surroundings (navigation) but mainly because walking, despite being a highly automated skill, in itself requires some amount of attention. Only reflexive action, like breathing, can go on without attention. A better definition is therefore that skill-based behaviour only requires a limited amount of attention, either continuously or at intervals. In contrast, both rule-based behaviour and knowledge-based behaviour require considerable attention.

In the TOTE paradigm one might assume that the T(est) would roughly correspond to rule-based behaviour and knowledge-based behaviour while the O(perate) would correspond to skill-based behaviour. The TOTE, however, emphasises that all performance is a mixture of T(est) and O(perate) - or a mixture of skills and procedures. Thus even on the level of skill-based behaviour, the performance is a mixture of T(est) and O(perate) - although highly automated so that the proportion has been changed to a small amount of T(est) and a large amount of O(perate), and with relatively few demands to attention. The difference again to rule-based behaviour or knowledge-based behaviour is simply that this can be described as TOTE with a high degree of attention in both phases, and with a high proportion of T(est) in relation to O(perate). We can thus characterise performance by noting the ratio of T(est):O(perate) as well as the ratio of attention required in either T(est) or O(perate). This provides a continuum of performance categories while maintaining that performance always is a mixture - rather than being pure skill-based behaviour or pure rule-based behaviour. This first of all eliminates the mistake of trying to force a person to work on, for instance, the level of rule-based behaviour (e.g. Rasmussen & Vicente, 1987). It furthermore also explains the difficulty in finding shifts between performance levels; there are simply no shifts but only gradual changes!

3. Contextual Control Models of Cognition

The prevalent point of view in many models of cognition is that human behaviour is constrained or governed by prototypical procedures, inherent strategies or dominant ways of doing things - such as diagnosis, decision making, scheduling, controlling, etc. These constraints in essence prescribe what the performance ought to be. Much effort is therefore spent on determining the extent to which actual performance complies with "prescribed" performance, and how far performance can be shaped or forced to conform to the normative description, e.g. through design of interfaces, working conditions, procedures, training, organisations, etc.

Although the procedural prototypes have so far dominated theories of human cognition, their weaknesses have been duly noted. The notion of a procedural prototype was discussed by Hollnagel et al, (1981), but only as a question for research; it was pointed out that the validity of prototypical performance could be tested experimentally by comparing it with actual performance. During the late 1980s a growing number of scientists made the observation that the procedural prototype models did not properly reflect how people performed; the influence of the context was obvious from many studies but was difficult to include in the models except as "events" or "stimuli": the coupling between the person and the environment was conspicuously absent.

In a concise analysis of a number of empirical studies Bainbridge (1991) has suggested that people should be modelled in terms of cognitive "modules" of processing, where a "module" refers to a specific cognitive goal (in distinction to task goals). In this way the control of the activity is determined by the sequence of cognitive goals rather than by the inherent structure of the activity; the sequence of cognitive goals is, in turn, determined by the context: the environment and the previous development. The notion of the cognitive processing module is illustrated in Figure 4-8, and is given the following description:

> "1. The aim of each processing module is to meet a particular cognitive goal, such as to find what is the present temperature, or to choose an action.
> 2. The 'answer' is found (stepped arrow) by referring to the environment or to knowledge, or by further cognitive processing...
> 3. The modules actively search for the information they need, rather than passively responding to information as it arrives.
> 4. Finding the relevant answer is done within the context provided by existing working storage, represented by the left of the two boxes.
> 5. The answer, represented by the right hand box, itself becomes part of working storage. It either provides the answer for a superior goal, or it becomes part of the data used in later cognitive processing ...

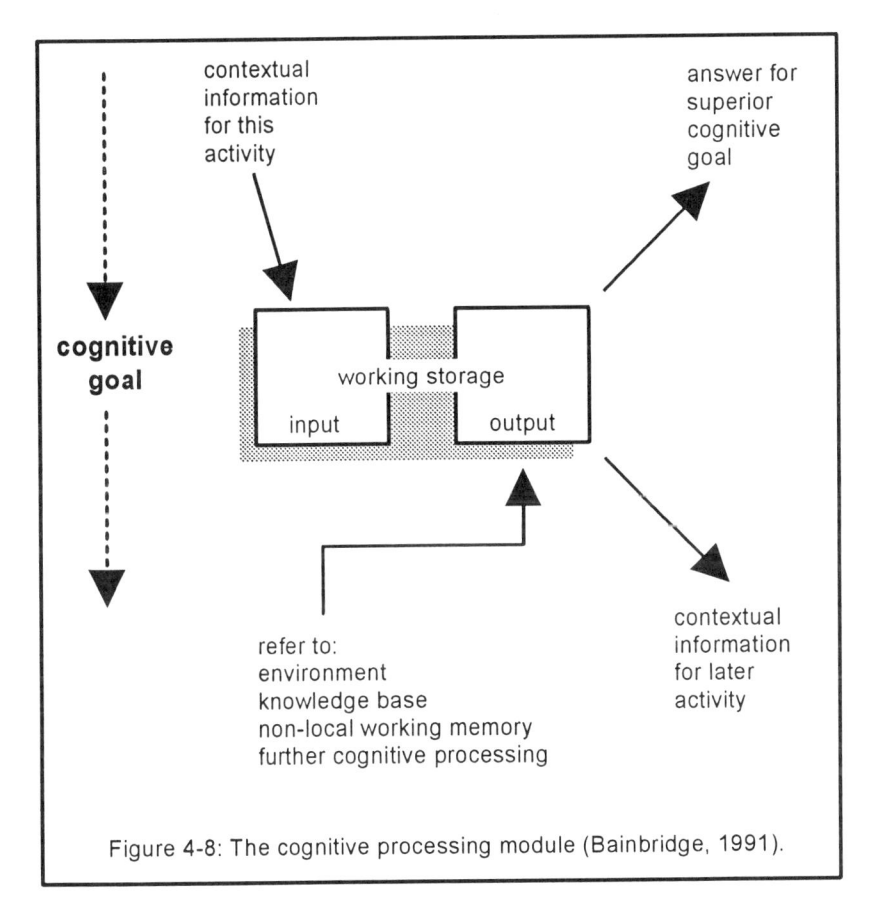

Figure 4-8: The cognitive processing module (Bainbridge, 1991).

6. The modules communicate with each other via working storage, and are sequenced via working storage."
(Bainbridge, 1991.)

This description clearly offers a view that is orthogonal to the procedural prototypes. In fact, the description is based on modules of action (cognition) that are bound by what actually happens rather than by the "mechanisms" of human cognition or information processing. Although this approach is not entirely complete (notably the concepts of working storage and sequencing) it does show that important aspects of performance are missing if the context is not taken into account.

The contextual control view regards human performance as determined, by and large, by the context. A person is, by definition, capable of doing a large number of things, i.e., has a large number of possibilities or functions available. The selection from these possible actions is not determined by normative characteristics of the elementary or component actions, but by the existing needs and constraints. There is therefore no pre-defined organisation of component actions, although there may be frequently recurring patterns or configurations, as described above. The scientific challenge is to provide a reasonable account of how this can be the case, without making any unnecessary assumptions about human cognition - whether it is disguised as an internal information processing system or as something else.

3.1 Competence and Control

A contextual control model of cognition consists of a **model of competence** together with a **model of control.** The arguments for that have been developed in the preceding sections and are shown graphically in Figure 4-9. They are simply that as the number of paths through a procedural prototype model grow, the more difficult it becomes to describe which path a person will follow in a specific situation. The problem is not so much **what** the person does, or what he can do, but **why** he does it. It therefore seems reasonable to separate the issue of control from the issue of competence - not only in terms of the focus of analysis but in terms of the underlying concepts and the corresponding model.

The model of competence is necessary because the person has certain competences, corresponding, for example, to the basic cognitive functions in decision making, which are applied according to the current needs and demands. The number of possible actions mainly depends on the level of detail or the granularity of the analysis. The set of possible actions should not be normative; the basic notion is that a person is capable of doing many things, but that some of these possible actions may be more readily available than others due to either training, experience, situation cues, etc. In some situations the set of possible actions may be very large, in others it may be quite small; it may vary considerably both between and within persons and situations.

In the model of competence the possible actions provide a set to choose from, which means that the person cannot do something which either is not available as a possible action or which cannot be constructed or aggregated from the available possible actions. Possible actions may be found on different levels of aggregation, i.e., a mixture of simple and

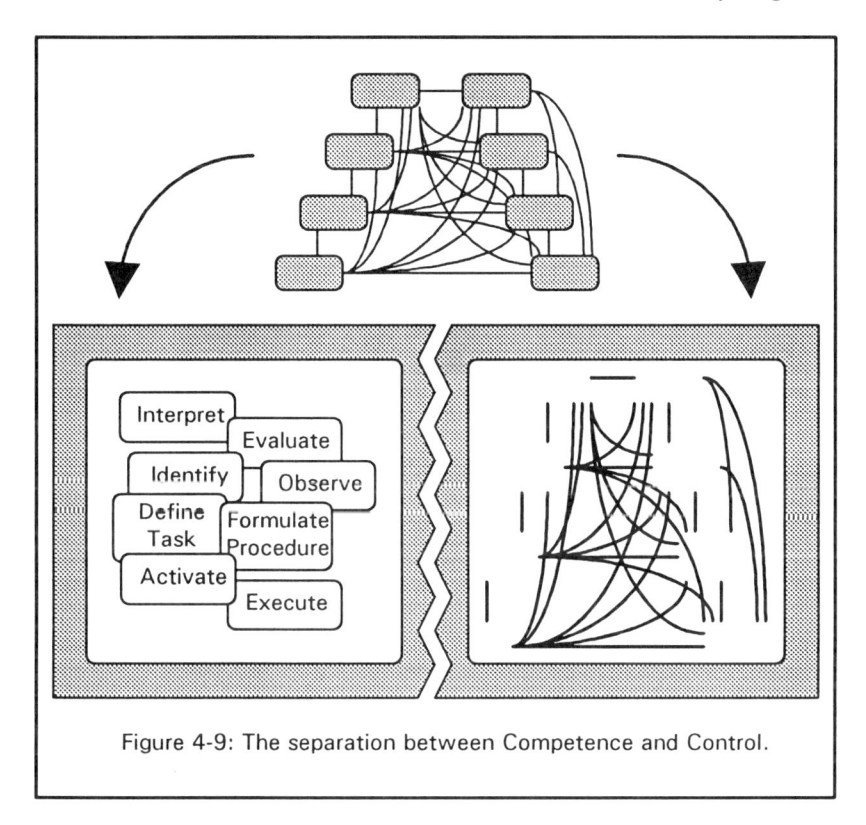

Figure 4-9: The separation between Competence and Control.

composite actions. Common to them is only that they are ready to use in a given context. One may speculate that the competences are hierarchically organised, so that higher level competences can be traced back to a smaller set of more fundamental, lower level competences. This may even in turn be used as an argument for a limited set of really basic cognitive functions. Pursuing that is, however, a theoretical exercise which takes us too far away from the present goal. For the current purpose it is simply assumed that for each given task and/or working context it is possible to define a small set of frequently recurring actions.

The control function can be defined and described in a number of ways. An important issue is the granularity of the description and the mode of functioning. Here the Law of Requisite Variety (Ashby, 1956) can be of assistance. The modelling of cognition must refer to a pragmatic definition of the variability that needs to be modelled. The nature and range of this variability - but not its origin - will provide guidance for how the control function is accomplished and for how much the control function should be able to do. From the practical point of view, it is sufficient to match only the required variety. In other words, it is enough to model what actually happens rather than what a specific theory says should happen.

In the contextual control model the pre-conditions, which are part of how actions are described, constitute the basis for tactical or strategic planning of actions. It is because we **know** that there are certain pre-conditions, and because we use them as a basis for planning that there is a certain sequentiality or ordering. Actions are only meaningful in a context, and an important part of every human undertaking is to establish the context and ensure that the pre-conditions obtain. But the resulting sequentiality is brought about by explicit control (and planning) rather than by the inherent characteristics of the elementary actions. Thus, if a person is inexperienced or confused or for other reasons does not know what to do, improper planning may ensue and actions may be carried out in a counterproductive way. That is, however, in itself ample evidence that sequentiality is not an inherent feature of the elementary events or the model, but rather comes from the way control is exercised.

3.2 The Model of Competence

The separation of the contextual control model into a model of competence and a model of control serves a pragmatic purpose, akin to the separation between knowledge and reasoning in AI models of intelligence. The main advantage is that it makes it possible to consider each aspect by itself, hence to avoid some of the traditional problems in understanding cognition and action. In particular, it provides an attractive possibility for actually implementing a software model of how actions can be controlled (Hollnagel *et al.*, 1992). The separation between the two models does not in any way imply or prove that something similar exists in the human mind. The model of competence itself may have several parts.

3.2.1 The Activity Set

First a model of competence must include the various competences (the "ready" actions). These need not be elementary actions, such as one would find them at the bottom of an action hierarchy. They will rather be the actions which the person is capable of carrying out and which are meaningful in the existing context. In computational terms they can be thought of as pointers to actions which belong together or which in the given context constitute a group. To take a prosaic example, changing oil in a car engine will be a single action for a skilled mechanic but a series or complex of multiple actions for the common driver. The definition of what constitutes a single or a complex action in the model thus depends on the situation and the person. This part of the model could be called the **activity set**.

Consider, for instance, the steps in the Approach To Landing (ATL) procedure described by Cacciabue *et al.*, (1992).

(A) Sub-phase: **Control the descent**.

   (1) **Extend flaps to position 1**: as the indicated air speed reduces below 250 Knots, the flaps are extended to position 1.

   (2) **Extend flaps to position 5**: as the indicated air speed reduces below 230 Knots, the flaps are extended to position 5.

   (3) **Extend flaps to position 10**: as the indicated air speed reduces below 215 Knots, the flaps are extended to position 10.

   (4) **Calibration of altimeter on QNH**: as the indicated altitude reduces below 7.000 ft, the pilot must calibrate the altimeter on the actual air pressure at sea level (QNH).

   (5) **Level flight on 4.000 ft**: as the indicated altitude reduces below 5.000 ft, the pilot changes his goal (and actions) from descent to levelled flight.

(B) Sub-phase levelled flight => **Adjustment to the ILS**.

   (6) **Execute approach check list**: as the indicated altitude reduces below 4.050 ft, the pilot executes the approach check list and, in particular, he controls whether the altimeter has, or has not, been correctly calibrated on the QNH.

Depending on the skill and experience of the pilot, the six steps can all be considered as a single action (called "ATL at airport X"), as six separate actions, or as a smaller number of actions grouped in various ways - for instance (1, 2, 3) + 4 + (5, 6). It is further reasonable to assume that a skilled person is able to unpack the actions to some extent, so that,

Fundamentals of the Model   166

for example, the model of competence could include both the joint of the six steps as well as each step by itself.

Another example is provided by the Human Factors Reliability Benchmark Exercise (Poucet, 1989) which has been described in Chapter 3. In the qualitative analysis of the second study case, the operational transient, only two of the 15 teams used the same breakdown of the actions:

> "The breakdown of the sequence into different actions and the modelling of the actions involving lumping identified actions together or decomposing some actions into even more detail, shows considerable differences from team to team."
> (Poucet, 1989, p. 113).

The definition of what the activity set of the person (or a team) is will clearly depend on how actions are grouped and/or decomposed. Regardless of whether the initial description is taken from written procedures or from observations, the "natural" activity set depends on the specific context - and on the assumptions that are used by the analyst. This may have serious consequences for the following steps of the analysis, particularly when the grouping/decomposition of the activity set serves as the starting point for a further quantitative reliability analysis! There is no objectively correct way of grouping actions and no ideal activity set that can be used for comparison. The determination of whether a specific structuring of the competences is reasonable must therefore be based on whether it is consistent with the rest of the analysis and with what is known about the case in question.

3.2.2   The Template Set

Another part of the model of competence will be a set of templates or patterns for carrying out actions - cognitive as well as physical. A template denotes a specific relation between two or more actions which in a given situation can determine the order in which they are carried out. The templates may be plans (pre-defined or produced during the task), procedures, rules, guidelines (heuristics), strong associations, or in fact anything else that may serve as a guide for performance. In the contextual control model the **control mode** will determine which particular template is used. A person may, for instance, follow a plan more or less rigidly. One assumption could be that compliance with the plan will be rather strict in the tactical mode, whereas more flexibility is displayed in the strategic mode (cf. below). This part of the model of competence could be called the **template set**.

Seen by themselves, templates may appear to be very similar to procedures. This is quite correct in the sense that procedures can be represented as templates. If, furthermore, the actions follow the template strictly, the resulting performance will be similar to a pass through a procedural prototype without shortcuts. The difference is, however, that the template does not itself accomplish the control. It is the person who decides what to do, whether to follow the template to the letter, whether to modify it, or whether to do something entirely different. The resulting performance is therefore not described in terms of whether or not it complies with the template (procedure) but in terms of how actions are chosen.

The separation of templates from activities serves several purposes. Firstly, it becomes easier to model how a person can be involved in several lines of actions at the same time, and how a new goal can be established (e.g. as a sub-goal). The person could be considering several templates (or plans), corresponding to several goals, at the same time. Actions would be chosen according to the most important goal, which might involve a switching of templates. If the choice could not be made, a new goal might be established, for example, to resolve a conflict or bring about a specific condition in the system.

Secondly, it becomes possible to model a number of characteristic phenomena, for instance:

(1) rigid performance, which can be described and constructed as overly strict compliance to a template,

(2) mistakes, in an everyday meaning of the term, which can be described as strong associations between actions which may override a plan or procedure, and

(3) random or scrambled performance, which can be seen as the result of an unsystematic choice among plans or even as choice among actions without consideration of plans.

The separation between templates and activities provides a way of modelling very diverse phenomena with a few and simple principles. Even though the model does not claim to be a correct scientific explanation it does make it easier to get a comprehensive view of how human action may be controlled.

4. Control Modes

Control is necessary to organise the actions in the short term, i.e., within the person's event horizon. The essence of control is planning what to do: this is influenced by the context as it is experienced by the person (e.g. the cognitive goals described by Bainbridge, 1991), by knowledge or experience of dependencies between actions (pre-conditions, goals-means dependencies), and by expectations about how the situation is going to develop, in particular about what resources are and will be available to the person.

This outcome of the planning prescribes a certain arrangement of the competences, hence a certain sequence of the actions. But it is important to note that the sequence is **constructed** rather than **pre-defined**. As argued above a frequently observed pattern or a characteristic distribution of actions may reflect the relative stability (repetitions, regularity) of the environment as well as a constituent feature of human cognition. But it does not in any way indicate that sequentiality must be a basic feature of the model itself or of the related theories of cognition.

The degree of control a person will have over a situation can vary. It seems reasonable to think of control as a continuous dimension where at one end there will be a high degree of control and at the other there will be little or no control. For practical purposes it is useful to make a distinction between four different control modes (without ruling out that others are possible). They will be called: scrambled, opportunistic, tactical, and strategic control.

4.1 Scrambled Control

Scrambled control denotes the case where the choice of next action is completely unpredictable or random. The corresponding situation is one in which the person is in a state of mind where there is no reflection or cognition involved, but only a blind trial-and-error. This type of performance is thus, paradoxically, characterised by the lack or **absence** of any control. This is typically the case when people act in panic, when all cognition is paralysed and there accordingly is little or no correspondence between the situation and the actions. The scrambled control constitutes the extreme situation of zero control. It may not happen very often, but in terms of modelling it is necessary to include as an anchor point.

4.2 Opportunistic Control

Opportunistic control corresponds to the case when the next action is chosen from the current context alone, and mainly based on the salient features rather than on more durable intentions or goals. It is opportunistic in the sense that the person takes a chance, not because he is deliberately exploring an alternative but because there is no time or possibility to do anything better. There is therefore very little planning or anticipation on the part of the person, perhaps because the context is not clearly understood or because the situation is chaotic (cf. the related phenomenon called **focus gambling** by Bruner et al, 1956).

Opportunistic control corresponds to a heuristic that is applied when the knowledge mismatch is large, either due to the inexperience or lack of knowledge of the person or due to the unusual state of the environment. As a result the choice of actions may not be very efficient and many useless attempts at finding a solution may be made. In this type of situation the person will often be driven either by the perceptually dominant features of the interface, e.g. those which most easily attract attention (loud sounds, blinking lights, patterns (Gestalts), etc.) or by those which due to experience or habit are the most frequently used, e.g. similarity matching or frequency gambling heuristics (Reason, 1990). Conversely, interface design may be crucial for how such situations are handled (cf. the drifting situation described by De Keyser, 1986). Bad interface design may serve as a trap which either brings about or prolongs the duration of opportunistic control. Conversely, good interface design may make it easier for the person to change to a higher control mode.

As described above, opportunistic control means less than adequate control. The term opportunistic is, however, often used with a different meaning, as in opportunistic planning. This characterises situations where the person is deliberately exploring new paths and testing new alternatives – either because the conditions allow that or because it is the only way out. Either way the person will be deliberately opportunistic or exploring, because there are no other options. I will therefore call this for **explorative control**. In opportunistic control as the term is used here the

Fundamentals of the Model   170

person could actually do better, but is unable to do so for a number of compelling, subjective or objective, reasons.

4.3 Tactical Control

Tactical control is characteristic of situations where the person's performance is based on some kind of planning, hence more or less follows a known procedure or rule. The person's event horizon goes beyond the dominant needs of the present, but the possible actions considered are still very much related to the immediate extrapolations from the context. The planning is therefore of limited scope or limited range, and the needs taken into account may sometimes be *ad hoc*.

If the plan is a frequently used one, performance corresponding to tactical control may seem as if it was based on a procedural prototype - corresponding to, for example, rule-based behaviour (Rasmussen, 1986). Yet the underlying basis is completely different. The conditions are such that the person can find an appropriate template for action and furthermore that he decides to follow it closely. The resulting performance is therefore caused by the similarity of the context or performance conditions, rather than by the inherent "nature" of performance. Performance in the tactical control mode will be more efficient and with fewer errors than in the opportunistic mode, but still not as good as in the strategic mode. The person is driven by the built-in functional logic of the context and therefore also heavily influenced by the interface, although not dominated by it as in opportunistic control.

4.4 Strategic Control

Strategic control means that the person is considering the global context, i.e., using a wider event horizon and looking ahead at higher level goals - either those which have been suspended and have to be resumed or those which, according to experience or expectations, may appear in the near future. The person will be less influenced by the dominant features of the interface, but may still be hampered by inadequate system support, e.g. for information retrieval. The design of the interface is therefore still important. The strategic control mode should provide a more efficient and robust performance, and thus be the ideal to strive for. The attainment of strategic control is obviously influenced by the knowledge and skills of the person, i.e., the level of competence; although all competence (basic functions) can be assumed to be available, the degree of accessibility or the ease with which they can be applied may greatly vary between persons, hence be a determinant of their performance.

In the strategic control mode the functional dependencies between task steps (pre-conditions) will be very important, simply because they are taken into account in planning. In the opportunistic control mode the person might, for instance, arrive at the comparison without having found the alternatives, hence have to backtrack. In the strategic control mode this would happen less often; backtracking or suspension of tasks may therefore be more dominant in the opportunistic and tactical control modes than in the strategic mode. This characteristic might be used by a plan recognition system as an indicator of the mode in which the person is working, hence on the type of support that is needed.

4.5 Control Mode and Subjectively Available Time

The four control modes describe characteristic regions of the continuum of control. An obvious next question is what determines the degree of control a person has of a situation and how that control can change, e.g. how the person possibly can shift from one control mode to another.

It is reasonable to assume that several factors will determine the control mode. One of the obvious candidates is the amount of subjectively available time. If the subjectively available time is short, the person will not be able to plan his actions properly; there will be little or no time to look for alternatives, to predict developments and likely consequences of actions, and to choose the best candidate. The person will be in a situation of high workload or information overload, and will have to resort to a number of strategies that can relieve the pressure (e.g. Miller, 1960). He will typically be in the opportunistic control mode (cf. Figure 4-10).

On the other hand, if subjectively available time is sufficient, the person can spend more time to plan his actions and to choose the proper alternative. If subjectively available time is unlimited, the person will be able to explore the situation and the possibilities for action as much as he can. The degree of control will accordingly be high, and the person will typically be in the tactical or strategic control mode (cf. Figure 4-10).

Figure 4-10 shows the correspondence between subjectively available time and control mode, but does not indicate what is cause and what is consequence. There is certainly a close coupling between the two. For instance, if a person is in an opportunistic control mode it is very likely that his choice of action will be inappropriate, hence that he will not achieve the desired goal(s). The situation is therefore that time is being

Fundamentals of the Model

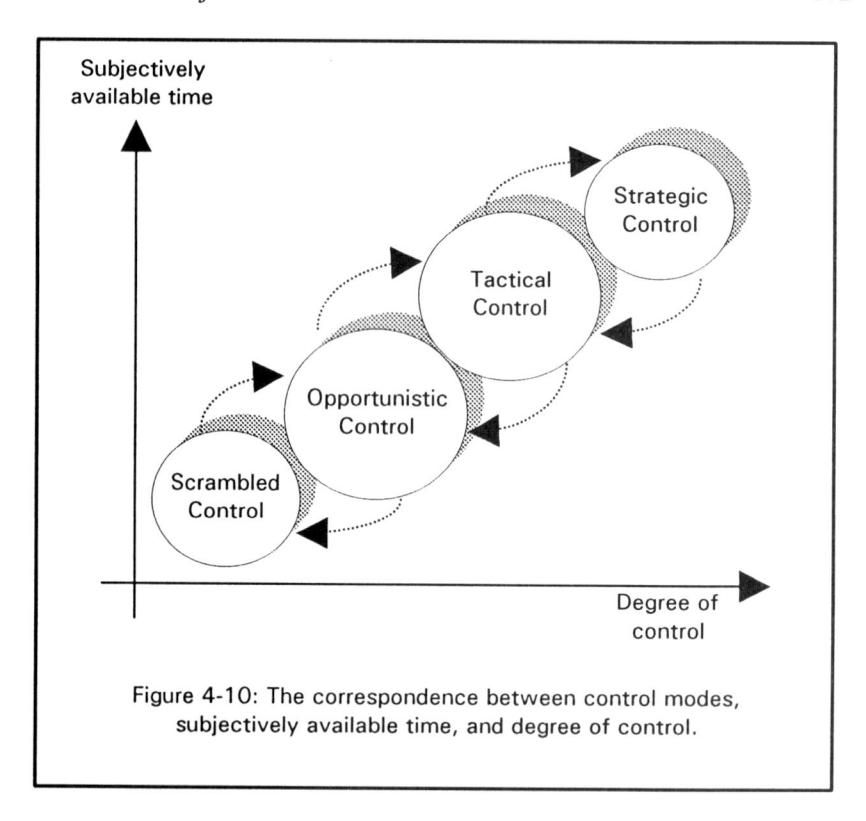

Figure 4-10: The correspondence between control modes, subjectively available time, and degree of control.

used but that the distance to the goals has not been reduced - and that it may even have been increased. The immediate consequence is that the subjectively available time may have been further reduced, and that the situation has possibly has become worse. Conversely, if the action is successful, then the distance to the goal may be reduced. This means that the subjectively available time may be increased, and that the degree of control therefore is improved.

While subjectively available time without doubt is an important variable, it is **not** an independent variable. The subjectively available time depends on the objectively available time, the rate of change of the process, the knowledge and experience of the person (expressed, for example, by the contents of the template set and the activity set), the cognitive style of the person and his level of aspiration, the procedural and organisational support, the physiological conditions, and so on. Although it may be tempting to reduce all these to a few variables - such as workload or information capacity - this will probably do more harm than good. The short discussion here has hopefully served to make clear that the description easily can become very complex. I will not pursue it further in terms of the possible cognitive "mechanisms" involved, but will elaborate it later in this chapter, and in the following, from the point of view of assessing the Reliability of Cognition.

Figure 4-10 can also be used to consider the issue of explorative control mentioned above. It may be assumed that explorative control depends strongly on subjectively available time; exploring new opportunities, testing - and possibly failing - and learning, requires that there is no powerful demand for action. In that sense there will be a limit below which it is impossible for explorative control to exist. This may mean that explorative control only can occur when the person is in the tactical or strategic control modes; yet explorative control is perhaps better thought of as another dimension of the control continuum.

4.6 Interaction between Competence and Control

There is clearly a strong interaction between the different control modes and the level of competence, i.e., the readiness of appropriate skills or abilities. Consider, for instance, what could happen under the following conditions. A driver is waiting at an intersection in the city centre for the traffic signal to change from red to green; it is in the middle of the rush hour; and raining cats and dogs; then the engine suddenly dies. The immediate reaction is to try to restart the engine. If the attempt fails, then several developments are possible. If the driver is inexperienced or has little understanding of how a car functions, it is very likely that he will simply repeat the attempts to start the engine. If the driver is more experienced, he may check the instruments to see if a cause for the engine failure can be found (e.g. out of petrol). Or he may accept the inevitable, get out into the rain and start to push the car towards the kerb. Thus, depending on his competence he may - in terms of the mode - end up in a scramble, opportunistic or tactical control mode.

In the interaction between competence and control concepts such as feedback driven and feedforward driven performance or behaviour may be important. Opportunistic control is typically feedback driven, hence the design of appropriate feedback indications is crucial. The guiding principle must be that interface design should sustain the prevailing mode of action while at the same time gently prompt the user towards a more strategic performance - assuming, of course, that it is possible to demonstrate that strategic performance on the whole is more efficient. This can, however,

Fundamentals of the Model 174

only be achieved if it is possible to detect or identify the current control mode. One suggestion for how this can be done is to rely on inference mechanisms and inferred measures (Hollnagel, 1992d). It is possible that more direct measures may also can become available. Assuming this to be the case, one can design a closely coupled system which will amplify the person's cognition and gradually achieve better performance. This will differ from the notion of forcing the person to work on a specific level (e.g. rule-based behaviour), and instead emphasise that the right or proper context is provided. The right context is one where the significant goals of the system are easy to see, and where the facilities and possibilities to pursue these goals are provided, i.e., in finding the means to solve them, resolve them into sub-goals, evaluate the consequences, plan on a long range, etc., without assuming *a priori* that a specific performance mode (level of behaviour) is to be preferred or even enforced. Unsatisfactory performance is often simply a consequence of not being able to comprehend the situation in is entirety.

When carrying out a plan it is assumed that only local (tactical, operational) modifications can be made. The making of a new plan, corresponding to strategic planning, is an activity in itself which cannot easily be merged or shared with others. The overall plan must therefore provide the conditions (and the performance modes) for the actual execution. One way of describing this is to refer to **schemata** in the sense of operational or tactical plans (Amalberti & Deblon, 1991). Schemata are pre-planned reactions, which usually are carried out when the corresponding conditions obtain. However, for some situations the reaction may be to defer action.[2] Setting up schemata requires considerable expertise and represents an expert trade-off between speed and efficiency; if the schemata work they provide a good and efficient simplification to situation control; if the simplification is incorrect performance will suffer.

If the conditions do not match the needs of a plan or schema, performance either has to revert to a default plan (default actions) or it will break down completely (opportunistic control). In this case an overriding goal is to regain control, i.e., to get back to another mode (cf. Amalberti & Deblon, 1991). Planning corresponds to defining a set of pre-conditions and actions. If the match is appropriate, then everything is OK. But if not, serious consequences will follow. Replanning (e.g. as part

---

[2] Depending on the model being used, this may lead to stack problems, hence information overload and a host of other interesting phenomena.

of explorative control) cannot usually be undertaken, because it is too time and resource consuming. If done, it will delay actions and lead to time compression problems.

One could assume that the basic limitation is that a person can only really follow one plan at the time (e.g. Moray & Rotenberg, 1989). In order to cope with the complexity of the environment, he therefore has to merge a number of smaller plans in a hierarchical fashion into a single higher-level plan. As long as compliance with that can be sustained, the situation is under control. If the compliance or synchronisation is missed the plan no longer fits the circumstances and the control breaks down. The alternative to planning is constant and undivided attention which is possible, but extremely resource demanding. If a person is confined to that the general flexibility and responsiveness to varying environmental conditions is definitely lost.

5. The Contextual Control Model (COCOM)

In order to be able to ascertain whether the approach advocated here is a useful basis for the practical modelling of cognition, it is necessary to examine a concrete proposal for modelling. The proposal must address both the modelling of competence and the modelling of control. At present we shall concentrate on the modelling of control and leave the modelling of competence untouched. Competence can, in principle, be modelled as a set of prepared procedures or routines (functions) which can be executed when the calling conditions are fulfilled. Each such procedure will, of course, be a mixture of actual procedural steps and knowledge. A potential practical candidate for this could be found in the notions of frames, scripts, or production rules. Control is, however, something which hitherto has not received the same attention as competence, and there is therefore little in the way of existing proposals for solutions. This is not to say that modelling competence will be easy - although it unquestionably will be easier than modelling control!

There are two important aspects of the control modes as part of the COCOM. One has to do with the conditions under which a person changes from one mode to another (keeping in mind that they only are points on a continuum); the other has to do with the characteristic performance in a given mode, i.e., what determines how actions are chosen and carried out? The first aspect is important for attempts to design and implement the COCOM as a dynamic tool, i.e., to make it operational. The second aspect is important for providing an account of the reliability of performance *vis-a-vis* the characteristic control modes. The reader must be warned, however, that the ideas presented in the following are suggestions only, rather than proven facts. Time and experience will show to what extent these suggestions have lasting merit. The present proposal, in particular, does not make any claim of actually representing the control function as it really is. The notion of control is a hypothetical construct which facilitates the analysis and understanding of certain characteristics of cognition (the observed variance). This, however, does not in any way prove that the construct is correct.

The modelling of control in a contextual control model of cognition must be able to reproduce the different control modes that are assumed necessary. In our case there are four: scrambled, opportunistic, tactic, and strategic, as previously outlined. These four named control modes are, however, only points on a continuum which ranges from no control at all to completely deterministic performance. The model must therefore be able to account for, and possibly generate, not just the four distinct modes but the whole range. One way to accomplish this is to identify a set of main control parameters which, on the one hand, provide a vehicle for constructing a model and, on the other, make it possible to reproduce the significant phenomena. The parameters must both be psychologically plausible and operational in a computational sense. **Psychological plausibility** is necessary because there must be a reasonable degree of compliance with our common way of describing human action, including how we experience that in ourselves. Achieving psychological plausibility is a case of arguments and inferences based on commonly accepted psychological "facts". The parameters must be **operational** because it thereby becomes possible to express the notions and assumptions behind the modelling in a testable way. It is not quite the same as requiring that a psychological or cognitive theory must also be a program (theories as programs, e.g. Boden, 1988) but rather a way of making sure that the theory is testable by means of explicit manipulation. It is thus a practical rather than a theoretical requirement. If the modelling does not work, then the underlying assumptions have been falsified; if it does work then the theory may not have been proved, but the assumptions have at least been made more credible.

5.1 Main Control Parameters

In order not to complicate matters more than absolutely necessary, only two main parameters will be considered to describe how a person may change from one control mode to another: the outcome of the previous action (in terms of success or failure) and the subjectively available time. Neither of these are, however, in any way simple and both depend on each other.

The **determination of outcome** (of the previous action) is not just a matter of ascertaining whether it succeeded or failed. On the contrary, the determination of the outcome is different for each control mode: in the strategic mode it may be a detailed evaluation of the feedback in relation to previous developments, expected outcomes, and delayed or otherwise removed rather than immediate consequences (e.g. side-effects), while in the scrambled mode it may be a rudimentary detection of whether a noticeable change in the right direction has occurred - or indeed if any change at all has taken place. Other complicating factors are the possible delays in the outcome (for systems with large time constants), ambiguous or incomplete state indications, equivocal rules for interpretation, etc. In both cases the determination may furthermore be objectively wrong, i.e., the outcome may be deemed a success although objectively it was a failure.

The **estimation of subjectively available time** may be similarly complex. It may involve a consideration of the number of activities that remain to be carried out, e.g. as suspended actions, the number of simultaneous goals, the predicted changes and developments in the process and the environment (hence "objective" time), the level of arousal, the level of familiarity of the situation, etc., as discussed already. The estimation may again range from being quite detailed and, presumably, precise to being a rough guess or gut feeling.

The two main parameters refer to aspects of the person's cognition (cognitive processes) - or rather to the outcome of the cognitive processes. In both cases an attempt must be made to describe how the two parameters may be influenced by other parameters which can be independently assessed. It would definitely be a naive oversimplification to assume that a simple relation could be found between a few observable (measurable) parameters and the main cognitive parameters defined here. Human action and human performance is far too complex to yield to simple-minded attempts at modelling. It may nevertheless be possible to find a number of observable parameters or indicators which will account

Fundamentals of the Model 178

for a significant part of the variance of performance. That would be sufficient basis for building a model and for establishing a working hypothesis about the causes of human reliability.

In both cases, however, is it necessary to introduce additional and independent parameters which may influence the preceding ones. One parameter is whether or not there are any **sudden changes** in the situation; a change could be the onset of an alarm or an avalanche of alarms, a violent change in the physical conditions (earthquake, fire), an abrupt change in the psychological conditions either individually or for the team (momentary incapacitation, pain), etc. Such sudden changes can alter the situation from being comprehensible to being incomprehensible and/or dramatically reduce the available time, hence have an effect on the two cognitive parameters: determination of outcome and estimation of subjectively available time.

5.2 Other Dimensions of Control

In addition to the two main control parameters it is necessary to consider a few others. The guiding principle has been the Minimal Modelling Manifesto. The Minimal Modelling Manifesto is the basis for a consistent approach to the modelling of humans in interactive systems, and can be expressed as follows:

> A Minimal Model is a representation of the main principles of control and regulation that are established for a domain - as well as of the capabilities and limitations of the controlling system.

A minimal model is concerned with control and regulation, hence with cognition at work. The model is minimal because it tries to make as few assumptions as possible, not because the phenomena it addresses are diminutive. For a further discussion, see Hollnagel (1992e).

5.2.1 Number of Simultaneous Goals

The first additional parameter is the **number of simultaneous goals**, which describes whether the person considers only a single goal at a time or whether possible actions from multiple goals are considered. This, in turn has an effect on the determination of outcomes and the estimation of time. It is, however, not the same as considering multiple choices or actions that may lead to the same goal (i.e., evaluating the effects of several possible lines of action, such as different ways of getting to work in the morning) because in this case there will not be a conflict between multiple goals; whether one line of action or another is chosen does not make any difference in terms of achieving the goal. It may make a difference in terms of producing different post-conditions (and possibly also latent conditions), for instance, one solution being more costly or time consuming than another, but this will only be important if other goals are considered within the same context or at the same time. Otherwise the selection of one route over another simply reflects the current value of resources and performance conditions.

If multiple goals are considered the situation is quite different. It now corresponds to having multiple active threads (lines) of action, and to choose the best one at each time, knowing that this may temporarily or permanently prevent other actions from taking place. It is still assumed that it is possible to be engaged in only one line of action at the time; cases of seemingly parallel activities (for instance, for highly skilled tasks) are assumed to involve distributed rather than parallel attention. The goals may either be independent top level goals, or multiple sub-goals from the same top level goal. The goals can either be competing (partly incompatible) or conflicting (mutually exclusive). Working with a number of simultaneous goals requires a good sense of timing and the ability to integrate or merge multiple plans, and is thus characteristic of tactical and strategic control modes.

5.2.2 Availability of Plans

A second additional parameter is the **availability of plans**, which means whether the person can refer to pre-defined or pre-existing plans (action templates) as a basis for choosing the next action. A plan can either be made on the spot, have been learned by experience, or have been defined explicitly in advance, e.g. as a written procedure. In either case the availability of plans requires that the situation is familiar. If a plan has come about by experience or is available as a written procedure the person must clearly recognise the situation before being able to use the plan. If the plan is made on the spot, it also requires a certain level of familiarity; when the situation is completely new and unanticipated, it is impossible for the person to prepare what to do. There must be some familiar features before a plan can be generated; the first step in any

Fundamentals of the Model 180

diagnosis and problem solving is to recognise the situation in whole or in part.

A plan can either be followed rigidly or serve as a guideline for actions to be taken. In many cases plans serve as incomplete frameworks for actions, which have to be modified as the situation develops. Examples are found in teaching, production scheduling, process supervision, on-line maintenance, or controlling an aircraft (cf. Amalberti & Deblon, 1991, or Sanderson, 1989). The existence and availability of a plan does not mean that the event horizon is automatically defined (cf. below); a plan can be carried out step by step without thinking ahead, or it can be used in its entirety.

5.2.3 Event Horizon

There are two further parameters which are important for the contextual control model. They both have more to do with how performance is shaped in a given control mode than with the change between control modes. The first of these is the **event horizon**. By this is meant how much of the past and how much of the future are taken into consideration when a choice of action is made. The event horizon is described in terms of the number of steps, moves or items that are considered, rather than in subjective or objective time. The extent of the past can be referred to as history size while the extent of the future can be referred to as prediction length. (This terminology is borrowed from game theory; the notion of distinct moves, actions and reactions, may be useful as a means of making the characterisation of the situation manageable.)

(1) **History size**. One issue is clearly whether the person makes use of any history, i.e., whether preceding events are taken into consideration when the outcome of previous actions is determined and when new actions are chosen. The past very much defines the context for the present; examples of that are many in for instance psychology, linguistics (natural language understanding), temporal logic, etc. The past provides the basis for resolving ambiguities in the interpretation of the present, hence also contributes to the choice of future actions.

In steady-state environments the influence of preceding events can be accounted for by using a Markov process which constitutes a very simple mechanical way of modelling. It is, however, clear that human action is not determined simply by Markovian transition probabilities. Other solutions should therefore be found, for instance using high level heuristics (conditions), plans (which effectively serve as a context frame, cf. below), possible worlds, etc.

(2) **Prediction length**. The prediction is essentially a prediction of the consequences if a given possible action is chosen. If there is no evaluation of possible actions, then the prediction length is zero. This means that the possible actions are taken at face value, i.e., the expected outcome of a possible action is assumed to be identical to its nominal purpose. This is therefore an occasion where memory heuristics such as "strong but wrong" (Reason, 1990) may play a role by (mis)leading the agent to assume that a given consequence will obtain - often without considering that option explicitly. The prediction length increases as the evaluation of possible outcomes is extended. The evaluation means that the possible effects (and in particular, the side effects) for each possible action are taken into consideration. In the case of dynamic environments, the evaluation may also consider the possible reactions, the choices that can follow from them, etc. In other words, the prediction length can possibly be expressed in terms of the number of steps or moves that are considered. Another way of describing this is whether the effects of the action are considered in detail, or whether only the default values are used. This may also correspond to whether predictions about outcomes, hence anticipated re-planning, is made.

The prediction can more precisely be described as the evaluation of the pre-conditions and post-conditions that apply for each possible action. If the pre-conditions are not considered, then the risk of wasting effort on a useless attempt is very great. Conversely, if the post-conditions (effects and side-effects) are not considered, the outcome may deteriorate the situation rather than improve it.

Although the event horizon stretches out from the present, it is not necessarily symmetrical for history (the past) and prediction (the future). A person may choose his actions with due consideration of preceding events (the context) but little attention to the possible uncertainty of expected outcomes. Or he may carefully consider the possible side-effects of a choice without recognising how previous actions may have changed the conditions, i.e., without considering the so-called latent conditions.

Although the event horizon is different from the notion of subjectively available time it is nevertheless closely coupled to it. Adequate subjectively available time is a necessary condition for having a large event horizon and considering many steps, but it is not sufficient; yet inadequate available time will clearly prevent a large event horizon. On the other hand, being able to consider past and future events may be conducive to maintaining an overview of the situation, hence sustaining adequate available time. Subjectively available time plays a role in all cases and for all control modes. In particular the lack of sufficient time can be assumed to lead to very rigid behaviour. Time is therefore a parameter which exerts its influence over and above the other parameters.

5.2.4 Mode of Execution

The final additional parameter is the **mode of execution** where a distinction can be made between **subsumed** and **explicit** actions. The mode of execution can be either ballistic/automatic or feedback controlled. In the ballistic mode the person automatically carries out the steps of a chosen plan until either a pre-defined checkpoint is reached or until external conditions force an interrupt.[3] In feedback driven execution each step (or group of steps) is followed by the evaluation of the feedback before the plan is continued. (In some sense the distinction between the two relates to how many steps there are between feedback points, i.e., how the large the span of automatic execution is in relation to the prediction length. Even in feedback control mode there will be sets of actions that are carried out automatically. Referring to the TOTE paradigm, the distinction can be seen as the size of the O(perate) part *versus* the T(est) part.) The ballistic (subsumed) and feedback (explicit) modes define two ends of a continuum, and actual performance falls somewhat in between them.

(A case might also be made for feedforward control, which involves planned reactions to predicted changes. However, feedforward is, in a sense, ballistic for the interval that is predicted - since it does not make use of feedback.)

---

[3] The term ballistic refers to the analogy of a cannonball which, once fired, continues until it hits something. An example of the opposite is a guided missile. A less martial example is the throwing of a javelin.

The Contextual Control Model (COCOM)

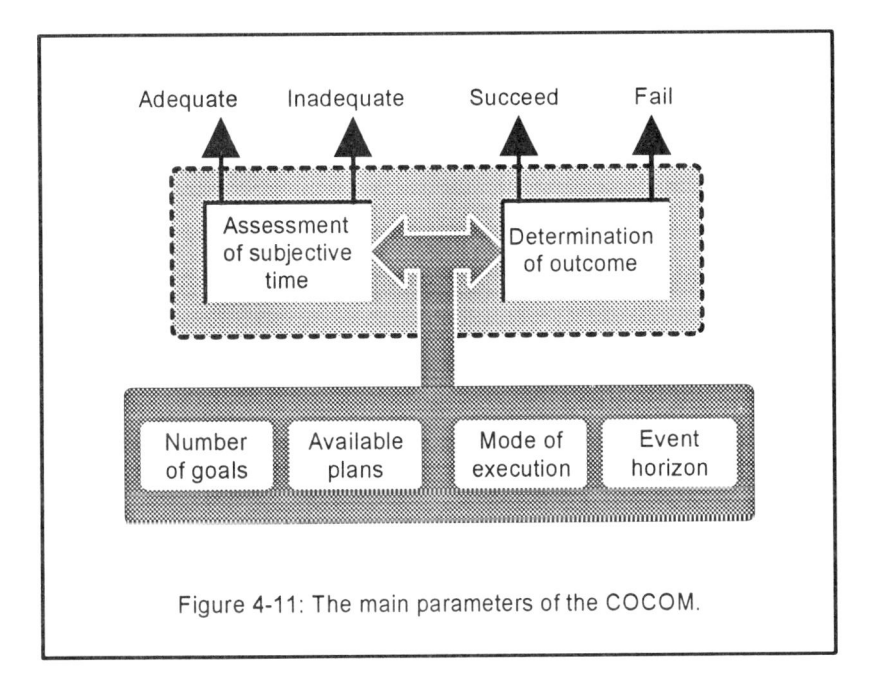

Figure 4-11: The main parameters of the COCOM.

The relations between the parameters proposed so far, two main ones plus five additional ones, can be shown as in Figure 4-11. Clearly, most of these parameters will be linked in some sense; this may either be described in terms of an actual correlation or be attributed to a common causal factor. It is also reasonable to assume that the tendency to perform in a specific way, e.g. that the control parameters have a specific value, can be related to the context or the state of the environment. Such relations must in themselves be fundamental, in order to avoid a potentially infinite regression of the explanation. Specific causal factors might be found in constraints of the context, limitations and traits of human thinking, or both combined. But for the theory or model it is sufficient to acknowledge this correlation; there is no great need to explain it, unless one is searching for the ultimate (or just proximal) causes.

Although the parameters proposed above are obviously not independent, the dependence or degree of co-variation is not easy to define *a priori*. It seems safe to assume that the degree of co-variation depends on the person as well as on the nature of the work and the working conditions. The nature of the parameters and their relations makes it impossible to draw a simple diagram of the control space or even to provide a table. On the other hand this may be quite fortunate, since it avoids the obvious limitations that come from using a two-dimensional schematic. Instead, each of the four control modes will be discussed *vis-a-vis* the parameters. This should provide an initial characterisation that may be used to set up a version of the model.

6. Control Modes and Performance Characteristics

There are two important questions that must be answered for each control mode: (1) how a person can change from one mode to another (control mode transitions) and (2) what the performance characteristics of a control mode. Both questions will in the following be considered for each of the four control modes. The description will refer to the set of parameters proposed above and try to indicate how an actual operational model of COCOM could be specified. By providing this degree of detail it will also be easier to describe how the Reliability of Cognition may manifest itself for a specific control mode.

Since the control modes represent characteristic points on a continuum rather than distinct states, it is reasonable to assume that the transition between control modes is from mode to mode, i.e., between "neighbours" - where the sequence is given by the order in which they have been described, as suggested in Figure 4-10. The only exception to that is that in some conditions it is possible to go directly between the tactical and the scrambled modes. The jump to the scrambled mode could be caused by a severe change in the situation, literally the roof caving in. The jump to the tactical mode is conceivable if the person manages to get away from a confined situation and regain previously lost control. In addition it is, of course, possible to go from a state to itself, i.e., to sustain a given control mode. The possible transitions are shown in Figure 4-12.

The following sections provide a characterisation of how control is carried out in each of the four distinctive modes. The descriptions do not refer to identifiable empirical investigations, but rather summarise the features that a contextual model of cognition should have in order to behave in accordance with the control modes. The descriptions should therefore be seen as a proposal or a basis for a more detailed specification of a model.

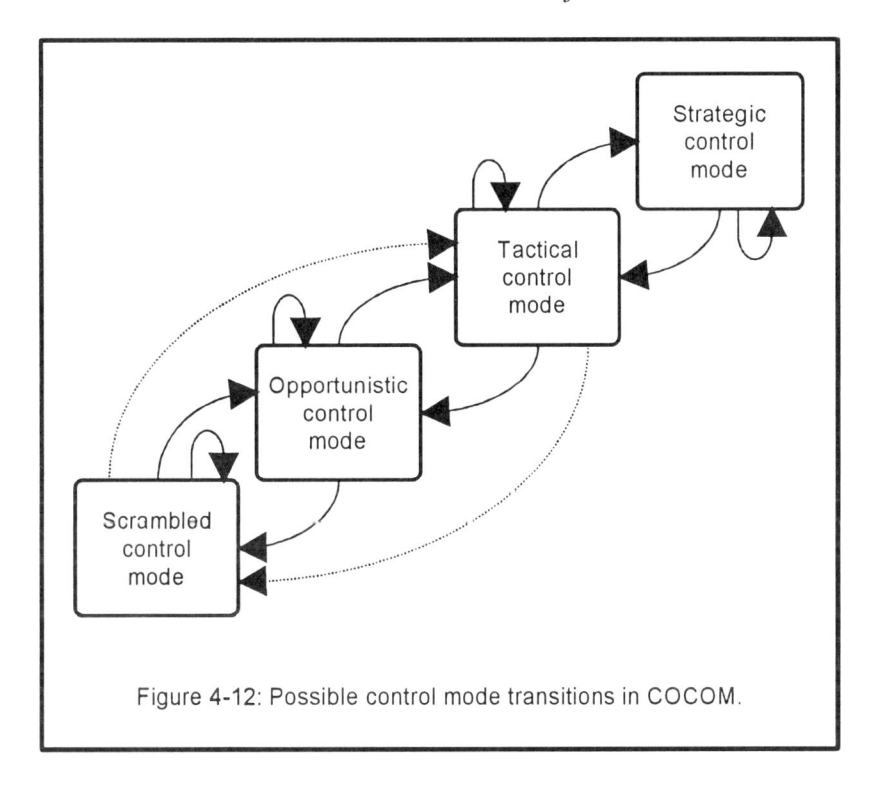

Figure 4-12: Possible control mode transitions in COCOM.

6.1 Scrambled Control

Scrambled control is the simplest of the four modes. In scrambled control the event horizon is confined to the present; there is no consideration of preceding events, nor any prediction of the outcome of future events. The choice of next action is seemingly random, as by blind trial-and-error, and plans are not used even if they nominally are available. Only one goal is considered at a time and it typically dominates the situation to the extent of excluding any other activities - planning, evaluation of feedback, etc. - as in a state of near panic. The goal is usually a very basic one, such as maintaining physical or psychical integrity, escaping from a threatening environment, obtaining a basic substance (such as oxygen), etc., rather than the goal that was current in the preceding control mode.

The execution of the action is completely subsumed or automatic, i.e., it is a rigid or stereotyped carrying out of the nominal action. The evaluation is reduced to noting whether the action succeeded or failed - there is no graduated evaluation of the effects. If the action did not succeed, another action (or the same action again) is tried, thereby possibly aggravating the situation. Finally, the time constraints are typically severe, since this is what may have brought about the scrambled control in the first place. Another possible cause is a complete lack of understanding of the situation coupled with a strong need to get away from it. The limited time may be real or perceived; what matters is that the context cannot be properly interpreted and that the need to get away from it looms large, effectively dominating any other consideration.

The scrambled control mode is typically entered from the opportunistic control mode. The specific causes can be that the previous action failed, that the available time is reduced (which, for example, could be due to repeated failures), or that there was a sudden change, e.g. a failure in the environment or process. If the person was in an opportunistic mode an unexpected deterioration of the situation might conceivably cause a change to the scrambled mode, for instance if it meant that the situation became completely incomprehensible. Similarly, severe changes of conditions may cause a shift from tactical control mode to scrambled. This could be the case if a major accident happened, for example, leaving the person with a completely unknown situation and little time to do anything about it.

Since the behaviour in the scrambled mode is characterised by random trial and error, the way to get out of it is that the person succeeds; this means that the dominating physical or cognitive restrictions are dissolved. If the attempted action fails, the person remains in the scrambled mode. When the action succeeds a physical danger might disappear or a mental/cognitive block may vanish. From the person's point of view this must be viewed more or less as fortuitous, because the actions were not part of any plan.

The cognitive functioning in the scrambled mode is characteristically shallow. There is only one goal which completely dominates the situation; there is furthermore no plans available for that goal, i.e., the person is faced with a strong need to achieve something but without any clear means of doing it. The choice of actions is therefore random in the sense of being unrelated to the goal constraints. The determination of the outcome of the previous action is rudimentary; it is more in the nature of noting whether there was a significant change in the situation, i.e., whether the threat or obstacle disappeared. The subjectively available time is very limited; if the person was able to look even one step ahead, he would no longer be in the scrambled control mode. Finally, arousal will typically be very high and attention therefore correspondingly low.

Altogether the scrambled control mode represents an extreme situation, best exemplified by a person in a state of panic. A person is unlikely to be in a scrambled control mode very often or for very long; if he was, there would surely be something terribly wrong with the working environment. But the scrambled control mode is necessary as an anchor at one end of the continuum.

6.2 Opportunistic Control

In the case of opportunistic control the action is chosen to match the current context with little thought of what effects it will have. In other words, the chosen action is one that matches the previous one or which corresponds to the dominant features of the situation, rather than an action which fits into a larger pattern and anticipates future changes. The event horizon is clearly asymmetrical: there is some effect of preceding actions but little or no prediction of possible outcomes. An example would be a situation where a powerful indicator, e.g. an alarm, appears. This will effectively attract the person's attention and the action will be chosen in accordance with that, without considering whether this might interrupt ongoing plans. Another possibility is that the next action is chosen to fit the previous one (e.g. a frequently occurring pair, described by a 1-order Markov chain) without following a deliberate line of action or plan. Accordingly, the use of plans in opportunistic control is negligible. The person may possible know of or have access to plans, but does not apply them. It is as if the attention span is too short to follow a plan.

Opportunistic control may either occur in the pursuit of a single goal or in the case of multiple goals. If the latter is the case, the situation is typically aggravated by the person vacillating between two goals. The choice is determined by whichever goal attracts the greatest attention, whichever action at the moment seems the most appropriate, either because the situation has changed or because of the association with the preceding action. There is little concern for the continuity with preceding actions or the possibly detrimental effects of shifting from one action to another. Such performance will clearly be very inefficient.

The execution of the action is typically explicit rather than subsumed, yet with the qualification that the feedback is not properly utilised. It is the action's manifest effects on the environment that are noted, rather than whether the effects of the action were those that were intended or anticipated - although the person surely will notice whether the desired effect was achieved. The (subjectively) available time may be small, although this is not necessarily the case. The opportunistic control may be an effect of the person's force of habit (or cognitive style) as well as of the context. But having limited time will obviously not be advantageous for improving the control mode, i.e., changing to tactical or strategic control.

A person can get into the opportunistic control mode from either the tactical control mode or the scrambled control mode; the latter change was described in the preceding section. The precipitating conditions are that the previous action failed, that available time has been reduced, and/or that there has been a sudden change in the working conditions or the state of the process. The transition from the opportunistic control mode to the tactical control mode is, again, due to either an increase in available time or that the previous action succeeded. Although the causes nominally are the same as for the scrambled mode, they differ in their details due to the different characteristics of cognition in the two modes. In a sense the criteria for success and failure are different.

The cognitive functioning in the opportunistic control mode is elaborate but not quite adequate. The person may have a single goal that he tries to achieve or two goals that compete for his attention; in the latter case the person is typically incapable of combining the two goals, thereby resolving the conflict between them. The person will typically know of a plan which can be applied to achieve the goal - although the plan may either be inappropriate for the specific conditions or not carried out in an ordered manner. The determination of the outcome of the previous action is very much tied to the concrete and immediate results, but may consider the action in relation to the immediate past (preceding attempts). This may also seriously influence the estimation of available time; a continued series of failures is likely to reduce subjectively available time and make the situation more tense while a continued series of successes is likely to improve the situation in every way. Subjective time will typically be just about adequate; the person may feel that there is sufficient time to make another attempt but other factors effectively prevents him from reasoning properly about it, i.e., the event horizon is very limited.

The mode of opportunistic control is a realistic scenario characteristic of, for instance, a situation of functional fixation. It may be found both during diagnosis, when candidate explanations are generated in a seemingly haphasard or unrelated manner, and in remedial action or recovery where alternatives are tried without any greater consistency. Performance in the opportunistic control mode may therefore be rather inefficient, because once failed solutions may be tried anew and because full advantage may not be made of successful actions.

6.3 Tactical Control

In the tactical control mode the person is less bound by the immediate needs of the task and the situation, and is able to view the action in a larger context. The event horizon is expanded, and the effects of an action are viewed in light of what went before. Similarly, choices for the next action are considered in some detail, and their potential effects (given the circumstances) are taken into account before the choice is made, although a considerable degree of "muddling through" may be observed. Plans are used as a basis for choosing an action, and quite often the actions follow some predefined or prescribed plan. The observance of a plan may even be rather rigid. This all depends on how experienced the person is and how much he feels in control of the situation.

The person can pursue more than one goal at the time and try to balance the choice of actions to reflect this. It is, however, rather more typical that one goal has priority over others, at least until a natural breakpoint occurs. The execution of actions is explicit, either in the sense that the plan may call for explicit evaluations or feedback, or in the sense that the evaluation of the outcome of an action is important for situation assessment. Actions are not merely chosen because of their dominant features or nominal effects, but based on a deliberate evaluation of their consequences (although possibly a shallow one). It follows, that there is reasonable time available for the person, or that at least there is no subjective time pressure. If there were, the evaluation would deteriorate and the control might move towards the opportunistic part of the control space (which is less demanding in terms of cognitive effort).

The change from the strategic control mode to the tactical control mode comes about if a plan fails to achieve its goals, if the person finds that available time is decreasing, or if a sudden change in the situation occurs. Conversely, the change back to the strategic control mode takes place when the actions achieve their goals, when everything goes as planned, and when time is abundant. The changes between the tactical mode and the opportunistic/scrambled modes have been described in the preceding.

In the tactical control mode the person is capable of considering more than one goal when the next action is chosen, though probably not very many. There will be several plans available, either known by heart or provided by external sources; the person is furthermore capable of looking for and/or modifying existing plans. The determination of the outcome of the previous action is based on the normal main effects, but also may consider major side-effects, deviations from expected effects, delayed effects, etc. The subjectively available time is adequate. It is typical that a person will be in the tactical control mode for long periods of time. In terms of the model this means that the limits for acceptable variations in outcome or available time are quite wide. The tactical control mode corresponds to "normal" performance in the sense that it is assumed to be the mode that the person is most frequently in.

6.4 Strategic Control

Strategic control represents the opposite of scrambled control. The person is fully aware of what is going on, plans deliberately, and chooses the action which best seems to fit the needs and demands. While not quite reaching the heights of rationality ascribed to a *homo economicus*, strategic control is as good as it is humanly possible under normal circumstances. (Conversely, scrambled control is as bad as it can be. The two are extremes, and therefore neither frequently occurring nor typical of specific scenarios.)

In strategic control the event horizon is extended, both for preceding events and with regard to predicting future developments, but the history size will typically be much larger than the prediction length. People find it far easier to remember what has happened - although with the possibility of misremembering or re-interpreting past events through hindsight (Fischhoff, 1975) - than to plan ahead. The number of steps that one can plan ahead is rather small even for experts (e.g. as in chess). The exception is probably the use of feedforward by experts, which seems to overcome the mental limitations by anticipating future developments; this obviously only works if the environment is stable and develops as expected. The person will rely heavily on plans, but will use them in a flexible manner rather than follow them rigidly. Plans may either have been defined or prepared in advance or be generated in the situation. Thus explorative control may temporarily supersede strategic control. Plans may serve as guidelines rather than as direct behavioural templates. The person can also consider multiple goals and is able, to some extent, to balance the requirements (pre-conditions) and expected outcomes.

The execution of the actions is a mixture of subsumed/automatic and explicit. The pursuit of multiple goals and the possible combination of several lines of actions will actually require that execution is partly subsumed. A plan may also specify a set of actions that can be carried out consecutively, or the action may refer to a well-learned routine or procedure. In this way concepts of nesting of actions, chunking into subroutines, etc. can be accounted for. The step size is, however, adapted to the situation and the needs, evaluations and feedback occur as planned, rather than haphasardly. There is no limitation on time, in particular there will be sufficient time to do the planning and evaluation - although this may be adapted to the demands of the situation. Typically, if the task is paced by a fast, dynamic process, the control mode is adapted to that - generally moving towards greater reliance on existing plans, hence a tactical control mode. Only when control is lost will opportunistic or scrambled control be likely to occur.

The strategic and tactical control modes are in many ways quite similar, and the changes between the two may be less abrupt than for the other modes. Typically, the change from the tactical to the strategic mode will come about if the person experiences sufficient control of the situation, i.e., that everything goes as planned, that predictions of developments are confirmed, that nothing untoward happens - and that motivation is sufficiently high. It is reasonable to assume that performance in the strategic mode is more demanding in terms of cognition than in the tactical mode; more planning, observation, reasoning, etc. is required. It is therefore not something the person will do unless there is a motivation for it. Conversely, the change back to the tactical control mode takes place if the complexity of the situation increases, shown for example by less than perfect predictions, by unexpected variations in the process, by increasing demands (hence reduced time) - or simply by a need to pay less attention to what happens or to share attention with something else, to relax a bit, etc.

The cognitive functioning in the strategic control mode represents the good end of the continuum - in contrast to the scrambled mode. The person is able to consider several goals at the same time and to balance the choice of actions between them. There will be a number of plans available for each goal, and it may even be possible to develop new plans or to explore new paths. The determination of the outcome of the

Fundamentals of the Model 192

previous event is detailed and balanced; previous as well as predicted developments are taken into consideration, delays are anticipated and accounted for, side-effects are carefully thought about, and information is carefully interpreted. Subjectively available time is adequate and assessed against what is expected to happen. Performance is strategic in the classical sense of the word (Schützenberger, 1954).

Combining the description of the four control modes and the assumptions about how changes between them occur, provide an overall notion of the internal structure of the COCOM. This is shown in Figure 4-13.

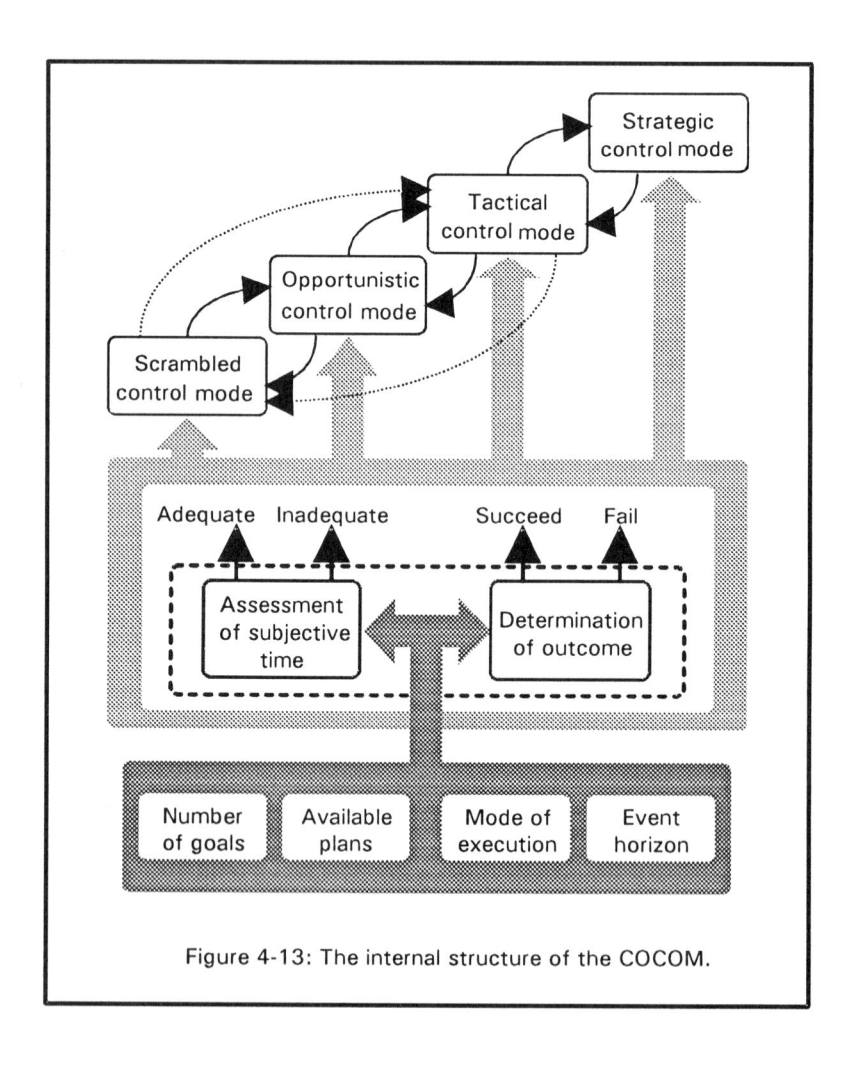

Figure 4-13: The internal structure of the COCOM.

The contents of the preceding sections can be summarised as shown in the following table. Note that this table introduces an additional category, called "choice of next action." It describes the way in which the next action is chosen, and in a sense summarises or combines the descriptions from the other categories. It could, of course, have been made a new category on line with the others; this is basically a matter of preference when the model description is defined.

Table 4-1: COCOM - Control Modes and Main Parameters

	Scrambled	Opportunistic	Tactical	Strategic
Determination of outcome (cognition)	Rudimentary	Concrete, limited by immediate effects	Normal, looking at main effects	Elaborate, considering history + predictions
Subjectively available time	Inadequate	Short or just adequate	Adequate	Adequate
Number of goals	One - not necessarily task relevant	One; or two competing goals	Several (limited)	Several
Plans available?	None	Negligible or limited	Available and used	Pre-defined or generated
Event horizon	Zero	Narrow (history > prediction)	Normal	Extended (history >> prediction)
Execution mode	Subsumed	Explicit	Explicit	Mixed, but controlled
Choice of action	Random	Association based (dominant effects)	Plan based	Prediction based

6.5 Changes between Control Modes

The conditions that may cause a change in control mode have already been described, but not in terms of a concrete algorithm or procedure - or

Fundamentals of the Model 194

even a "mechanism." It is, of course, entirely possible to do so, and it is even necessary to do it when the COCOM is to be implemented in software (see Hollnagel, 1992a). Yet going into the details of this will take us too far away from the topic of this book, which is the Reliability of Cognition. For that purpose it is sufficient to note that a contextual control model, like the COCOM, can serve as the basis for describing the Reliability of Cognition - and human reliability in general - without relying on either the atomistic nor the mechanistic assumptions.

An obvious question is how often changes between control modes will occur, for instance whether control modes could fluctuate between one action and the next or whether the changes will be between longer periods of relative stability. Looking at the general experience from studies of people at work, as well as from experimentally controlled situation - e.g. the classical literature on problem solving - it seems reasonable to assume that the changes will be transitions between longer periods of stability rather than rapid fluctuations. The important point in relation to COCOM is, however, that it will be able to model both one and the other. The change between control modes is determined by a combination of situational and personal (or internal) conditions - in other words by the existing context. If control modes fluctuate, then so be it. Because the change between control modes is not assumed to be deliberate, it does not require a specific higher order principle (or a meta-plan) to take place. In the contextual control model all changes are described on the same level and in the same fashion. If the changes are of a specific nature, e.g. rapid fluctuations or "mixed" control modes, then this is due to the existing conditions and to those only.

6.6 Relations to Other Descriptions

If the row headings of the table above are compared with, for example, the commonly used Performance Shaping Factors, it may be noted that a parameter which corresponds to workload or stress is missing. This is not an oversight but quite intentional. Although workload is a ubiquitous concept it is also an extremely ill-defined one. In terms of the COCOM one might say that workload is represented not as a single parameter but as the combined effect of a number of parameters (cf. also the previous discussion of the relation between subjectively available time and control mode). Thus, if subjectively available time is low, if there are sudden changes, and if number of goals is high, then workload can be assumed also to be high. The situation may deteriorate if the mode of execution is required to be explicit; conversely, it may be improved is execution can be automatic, if well-known plans are available, etc. There are practically no limits for the number of variations and hypothetical situations that one can imagine - which only goes to show that it is an advantage not to have workload as a primary parameter.

Another apparent deficiency of the COCOM is the absence of characteristic performance categories, such as the skill-based, rule-based, and, knowledge-based types of behaviour (Rasmussen, 1986). Yet neither is this an oversight. As described above, various characteristic types of performance (e.g. rigid, flexible, explorative, opportunistic, cautious) may come about depending on the state of the model and the control mode. The three characteristic categories referred to by Rasmussen only constitute a subset of these. They have traditionally been accounted for in terms of three levels of a specific deterministic (but unnamed) information processing model (Rasmussen, 1986, p. 101) or as characteristic ways of making short-cuts in the step-ladder model (SLM). Although a contextual control model cannot replicate this precise mechanism, the corresponding categories of performance can easily be produced by the COCOM, or a similar type of model, simply by assuming corresponding control mode characteristics (parameters). For example, skill-based behaviour corresponds to a situation where execution is automatic, where the situation is familiar, where there is a well-learned plan available, and where there is sufficient subjective time; the absence of either of these conditions would change the performance. It is equally easy to exemplify other characteristic performance categories. The important point is that a contextual control model is able to account for a broad spectrum of performance from a rather simple set of principles, hence that a procedural prototype model, like the SLM, can be considered as a special case of a contextual control model.

Although it is inappropriate to picture the COCOM as a classical information processing model, it may still be useful to show all the details described previously in a single drawing; if nothing else, it can serve as a mnemonic device. This has been attempted in Figure 4-14. This figure does not show any arrows internally in the model, but only the connections with the environment - illustrated by the influence from the context and the resulting performance. The fundamental internal functions have been described in the preceding and partly shown in Figure 4-13. When the COCOM is implemented in a specific version, it will be possible to show the relations for that particular rendering; work in this direction is

Fundamentals of the Model

Figure 4-14: The contextual control model.

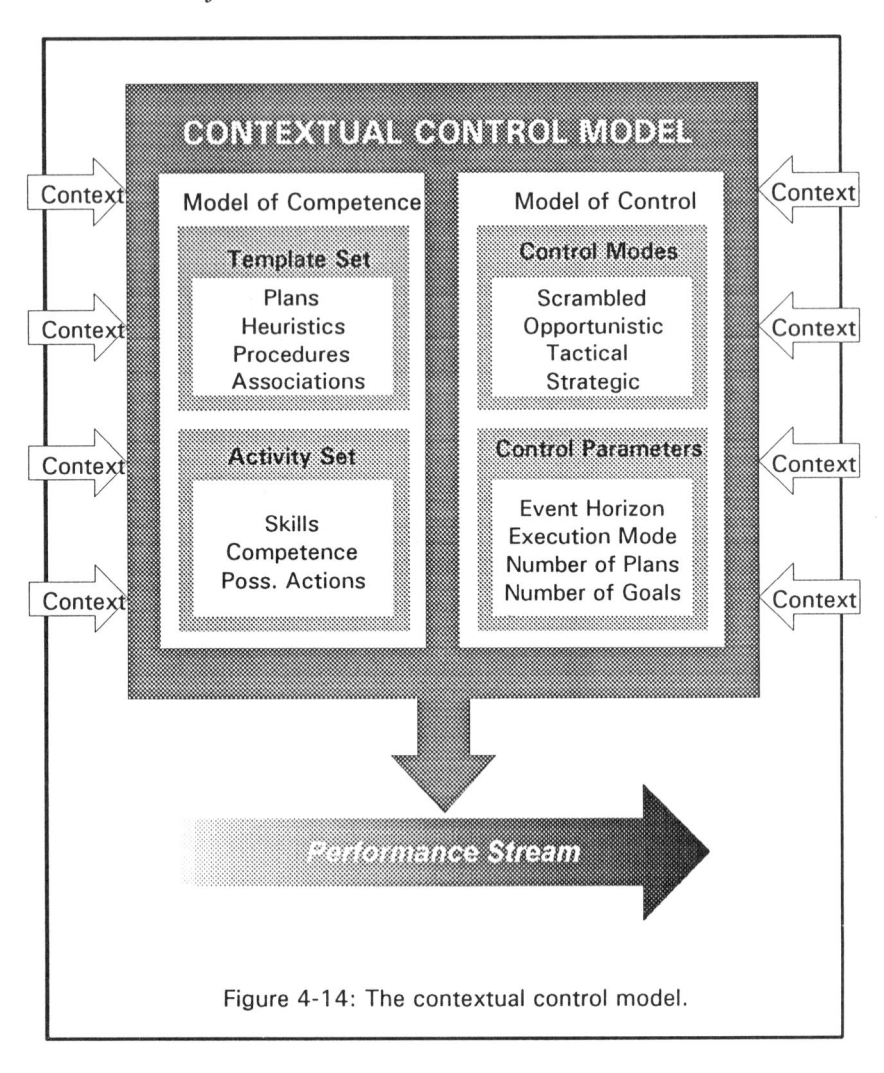

presently being done in the System Response Generator project described in the Appendix.

6.7 Control Modes and Reliability of Cognition

The COCOM was proposed as a way of modelling cognition in terms of contextual control rather than procedural prototypes. It is therefore necessary to consider whether it is possible to say anything about the

Reliability of Cognition on the basis of the COCOM as it has been described in the preceding sections.

As a starting point we may consider the Reliability of Cognition in relation to the four control modes: scrambled, opportunistic, tactical, and strategic - considering these as four characteristic points or regions in the continuum from no control to complete control. Referring to the characterisations given above, it is reasonable to conclude that the Reliability of Cognition must be low for both the scrambled and opportunistic control modes, although for different reasons.

In the scrambled control mode the person's performance is completely unreliable from the point of view of required performance - although it probably is reliable in the sense that if the performance conditions are known then it is fairly easy to predict what the person will do. In a state of panic people will resort to stereotypical actions, but this is usually not the kind of reliability that is appropriate for accomplishing the task. Since the person is neither considering the relevant system goals nor the appropriate or prescribed procedures, actual performance will diverge from required performance.

In the opportunistic control mode the person will to some extent consider relevant goals and relevant plans, but the choice of action will often not be the prescribed one. The Reliability of Cognition is low because the choice of action is determined by the gross features or surface characteristics of the available data (input information, events) rather than by intentions and cognitive (internalised) goals. This may disrupt the smoothness of performance. The match between the desired and the actual performance is therefore incomplete, and the overall result is reduced reliability.

In the tactical and strategic control modes the Reliability of Cognition is assumed to be relatively high. In both modes there is little or no time pressure, and there is an adequate comprehension of the situation and of the tasks. In neither case is the person likely to make major mistakes or miss events or decision points. In the tactical control mode the person may at times return to a type of performance which is driven by rules and procedures and where insufficient attention is paid to the ongoing variations in the environment. It is therefore possible that the performance will not be entirely adapted to the circumstances, hence that the outcome will not completely match the requirements. The Reliability of Cognition may as a result be lower than desirable.

Fundamentals of the Model 198

In the strategic control mode we must assume that the Reliability of Cognition is high - but not perfect. The limiting factor is the capacity of human cognition. In the strategic control mode the person will do his best to consider as many relevant aspects of the situation as possible and to choose an action that is a good compromise between needs and possibilities. But if the person tries to take too many details into consideration the ambition is self defeating: by trying to bite off more than he can chew (in terms of cognition!) the result will be of a lower quality due, for example, to misunderstandings, incomplete reasoning, mistakes, reliance on heuristics, etc.

The relation between COCOM and the Reliability of Cognition is important in two different ways. Firstly, if human performance is analysed in terms of cognition as described by the contextual control model, then it is possible to improve results that come from the conventional methods, e.g. in terms of consistency. This will be the topic for Chapter 5. Secondly, if we can describe or identify specific conditions (read: control modes) where the Reliability of Cognition is high, then this can be used as a basis for designing interactive systems. The ultimate goal of an analysis of human reliability is to identify the weaknesses in the system, not to provide numerical estimates of their size but to use them as a basis for improved design (cf. Moray, 1990). This issue will not be treated in any kind of detail in this book, but must await another opportunity.

6.8 Control Modes and User Modelling

The descriptions above serve to characterise four typical control modes. The four modes are assumed to be typical points in the control space without claiming that they are representative or absolute categories. It is reasonable to suggest that scrambled and opportunistic control modes should be avoided because they are inefficient, but the choice between tactical and strategic control is not that simple. Sometimes tactical control is to be preferred, sometimes strategic.

The four control modes described here are orthogonal to other descriptions of strategies, e.g. symptomatic and topographic diagnosis (Rasmussen, 1984). Both topographic and symptomatic diagnosis may be either tactical or strategic, depending on the situation, the resources available, the pressures, and the person. The common feature of the four modes is that they all depend on the context and that transitions between them - gradual or abrupt - happen because the context changes, rather than because of internal factors. A possible explanation or mechanism for that has been outlined previously (cf. also Hollnagel, 1992a); similarly, a specific software implementation of this kind of contextual model of control is taking place in the SRG project (Hollnagel & Cacciabue, 1991).

User modelling is an important topic in the field of man-machine interaction and dynamic process supervision. Models are needed both for system and interface design, for safety and risk analyses (PSA), for specification of supporting tools, for development of training, etc. Cacciabue (1991) has pointed to some characteristics of man-machine systems which should be reflected in the models:

(1) The systems evolve with time and future developments depend on past history, even if it is "hidden", i.e., not obvious from an assessment of the present state. Their development can be described as a series of bifurcations which reflect both the principles that govern system behaviour and the influences from the external world.

(2) The human controller is part of the system; this means that a description of the system itself is embedded in the system. A model of the system must therefore not only describe the system as such, i.e., from the viewpoint of an external observer, but also the way the system is represented internally, i.e., the model that the person may have. The person's model of the system is, of course, complemented by the system's image of the person (cf. Hollnagel & Woods, 1983). The system's image is to a large extent determined by the assumptions that system developers build into the design.

(3) The management and control of a man-machine system includes cooperation and communication among people, in addition to the specific man-machine interaction. This "additional" interaction in many ways constitute the context for the specific control tasks.

These characteristics might be used as the basis for a full requirement specification for models. Even without going that far they emphasise that the model must be able to replicate the dynamic selection of actions, the various modes in which this can take place, and the influence of internal and external information (knowledge states) such as the person's current interpretation of the state, previous actions (changed states, latent conditions), time and resources constraints, and external factors (communication). All models of human performance contain some distinction between the various ways in which a plan or an action can be carried out, i.e., something which corresponds to different levels of performance. Since this clearly is one of the essential phenomena that must be modelled, it seems sensible to achieve that by making a clear separation between control and competence in the models. This will provide a way of working with the **principles** which govern behaviour rather than with specific **exemplars** or prototypes of behaviour. A good model of human performance should not prescribe a sequence of actions but describe how sequences are planned and specific actions are chosen. In doing that it may fall short of being a scientific model which explains the phenomenon in question, but it may still serve the needs for modelling in man-machine systems and dynamic process control.

7. Summary

In order to assess the Reliability of Cognition - in terms of a qualitative analysis as well as a quantitative calculation - it is necessary to have an adequate description or model of human cognition. The model provides the framework for identifying and characterising the essential features of the situation, and for relating them to the Reliability of Cognition.

There are three main approaches to or metaphors for the modelling of human performance and of human cognition. The oldest one is the Stimulus-Response-Organism model; this is similar to a black-box approach or to viewing the human as a Finite State Automaton. The S-O-R model has a long tradition in psychology although it has been abandoned by the mainstream of cognitive psychology. It can nevertheless be found in most engineering approaches to the modelling of human performance, as in (supervisory) control models.

The S-O-R model has mainly been replaced by the view of the human as an information processing system. Although this appears to be more sophisticated than the S-O-R model, it is principally the same. The information processing mechanism is still a mechanism, despite its higher degree of complexity. The view of the human as a machine, although possibly a fallible one, pervades most cognitive approaches to human performance analysis and description, and is also the basis for the prevalent methods of human reliability analysis.

The third metaphor is the cognitive viewpoint; here cognition is viewed as **active** rather than **reactive** and the focus is on overall performance rather than the hypothetical mechanisms of performance. This difference is clearly expressed by the use of cyclical rather than sequential models. A further analysis of the difference can be used to identify two main classes of models, called **procedural prototype models** and **contextual control models**. A procedural prototype model implicitly expresses the view that there exists a characteristic or pre-defined structure of actions which represents a natural or preferred way of doings things. A contextual control model implies that actions are determined by the context rather than by any inherent relation between them. Both views recognise that there are strong recurrent sequences or patterns of actions. But while the former view focuses on the patterns and tries to explain the possible deviations from them, the latter view puts the focus on the control of actions, thereby making the patterns rather incidental.

The procedural prototype models are exemplified by the well-known Step-Ladder Model, which tries to explain performance as repeated passes through the same procedural prototype. Although this view is efficient when performance is regular, it does present some problems when deviations are to be explained because the control is implicit in the model. Procedural prototype models nicely match the Western cultural preference for determinism and order, but are not the only alternative; even a relaxed control paradigm such as the Test-Operate-Test-Exit (TOTE) provides a more powerful and less restraining representation of human action.

The alternative to the procedural prototype is found in the contextual control models, which emphasise how human action is shaped by the context, for instance in terms of the person's cognitive goals. Contextual control models are compatible with the cognitive viewpoint and therefore have a relatively long tradition - but with scanty recognition. A contextual control model is described in terms of two main parts: a model of competence and a model of control. The competence describes the possible actions that the person may call upon (the **activity set**) as well as the characteristic ways in which actions may be grouped (**templates**), for instance as pre-defined procedures. It is pointed out that a sequence of actions, such as a procedure, may be broken down in many different ways, depending on both the situation and the person. There is no objectively right way of decomposing actions; but given information about the context (the skill and experience of the person, the typical work situation, the task and support design, the process states, etc.) it is possible to define the principles for decomposition that should be applied. The separation of competence into the activity set and the templates makes it possible to model a number of characteristic performance phenomena (e.g. rigidity, mistakes) with a few and simple principles.

The other main part of the model is concerned with how the control of actions takes place, in terms of both choosing the next action and carrying it out. In the model presented here four control modes are proposed: scrambled, opportunistic, tactical, and strategic. Each of these are defined and described in some detail. Following that a specific contextual control model, COCOM, is defined. An important part of the model is the main control parameters which determine how the person may enter (or leave) a particular control mode and what the characteristics of performance are in a given mode. The control parameters are suggested to be the determination of the outcome (of the previous action) and the estimation of subjectively available time. Further parameters are the number of simultaneous goals (e.g. related to the occurrence of sudden changes, alarms, faults, disturbances), the availability of plans, the person's event horizon (in terms of both history size and prediction length), and the mode of execution. A description is given of how the main control parameters may interact with the four control modes to cause changes between modes and to produce the characteristic types of performance within a mode. The modes are in turn described in terms of the Reliability of Cognition and the expected reliability of performance. It is emphasised that the primary purpose of human reliability analysis is to improve design. The contextual control model must therefore serve not only as a foundation for reliability analyses but also as the basis for specific design decisions.

5.
The Dependent Differentiation Method

1. Framework for Assessing Human Reliability

The preceding chapters have served to define the background and the context for how we can study the Reliability of Cognition by developing a description of what the Reliability of Cognition is. In Chapter 3, a characterisation of current approaches to Human Reliability Assessment (HRA) were given. In the preceding chapter (Chapter 4), a specific model of cognition, the Contextual Control Model (COCOM), was presented as an alternative to the decomposition approach and as an improved basis for analysing the Reliability of Cognition. A better concept, theory, or model of the Reliability of Cognition does, however, not in itself solve the present problems. There is a practical need for better **methods** to assess human reliability. The acid test of any conceptual framework for the Reliability of Cognition is therefore whether it can be used as the basis for developing an assessment method, and whether that method works. Quite another concern is whether the results provided by the new method are more precise than the results which can be obtained from existing methods. This issue can only be resolved if a sufficient empirical basis for comparison can be found.

The purpose of the present chapter is therefore to define a clear set of guidelines, leading to an operational method which can be used to address the problems of human reliability assessment and achieve usable results (i.e., results which ultimately can be used in a PSA). The overriding criticism against the existing methods has been that they have all been built on the notion of decomposition and thereby have embraced the assumptions implied by the decomposition principle. Since these assumptions have been demonstrated to be wrong, the conclusion is that a method for analysing Reliability of Cognition should have a different basis. We must therefore carefully consider how that objective can be achieved, i.e., which assumptions one should use instead.

1.1 The Decomposition Principle

The decomposition principle was briefly introduced in Chapter 1 and discussed again in Chapter 3. To summarise, the assumptions underlying the use of quantification were the following:

(1) **The Atomistic Assumption**: Human performance can be adequately described by focusing on the individual elements or parts of the performance. A description of the total performance can be produced by aggregating the descriptions of the individual performance elements.

(2) **The Mechanistic Assumption**: The human mind can be adequately described as an information processing system, i.e., man can be described as a machine.

The main consequence drawn from the atomistic assumption was that the whole was equal to the sum of the parts, and that one therefore conveniently could replace a study of the whole by a study of the parts. The main consequence drawn from the mechanistic assumption was that any human performance could be decomposed into a number of elementary steps. Together this meant that human reliability assessments could be carried out as step-by-step analyses of fixed or static event sequences and that human performance was considered only on the level of the individual step or action.

(An argument against this view may be that it is possible in some techniques, such as THERP, to consider the dependence between steps, e.g. correcting an incorrect diagnosis. This dependence, however, still remains on the level of individual steps and does not look at performance as a whole. Interestingly enough, many of the comments that go together with such analyses reveal the discomfort in doing so and refer to the cliché that operators work in a context - although the analyses do not show it.)

An alternative set of assumptions can obviously be obtained simply by negating the original ones:

(1) Human performance cannot be adequately described by focusing on the individual elements or parts of the performance. A description of the total performance cannot be produced by aggregating the descriptions of the individual performance elements.

(2) The human mind cannot be adequately described as an information processing system, i.e., man cannot be described as a machine.

The main consequences from these new assumptions are (i) that the whole is **more than** the sum of the parts, and that a study of the whole therefore **cannot** be replaced by a study of the parts; further (ii) that human performance cannot be decomposed into a number of elementary steps. Another way of saying that the whole is more than the sum of the parts is by emphasising the interaction that takes place in any dynamic system - and which even is a defining characteristic. Similar to this interaction there is a dependency between the identifiable steps of an activity, which neither can be neglected. It is, however, necessary to be more constructive than this. The negated assumptions define what one **cannot** do; but this must be supplemented by descriptions of or guidance on what one **can**, and should, do.

Part of the answer has been provided by the COCOM described in the previous chapter. This model offers an alternative to the information processing view that is implicit in the procedural prototype models. Human cognition is still **described** using information processing terms, but is **not explained** as information processing. The COCOM is thus a departure from the strict physical symbol system hypothesis that lies behind the classical notion of the human information processor - which thereby changes from having an ontological to having an analogical status. The COCOM demonstrates that there is an alternative to the traditional view of how the human mind works, and that this alternative can be used to describe human cognition.

One way of proposing a constructive alternative to the atomistic assumption is to focus on performance as a whole. The basic position must be that the analysis should strive to be complete in the sense of including all actions as they would naturally occur. An initial stage must therefore be to establish the overall conditions under which performance will take place. (The term "stage" is used to denote the different phases in the analysis of the Reliability of Cognition in order to distinguish them from the "steps" that are part of the performance being analysed.) These conditions can then be used to define the questions and filter the information that is needed for the following stages. The analysis may still involve a gradual refinement or decomposition, but it will be context dependent rather than based on the constituent actions in isolation. By using high level concepts, such as the Common Performance Modes (CPM) to be introduced in a later section, it may not even be necessary in every case to go into detailed analyses of performance steps, since the expected benefit could be negligible. The quest for data is re-directed to the more aggregate performance levels or global situations. Data may be given in the form of empirical data from characteristic situations (real or simulated), or as subjective or conditional probabilities. Whether it is one or the other is of secondary importance as long as the principles are consistently applied.

1.2 The Need for Task Analysis

The proper basis for assessing human reliability must be a systematic description of the appropriate actions that the person is expected or required to carry out, including how they are related or structured. A description of this kind is commonly called a task analysis. Every form of task analysis implies a set of assumptions about operator capabilities and limitations, hence a model of human action - even though it is not always expressed explicitly as a model. This model, or reference description, must enable not only a *post hoc* explanation of a given set of observations, but also **the prior analysis of probable or likely ways in which an action sequence can develop**, hence the ways in which performance can be impaired. We shall use the COCOM as the reference description.

It was argued previously that one cannot assess how a task is carried out and how it may develop without considering the context. On the other hand, the context is partly determined by the task, thereby establishing a reciprocal dependency. It is clearly necessary to start somewhere, and since we usually have available either a description of the task (e.g. as a procedure) or a definition of the task objectives (goals, targets) it is reasonable to begin with the task analysis. The reciprocal dependency between task and context will appear again in the discussion of the Dependent Differentiation Method.

The task analysis is a prerequisite for the Dependent Differentiation Method although it is not a part of the COCOM itself, nor can be derived from it. There is nevertheless one particular type of task analysis which is consonant with the principles of the COCOM. This is the Goals-Means Task Analysis Method (GMTA) which is described in the following.

2. Task Analysis Principles

The term task analysis is commonly used as a generic label. Any survey of what usually is called task analysis methods will quickly show that they represent many different meanings of the term. A little closer inspection will, however, reveal that they fall into a few main categories:

(1) the analysis and description of tasks or working situations which do not yet exist or which are based on hypothetical events; examples are Hierarchical Task Analysis (Annett *et al.*, 1971) and the Goals-Means Task Analysis described below;

(2) the description and analysis of observations of work being carried out or of event reports (accident investigations); examples are event trees (Hall *et al.* 1982), and various observational techniques;

(3) the representation of the results of the above, in the sense of the notation used to capture the results; this is of recent interest due to the increasing use of computers to support task analysis, and examples are therefore few;

(4) the various ways of further analysis or refinement of data about tasks (from either of the above sources); examples are Failure Mode and Effects Analysis (FEMA) (Henley & Kumamoto, 1981) and Link Analysis; and finally

(5) the modes of presentation of results and the various ways of documenting the outcomes; examples are event trees, time-lines, and operational sequence diagrams (Kurke, 1961).

The relations between the main task analysis concepts are shown in Figure 5-1, and described in the following sections.

2.1 Task Analysis

The term **task analysis** will be used to refer to the analysis and description of tasks based on information about the system (specifications and detailed design data) rather than based upon observations. Task

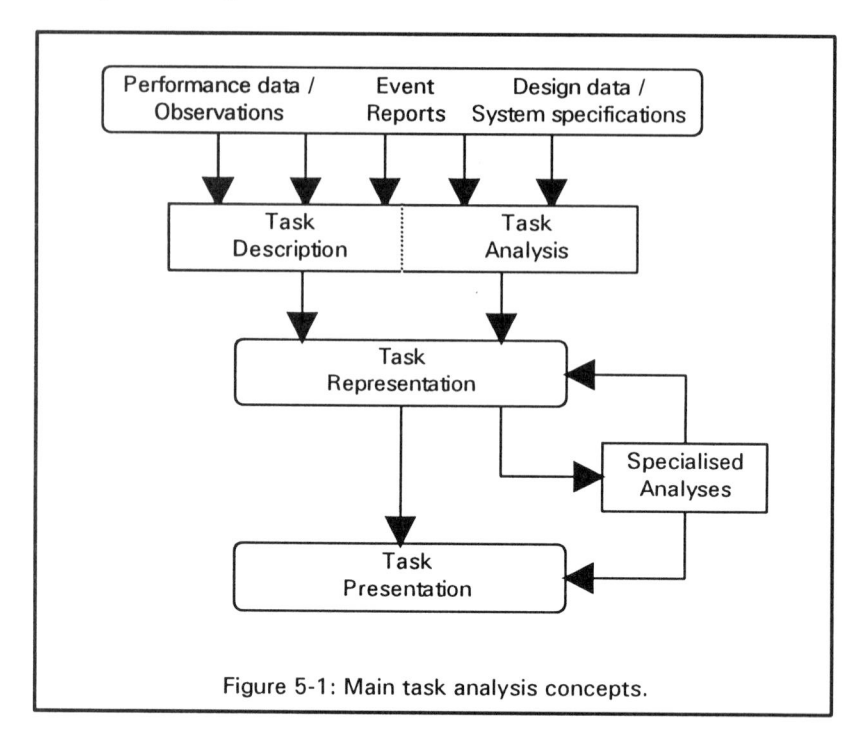

```
┌─────────────────────────────────────────────────────────┐
│  ┌──────────────┐  ┌───────┐  ┌──────────────────┐     │
│  │Performance   │  │Event  │  │Design data /     │     │
│  │data /        │  │Reports│  │System            │     │
│  │Observations  │  │       │  │specifications    │     │
│  └──────┬───┬───┘  └───┬───┘  └────┬──────┬──────┘     │
│         ▼   ▼          ▼           ▼      ▼            │
│       ┌────────────┐      ┌────────────┐               │
│       │   Task     │......│   Task     │               │
│       │Description │      │  Analysis  │               │
│       └─────┬──────┘      └─────┬──────┘               │
│             ▼                   ▼                      │
│           ┌──────────────────────┐                     │
│           │        Task          │◄──┐                 │
│           │   Representation     │   │                 │
│           └──────────┬───────────┘   │                 │
│                      │           ┌───┴────────┐        │
│                      │           │Specialised │        │
│                      │           │ Analyses   │        │
│                      │           └───┬────────┘        │
│                      ▼               │                 │
│           ┌──────────────────────┐   │                 │
│           │        Task          │◄──┘                 │
│           │    Presentation      │                     │
│           └──────────────────────┘                     │
│                                                         │
│        Figure 5-1: Main task analysis concepts.        │
└─────────────────────────────────────────────────────────┘
```

analysis in this sense is carried out as a part of the system design, and can serve several purposes such as preparation of procedures, detailed design of support systems or man-machine interface, task allocation, etc. In our case the purpose is to prepare the ground for a human reliability analysis. Task analysis describes what the tasks **should be** (or are expected to be) and is in this sense prescriptive. Task analysis is, however, not prescriptive in the meaning of being normative. (A possible exception is when a task analysis is used to define procedures, which normally are supposed to be followed to the letter.) Task analysis is a prescription of the general principles rather than of the minute details.

2.2 Task Description

Task analysis is seen in contrast to methods of **task description** or **performance analysis**. A task description produces a generalised account or summary of activities as they **have been** carried out. It is based on empirical data or observations rather than on design data and specifications. A typical example is link analysis or even Hierarchical Task

Analysis (Annett *et al.*, 1971). Properly speaking, task description or performance analysis deals with actions rather than with tasks. This, by the way, is comparable to the French tradition in cognitive psychology where a distinction is made between tasks and activities. The task is the objective to be attained while the activity is the instantiation of the task, i.e., the actual performance - which, of course, may be different from the prescribed task (Leplat, 1989). Another possible term would therefore be **activity analysis**.

A very special source of performance analysis data is found in event reports, whether of near-misses or accidents. Event reports differ from the main data source of performance analyses/task description by being an account of something that has actually occurred. But they differ from the typical empirical observations by referring to a unique occurrence, i.e., they are idiographic rather than nomothetic. Event reports are therefore considered both from the point of view of tasks analyses and of task descriptions.

(The terms "task analysis" and "task description", as used here, correspond to what customarily are called "prescriptive task analysis" and "descriptive task analysis". Since "descriptive task analysis" in the common usage is the analysis of performance to provide, among other things, the description of a task, I find that the term is ambiguous and therefore try to avoid it.)

2.3 Task Representation

Task representation refers to the formats that are used to describe the outcome of task analyses/task descriptions. These include tables, forms, matrices, etc. **Task representation** must be distinguished from **task presentation**. The latter refers to the actual mode of presentation which is used to convey the results. Thus, the same task representation may be expressed by different presentation forms (e.g. matrices, trees, time-lines).

2.4 Specialised Analyses

The final set of methods have to do with **specialised analyses** which either may refine the previous results or lead to different representations of the outcome. Examples of that are workload analysis, time-line analysis, operational sequence diagrams, critical path analysis, etc. The common characteristics of these methods are, firstly, that they are not really task analysis methods in the sense defined here and, secondly, that

The Dependent Differentiation Method

```
┌─────────────────────────────────────────────────────────────┐
│  Performance data /   Event       Design data /             │
│  Observations         Reports     System specifications     │
└─────────────────────────────────────────────────────────────┘
                              │
              ┌───────────────┴───────────────┐
              │ Task Description │ Task Analysis │
              └───────────────────────────────┘
                          │
              Link    Hierarchical   Goals-
              analysis    task       means
                        analysis     analysis
                          │
                   ┌──────────────┐            Cognitive
                   │     Task     │            task analysis
                   │Representation│            Signal flow
                   └──────────────┘            graph analysis
                          │        Specialised Decision/action
                          │        Analyses    flow analysis
                          │                    Confusion
                          │                    matrix analysis
                   ┌──────────────┐            Taxonomic
                   │     Task     │            decomposition
                   │ Presentation │            Operational
                   └──────────────┘            sequence analysis

                        Applications

         Staff      Reliability    Team
         selection  analysis       organisation

         Procedure  Interface      Workload
         design     design         analysis

         Risk                      Task
         analysis   Training       allocation
```

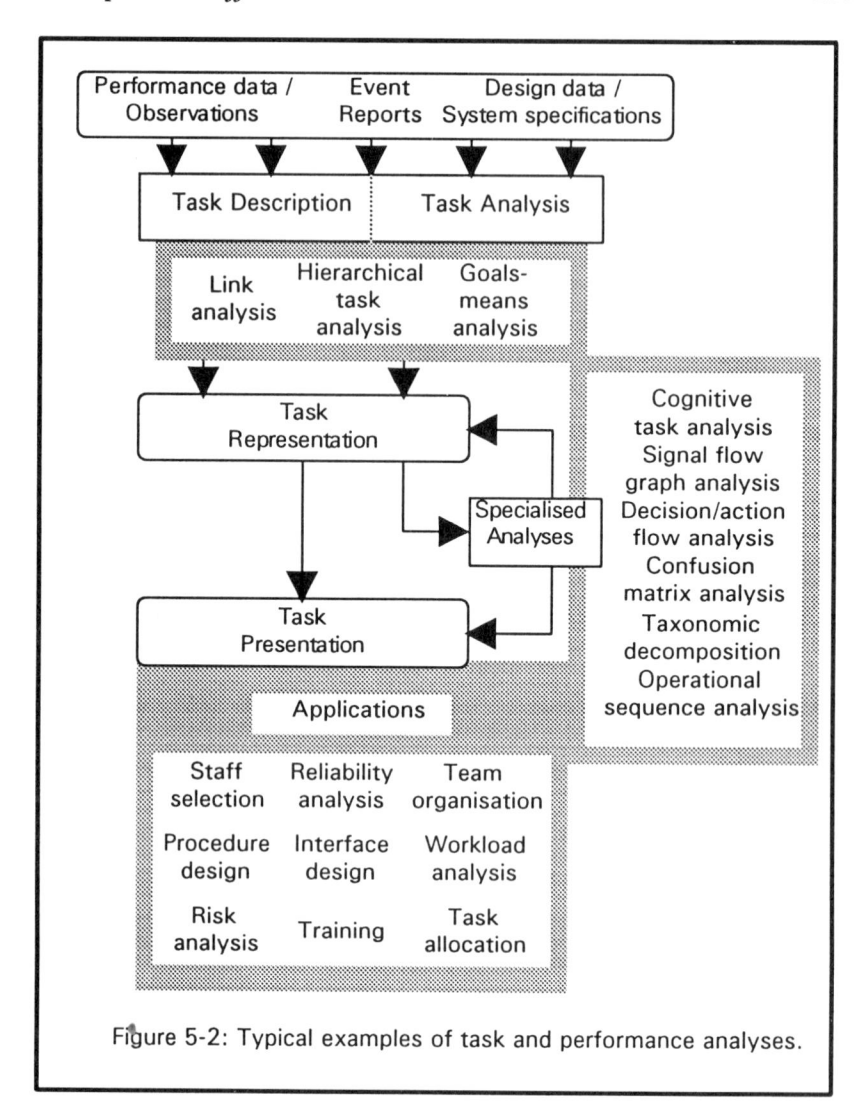

Figure 5-2: Typical examples of task and performance analyses.

they can be applied to data about tasks whether they are derived from design specifications or empirical observations. For instance, the Goals-Means Task Analysis described in the following does not in itself point to any specific way of representing or presenting the outcome, and need therefore be supplemented by a suitable way of task representation and task presentation.

Some typical instances of the categories described in the preceding sections are indicated in Figure 5-2; here the task presentations are seen as derived from the needs of specific applications.

3. The Logic of Task Analysis

The purpose of a task analysis is to define the essential contents of the task in order to describe the parts as well as the ways in which the task is organised. The basic organising principles must be that each step of the task either brings the person closer to the stated goal or achieves a (pre-) condition which is required if further progress is to be made. If that is not the case the task step is clearly inappropriate and should be avoided or not be prescribed. A task must consequently be described with reference to the goal of the task. The following definitions are applied:

(1) A **goal** is a specified condition or state of the system which signifies the accomplishment of a task or a task step. A goal is what the person intends to do or achieve. A description of the goal usually includes or implies the criteria of achievement, i.e., the conditions that define when the goal has been reached.

Goals are either defined as **derived goals** or **independent goals**. The difference is whether the goal is the result of the decomposition of another (parent) goal or whether it is defined independently. The former are called derived goals (or sub-goals) and the latter are called independent goals or top-goals. A pre-condition (cf. below) is often equivalent to a sub-goal. A sub-goal always refers to a parent goal (which itself may be a sub-goal), and when the sub-goal has been achieved control of the task returns to the parent goal; this means that the current goal changes. When the top goal has been achieved the task is for all practical purposes completed. The distinction between derived and independent goals is not absolute but pragmatic, and depends on the chosen level of description. Just as some task steps are considered elementary in a practical sense, so some goals will be considered as independent goals or top goals. This is partly done to limit the number of levels of recursion.

Semantically, a goal is described as a **system state** or a system condition. To take one of the examples used below, the goal would be "the roof has been painted" rather than "to paint the roof" because the latter describes an activity rather than a state. To achieve a goal it is necessary to carry out a task or a task step. (The terms "task" and "task step" will be used interchangeably in what follows.)

(2) A **task** is an organised collection of actions carried out by a person in order to achieve a goal or an objective. The actions are referred to as **task steps**. The set of task steps need have only a single member. The organisation of the actions may be derived from the fact that the individual actions together serve to accomplish the goal or may be prescribed by a procedure - either a predefined sequence of steps, a predefined organisation principle, or specified dependencies between steps.

A task step can be composed of other task steps and its name may serve as a label or reference for the set of task steps (a routine, a group, a procedure). This is usually done merely for notational convenience. Each set of task steps may be concatenated as a larger task or task step; conversely, each larger task step can be characterised by means of the constituent task steps. Task steps describe activities which can be carried out, i.e., where it is assumed - for the sake of the analysis - that the pre-conditions are true. This assumption can, of course, always be challenged and the analysis taken one step further. In this way a task step which has a pre-condition will create a sub-goal which expresses the pre-condition.

Semantically, a task is described as an activity which causes a transition between system states; using the same example, the activity "to paint the roof" causes a transition between the states "the roof is not painted" and "the roof has been painted."

A task analysis consequently denotes a method which can be used to produce a description of the task steps that will lead to a stated goal as well as the principles by which they are organised. Since each task step can be seen as part of the means to achieve the goal, the basic method must be a recursive refinement based on a goals-means relationship: a **Goals-Means Task Analysis (GMTA)**.

3.1 Pure Tasks

In a pure task the steps can be carried out without imposing further conditions on the system. The steps of the task can be described as a simple sequence, either in a strict linear sequence or with branch points (decision points), i.e., as a tree. If the task steps are single or elementary actions, the result is a **non-hierarchical** task description.

An elementary step is defined as a step that is not further refined at the current level of analysis. The category is thus not absolute. In practice, any given application will have some steps that commonly are considered as elementary, in the sense that one can safely assume that they can be performed no matter what the conditions are. These are usually the basic skills that every person has *qua* being a person or *qua* being qualified for the work in question. These elementary steps correspond to the action set in the COCOM.

If the task steps are composed of other task steps (aggregated steps), then the result will be a **hierarchical** task description; it simply means that the task description includes elements at different levels of elaboration or granularity, from global descriptions to detailed elementary task steps. Most task descriptions will be hierarchical due to the limitations we encounter (size of paper, span of memory), as well as the way in which we intuitively organise work.

The pure task thus describes or points to the means by which the goal can be achieved. This can be written as follows:

Goal: "The named goal or state which is to be achieved."

Task: "The named task which is sufficient to achieve the goal."

The principle of a pure task is shown in Figure 5-3.

3.2 Tasks with Pre-conditions

In other cases the task steps cannot be carried out unless certain conditions are met. A **pre-condition** describes the conditions that must be satisfied for a task step to be carried out. When a pre-condition is found, it is considered as a sub-goal (a derived goal). It is this feature which introduces the recursiveness of the GMTA. A pre-condition corresponds to a sub-goal and can be followed by one or more task steps which, when carried out, will achieve the pre-condition. The task analysis must therefore clarify under which conditions the task steps can be carried out and under which conditions they cannot. These conditions can either be the pre-conditions that must exist when the task step is started and/or the conditions that must prevail during the execution.

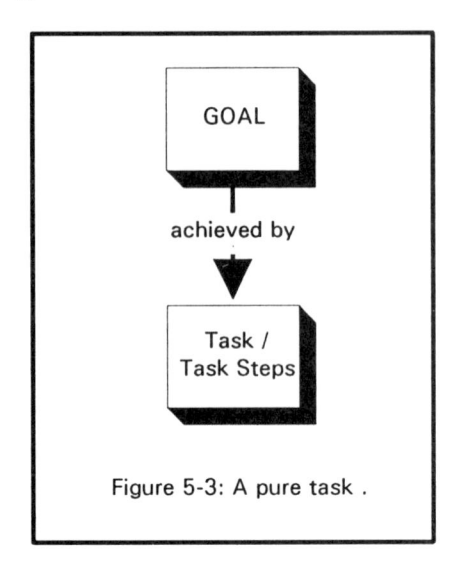

```
                    ┌─────────────┐
                    │    GOAL     │
                    └──────┬──────┘
                           │
                       achieved by
                           ▼
                    ┌─────────────┐
                    │   Task /    │
                    │ Task Steps  │
                    └─────────────┘

              Figure 5-3: A pure task .
```

Example: In order to repair the roof of a house, a ladder must be present (to get on to the roof). The ladder is, however, not required during the repair (although it may be useful for getting down again, hence completing the task). This can be written as follows (using a pseudo-programming language notation, where the indentations are used purely for readability):

Goal: The roof has been repaired.

    ...

    Goal: Person is on the roof.

        Task: Climb the ladder

        Pre-condition: Ladder is present.

If the pre-conditions are not true when the task step is to be carried out, they must be established; this leads to the definition of a new goal hence a new task, the purpose of which is to establish the pre-conditions and thereby enable the (preceding) task step. The new subsidiary task will not necessarily directly bring the person closer to the goal if seen in terms of an immediate distance measure, but rather serves as an **enabling** task. However, in so far as the goal of the enabling task is to establish the conditions that makes it possible to carry out the preceding task step, the enabling task does, in some sense, bring the actor closer to the goal. (As an example, it may be necessary to make a detour to buy additional petrol (gasoline). Although this may objectively increase the distance to the goal, it serves to establish the necessary pre-condition for ultimately getting there.)

The notion of a pre-condition can be expressed as follows:

Goal: "The named goal which is to be achieved."

    Task: "The named task which is sufficient to achieve the goal."

    Pre-condition: "The condition that must be fulfilled before the task can be carried out."

        Task: "The named task which is sufficient to establish the pre-conditions."

3.2.1 Timing Conditions

An additional and important concept is the **sequence** or **timing conditions** of a task step. Sequence conditions describe constraints on the sequencing (starting and stopping) of a task step. The sequence conditions can be expressed as relative timing conditions using, for example, the suitable terminology proposed by Allen (1983). According to this any event can be described in terms of four self-explanatory time indicators:

(1) **Earliest Starting Time (EST),**

(2) **Latest Starting Time (LST),**

(3) **Earliest Finishing Time (EFT), and**

(4) **Latest Finishing Time (LFT).**

It is clear that these timing conditions need only refer to an ordinal rather than an interval scale. Neither do they require estimates of duration nor references to measured time (absolute or elapsed). Such information is rarely available until the last stages of system development, but may, of course, be used if it is known. Even ordinal timing conditions are, however, extremely useful to provide further details in the description of pre-conditions or execution conditions. A timing condition could, for instance, be that the beginning of a task step must occur after the ending of another task step, or that two task steps must be carried out simultaneously:

...

Task: Do X

Task: Do Y

Pre-condition: $EST_{stepX} > LFT_{stepY}$

...

Indication of simultaneous execution of task steps is relevant mostly for steps belonging to different tasks (i.e., with different but possibly related goals). This is a useful way in which to express the requirements for parallel execution of tasks. Time indications may be given as absolute time (clock time) or relative to other events, if this information is available to the analyst. An example is:

...

Task: Do X

Task: Do Y

Pre-condition: $(LST_{this\ step} - LFT_{previous\ step}) < 2$ minutes

...

In this example the pre-condition directly describes how the condition is going to be determined and there is consequently no need for other task steps. Although a sequencing/timing pre-condition functionally is equivalent to a sub-goal, because it is a condition which must be fulfilled, it does not make sense actually to transform it into a sub-goal since there are usually no explicit task steps which will accomplish it.

In order to make the terminology unambiguous only task steps can have pre-conditions.

3.2.2 Execution Conditions

In addition to the pre-conditions there may also be conditions which must be maintained during the execution of the task step. An execution condition describes the conditions that must be satisfied **while** a task step is carried out. An execution conditions is thus always attached to a task step. In cases where the execution condition is **implied** by the description of the task step, it need not be defined explicitly. Most execution conditions will therefore refer to temporal conditions and relations to other task steps. The execution conditions will typically be of two types: either the availability of the resources necessary to carry out the task steps, or conditions that refer to other task steps. An example of the latter could, for instance, be that two tasks were completed at the same time or - even more stringently - that they were carried out during the same time interval or simultaneously. In the majority of cases execution conditions are not stated explicitly, but are rather assumed by the description of the task step. Thus if the task step is to paint the roof, an execution condition is that there is a constant supply of paint. But since it is very difficult to paint something without having paint, it is not strictly necessary to express the execution condition. However, if the task step is to open a valve, this does not indicate how the step can be performed; it may be by remote operation or by using a valve wrench.

Pre-conditions obviously need not always be present during execution. A related question is whether execution conditions must be present when a task is started. Although it may seem so in many cases, it is not difficult to find examples of the contrary. If, for instance, I want to drive in my car from Copenhagen to Paris, I need enough petrol to do so. But I do not need to have it all when I start the journey, since I can refuel on the way. (Refuelling thus becomes an intermediate goal.) An aircraft, on the other hand, usually needs to have all its fuel on board when it takes off. In another example, if the task is to photograph a solar eclipse, then the eclipse should clearly not be there when the task or mission is started. Despite these counter-examples it is probably more often than not the case that execution conditions must be present from the start.

The Goals-Means relation is a principle that requires the task analysis to be **recursive**. If the task steps can be described without pre-conditions as a linear sequence (a pure task), the task analysis need only be **repeated** rather than recursive, i.e., there is nesting of actions but no nesting of goals and sub-goals. The recursive analysis will, of course, automatically produce a hierarchical result. The relation between goals, task steps, and pre-conditions is shown in Figure 5-4.

Figure 5-4: A task with pre-conditions.

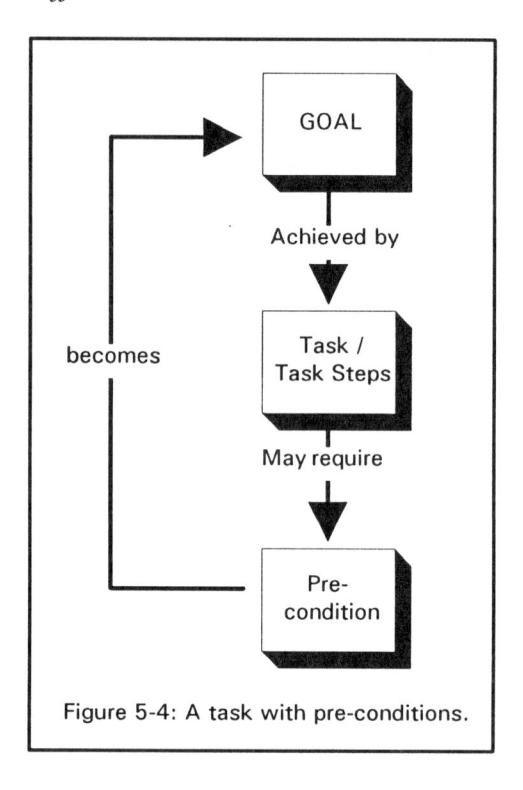

3.3 Post-conditions

Formally speaking, one can consider each task step as producing a **post-condition** (or rather a post execution condition). A pre-condition will always refer to a set of task steps which have to be carried out if the pre-condition is not met. The outcome of the task steps (their post-condition) is what makes the pre-condition true (i.e., the value of the post-condition of one step becomes the value of the pre-condition of the following step). Task steps can be described so that their post-condition will either be true or false according to some criterion. The task steps can therefore also be described in terms of these conditions, meaning that a task step is only considered completed if its post-conditions are true. Otherwise the analysis must be taken a step further (by additional refinement) in order to achieve the desired post-condition. (Note that this provides another definition of an **elementary task step**, as being one which always returns the condition true. In other words, it is a task step which, for the sake of the analysis, can be assumed always to be executable.) Clearly, if the post-condition is false, and there are no ways in which it can be changed, then the goal cannot be achieved and the task therefore has failed.

Another type of post-condition describes the conditions which are not defined or included by the goal or task step, but which may occur anyway. These post-conditions refer to the **side-effects** which are deemed to be important because they are known or assumed to have consequences for, as an example, other pre-conditions - either in a positive (synergistic) or negative (antagonistic) way. Practically all actions will have some side-effects, although most of these safely can be neglected. (That is the case, at least, for normal conditions. The art of risk and reliability analyses is, of course, to know when a side-effect should be neglected and when it should not.)

The relation between the basic building blocks of a GMTA is shown in Figure 5-5.

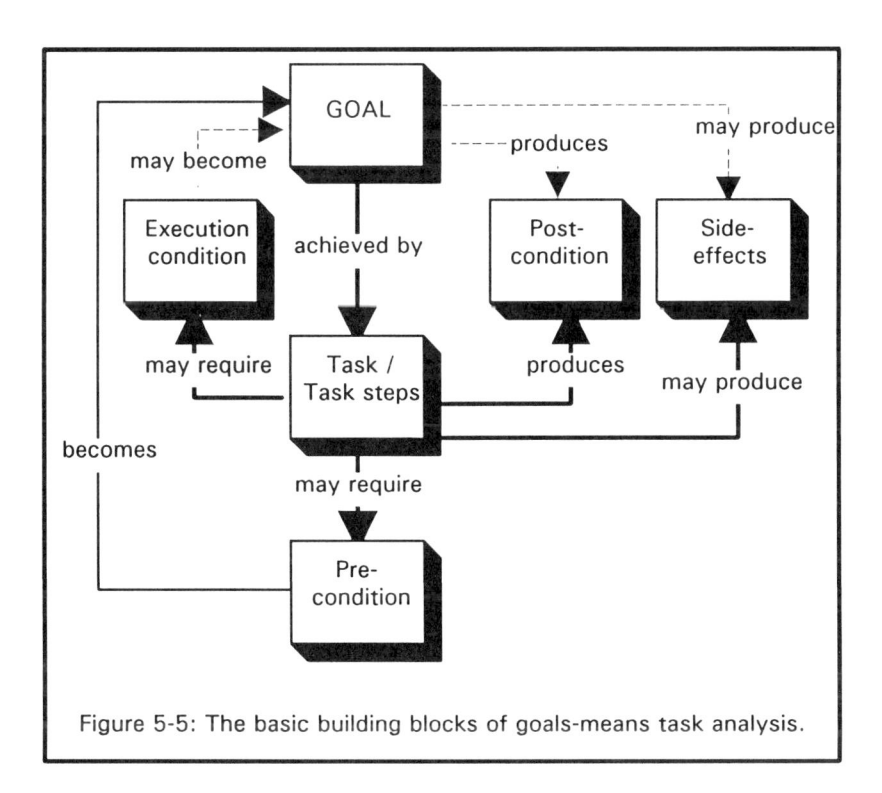

Figure 5-5: The basic building blocks of goals-means task analysis.

4. Goals-Means Task Analysis Method

One of the problems in describing the Goals-Means Task Analysis method is that the description must cover two levels at the same time. One level is a description aimed at the user, i.e., the person who is going to apply the task analysis method. This chapter tries to deduce how such a task description (or task phenomenology) should be structured. The other level is a description of the logical relations between tasks, i.e., the principles that define the underlying structure of the task representation. The Goals-Means Task Analysis method uses a powerful, yet simple, description which only makes use of a few concepts, but which relies heavily on the principle of recursion.

4.1 Example: "Feed and Bleed"

As an example I will use the general procedure for Feed And Bleed (FAB) as prescribed for Pressurised Water Reactors (PWR).[1] (The procedure used here is a generic one which therefore leaves out all plant specific details.) The FAB procedure describes a way of cooling the reactor when the normal cooling has failed.

As shown in the simplified schematic in Figure 5-6, cooling of the reactor core is normally accomplished by means of the steam generators where the primary cooling loop exchanges its heat with the secondary cooling loop. In the secondary loop the heat is absorbed through the condensers - after the steam has been used for its designated purpose, typically to drive the turbines. In the cases where the normal cooling is not available, i.e., when the steam generators for some reason do not function as required, a temporary cooling can be achieved by "feeding" through the high pressure safety injection system and "bleeding" through the pressure operated relief valve(s). Since this solution breaks a radiological barrier of the plant, by releasing radioactive coolant into the containment, it is only used in emergencies.

The initial condition which leads the operators to consider the possible use of FAB is that the main feedwater as well as the auxiliary feedwater have failed. The task analysis starts from the top-level goal of establishing core cooling, and proceeds as follows:

---

[1] I am indebted to Ed Dougherty for assisting me in developing this example. A relevant event is described in NUREG-1154.

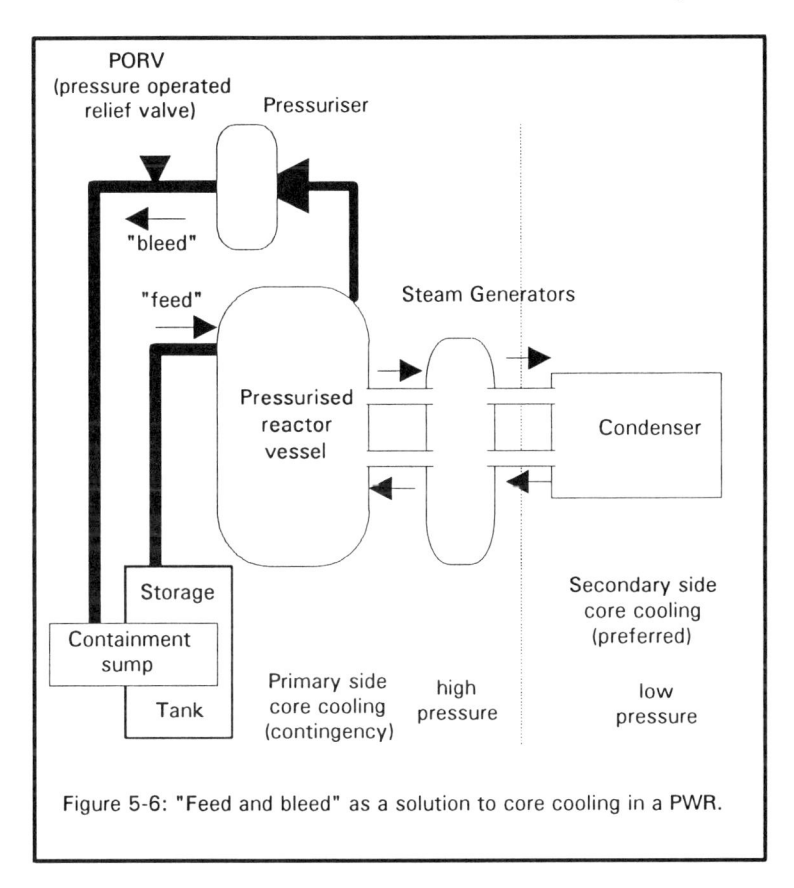

Figure 5-6: "Feed and bleed" as a solution to core cooling in a PWR.

Goal: <Core cooling has been established>

  Task: Establish core cooling.

The goal is the system state when core cooling has been (re)established; the corresponding task is therefore to establish core cooling. If the task had referred to a well-defined activity (a plan, a procedure, a set of steps) that could be carried out with high reliability, there would be no need to proceed further. In this case, however, it is not assumed to be so, and the analysis must therefore continue:

Goal: <Core cooling has been established>

  Task: Establish core cooling.

Pre-condition: <AFW has been re-established> OR <FAB has been established>

Further analysis shows that core cooling can be established in two different ways, either by re-establishing the Auxiliary Feedwater (AFW) supply to the secondary side or by establishing FAB. (The ordering reflects the priority of the two conditions; it is clearly preferable to establish core cooling by means of the AFW.) This pre-condition means that the task "establish core cooling" really refers to two different tasks as defined by the pre-conditions. Describing each pre-condition implies that the status of the system is checked to determine whether or not the pre-condition has been fulfilled. If this can be assumed to be a highly reliable action it need not be made more detailed; if not, the description must be expanded:

Goal: <Core cooling has been established>
   Task: Establish core cooling.
   Pre-condition: <AFW has been re-established> **OR** <FAB has been established>
   Goal: <AFW has been re-established>

The pre-condition for "establish core cooling" creates two new (sub)goals, the first of which is to re-establish AFW in the system. The semantics of the notation is that if the goal <AFW has been re-established> has been achieved (i.e., if it is true), then the (disjunctive) pre-condition is also true, which means that the top level goal has been achieved. On the other hand, if the goal <AFW has been re-established> has not been achieved (i.e., if it is false), then the pre-condition may or may not be true depending on the second part of it:

Goal: <Core cooling has been established>
   Task: Establish core cooling.
   Pre-condition:: <AFW has been re-established> **OR** <FAB has been established>
   Goal: <AFW has been re-established>
     Task: Restart AFW from control room.
     Pre-condition: <AFW is in working condition>

Goal: <AFW is in working condition>

Task: Send assistant operators to restore the AFW system.

Having identified the pre-condition <AFW has been re-established>, the corresponding task must be described and, if necessary, expended. If the operators succeed in re-establishing the AFW, then the situation is under control. In the example the goal can be achieved if the AFW can be restarted from the control room, which in turn requires that the AFW is in working condition. However, if this is not the case the operators must try to achieve the second part of the pre-condition - while simultaneously continuing to achieve the first part. The need to achieve both parts of the pre-condition simultaneously can be expressed by using timing conditions. This need for simultaneity is specific for the FAB procedure and is not a general trait of the GMTA:

Goal: <Core cooling has been established>

Task: Establish core cooling.

Pre-condition: <AFW has been re-established> **OR** <FAB has been established>

Goal: <AFW has been re-established>

Task: Restart AFW from control room.

Pre-condition: <AFW is in working condition>

Goal: <AFW is in working condition>

Task: Send assistant operators to restore the AFW system.

Goal: <FAB has been established>

Task: Establish FAB.

Pre-condition: <FAB criterion is met> **AND NOT** <AFW has been re-established>

If FAB has not been established, then the task is obviously to establish it. This new task has a composite pre-condition: FAB should not be established before the criterion for FAB has been met, nor should it be established if in the meantime AFW has been re-established. This could happen because the higher level pre-condition was disjunctive, hence implying that two goals are pursued at the same time. In a full GMTA this condition must, of course, also be made explicit. It has not been done here because it would make the description unnecessarily complicated for the present example:

Goal: <Core cooling has been established>

    Task: Establish core cooling.

        Pre-condition: <AFW has been re-established> **OR** <FAB has been established>

    Goal: <AFW has been re-established>

        Task: Restart AFW from control room.

        Pre-condition: <AFW is in working condition>

    Goal: <AFW is in working condition>

        Task: Send assistant operators to restore the AFW system.

    Goal: <FAB has been established>

        Task: Establish FAB.

        Pre-condition: <FAB criterion is met> **AND NOT** <AFW has been re-established>

    Goal: <FAB criterion is met>

        Task: Monitor FAB criterion. "The criterion is that steam generator level is below a critical limit."

The last pre-condition again creates a new (sub)goal. It only creates a single new (sub)goal because the second part, <AFW has been re-established>, already has been defined in a higher level pre-condition. The task related to the new goal is in this case simply to monitor the steam generator level and compare it to the predefined criteria for starting the FAB:

Goal: <Core cooling has been established>

    Task: Establish core cooling.

Pre-condition: <AFW has been re-established> **OR** <FAB has been established>

Goal: <AFW has been re-established>

    Task: Restart AFW from control room.

    Pre-condition: <AFW is in working condition>

    Goal: <AFW is in working condition>

        Task: Send assistant operators to restore the AFW system.

Goal: <FAB has been established>

    Task: Establish FAB.

    Pre-condition: <FAB criterion is met> **AND NOT** <AFW has been re-established>

    Goal: <FAB criterion is met>

        Task: Monitor FAB criterion. "The criterion is that steam generator level is below a critical limit."

    Pre-condition: <High Pressure Safety Injection has been turned on>

        Task: Manually actuate all HPI pumps.

    Pre-condition: <PORVs have been opened>

        Task: Open all pressuriser PORVs

When the FAB criterion occurs the goal <FAB criterion is met> becomes true. If the second part of the pre-condition also is true, i.e., if AFW has not been re-established, the next steps in the task "establish FAB" can be carried out. In this example FAB has been established when three pre-conditions have been met, i.e., <FAB criterion is met> **AND NOT** <AFW has been re-established>, <High Pressure Safety Injection has been turned on>, and <PORVs have been opened>. In the example it is assumed that the two last pre-conditions can reliably be achieved by their corresponding tasks. This therefore concludes the higher level task of establishing FAB.

The Dependent Differentiation Method

A graphical presentation of this task analysis is shown in Figure 5-7. The results from this rather detailed description of the GMTA will be used later in this chapter.

4.2 Formalisation of the GMTA Method

The relations between the main building blocks of a GMTA were shown

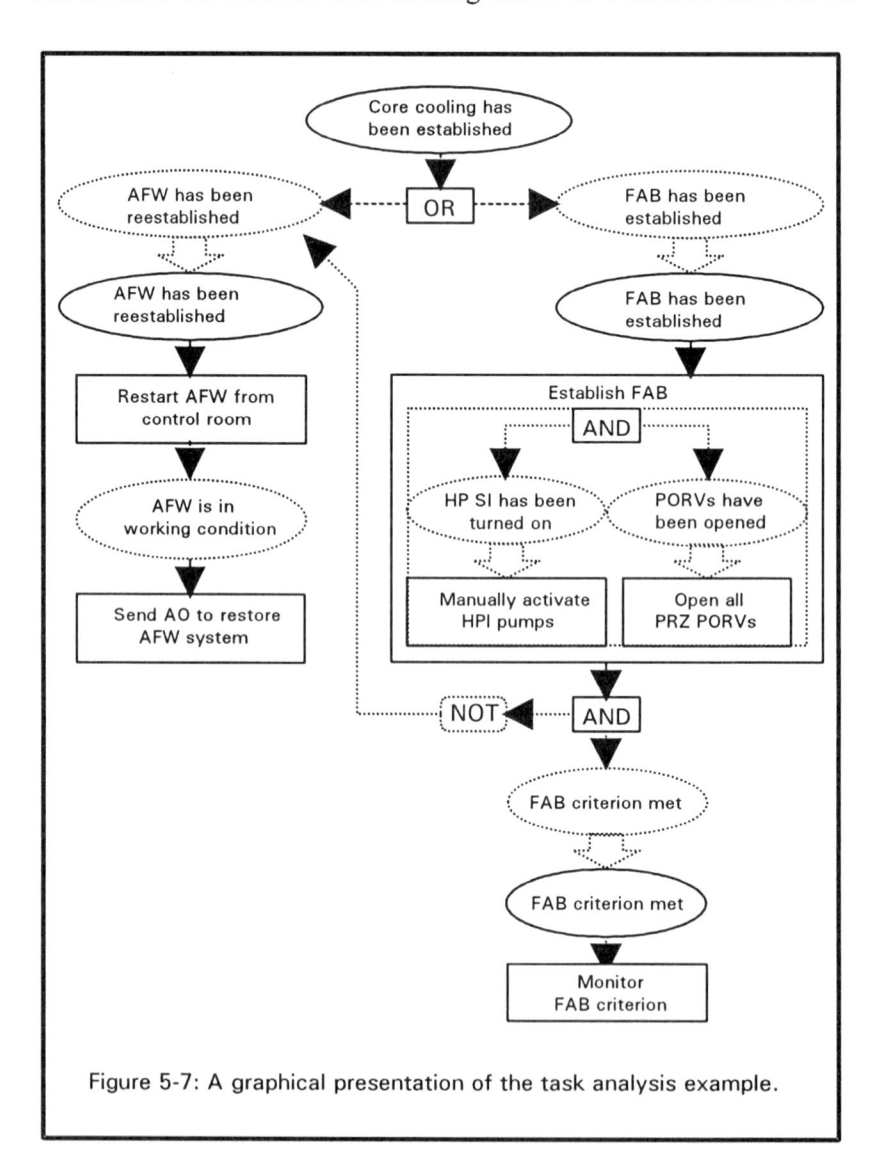

Figure 5-7: A graphical presentation of the task analysis example.

in Figure 5-5. (Note that this is **not** intended to depict the control structure of a task.) This will serve as a useful anchoring point in describing the method in a step-by-step fashion. The minimal description of a task must contain a goal and a task step, where the task step is an elementary step as defined above. An example taken from the Reference Manual to a standard DOS computer is:

Goal: <The computer has been enabled>

    Task: Start the Computer.

In most cases there will, however, be more than one task step needed to achieve the goal, as well as some pre-conditions for the goal. The above example would be:

Goal: <The computer has been started>

    Task: Start the Computer.

    Pre-condition: <Power Source Is Connected>

    Pre-condition: <Make Sure There Is No Diskette In Drive A>

        Task: Turn the Computer On.

        Task: Press Enter In Response To the Date and Time Prompts.

        Task: Confirm That the MS-DOS Prompt Is Correct.

Depending on what one wants to assume as an elementary step, this description may either be accepted as adequate or developed further. In many cases a task will only exist as a variable name for a set of task steps, rather as an entry in itself. This can be done according to convenience since it does not introduce any modifications to the underlying structure.

The possibility of refining the task description makes it extendible, as it must be. The important point is that at each iteration a consistent and coherent description is produced. This means that all goals or pre-conditions must be described, i.e., they must exist as names. Neither must there be tasks (decomposable, non-elementary) that are not described.

The steps in the GMTA can now be described as follows:

The Dependent Differentiation Method

(1) For the given task domain (application domain), identify the top goal, or top goals if there is more than one.

(2) For each top goal and for each goal, do the following:
   (a) Give the name of the goal.
   (b) Describe the goal.
   (c) Identify and name the attached set of task steps. For each task step mark if it is decomposable or elementary.

(3) For each task step, do the following:
   (a) Give the name of the task step.
   (b) Give the name(s) of the relevant goal(s) or higher order task step(s), i.e., the goal(s) or task step(s) that refer to this task step.
   (c) Identify and name the possible pre-conditions, including execution conditions.
   (d) Identify and name the possible post-conditions (and side-effects).
   (e) Describe the details of the task step, including the possible execution conditions.

(4) For each pre-condition, do the following:
   (a) Give the name of the pre-condition.
   (b) Give the name(s) of the relevant task(s), i.e., the tasks that have this pre-condition.
   (c) Define and describe the enabling set of task steps. For each task step mark if it is decomposable or elementary.

(5) For each post-condition, do the following:
   (a) Give the name of the post-condition.
   (b) Give the name(s) of the relevant task step(s), i.e., the task step(s) that refer to this post-condition. This also means that the post-conditions are attached to the goals that refer to these task step(s).
   (c) Describe the details of the post-conditions/side-effects.

The completeness of applying the GMTA can be checked in the following way:

(1) For each goal except the top goal(s) there must be a parent goal. For each goal there must be a goal description.

(2) For each task step there must be a task step description. Task steps are defined for goals.

(3) For each pre-condition there must be a pre-condition description. Pre-conditions are defined for task steps only.

(4) For each post-condition there must be a post-condition description. Post-conditions, including side-effects, are defined for task steps.

5. Task Description Requirements

The task analysis, in this case the GMTA, provides a definition of the tasks that need to be carried out. The resulting description of the tasks is the basis from which the reliability analysis proceeds, regardless of which type of human reliability analysis method is used. Since the aim is to focus on the performance as a whole, the step after the task analysis is to consider the context or the conditions under which the tasks are to be carried out. In order to describe the context it is necessary to have a predefined point of view, leading to a set of categories which can be used to separate the relevant from the irrelevant; otherwise the description of the context is potentially without limits (cf. the discussion of data analysis in Chapter 3). Since there, unfortunately, is no absolutely correct point of view to start from, the next best thing is to have a point of view which is consistent and which can be supported by arguments. In the present case the point of view is provided by the COCOM, complemented by general experience about human action.

Regardless of which model of human action or human cognition one chooses to use, there are a number of requirements which must be put to the resulting description of tasks or actions. These requirements, which characterise the requisite variety of human action, reflect the needs of a human reliability analysis in a realistic context.

5.1 Continuity

People are usually involved in continuous actions. That is, even if they bring a specific task to its conclusion they are still actively involved in other tasks. Each task, and each step in a task, is thus embedded in a context. The extent of the context, hence of the embeddedness of the tasks, may vary. In extreme cases, where the context is (subjectively) absent we may witness performance that corresponds to a scrambled control mode - which in effect means that the continuity of actions is lost. It is, however, more common that performance corresponds to tactical or strategic control which involves an appreciable amount of continuity between tasks. The subjective context may, of course, also be incorrect. In such cases the person will be unlikely to achieve his objectives, although there will still be a continuous set of actions.

The degree of continuity can be expressed in various ways. Two obvious indicators are parallelism and hierarchisation:

(1) **Parallelism**: People are usually involved in more than one task or set of actions at the same time. This does not imply that they do more than one thing at any given moment, but only that they will be involved in the pursuit of more than one goal seen over a time interval, for instance corresponding to the duration of a task.

(2) **Hierarchisation**: People may be executing one task as a part of another (e.g. trying to fulfil a pre-condition). In this case actions will be nested or recursive. Hierarchisation describes a form of parallelism where there is a dependence between the tasks, e.g. between goals and sub-goals. In parallelism as such the tasks merely co-exist within the same period of time - possible guided by a more general purpose, e.g. better utilisation of resources (first of all, time itself).

5.2 Performance Variability

People rarely follow prescribed procedures to the letter. For one reason or another they may deviate from the given instructions, or simply vary in their performance from time to time. The reasons can be many: an incomplete view or comprehension of the situation; random variations in capacity or support availability; actual differences between two situations (or that the situation does not match the nominal description completely); boredom; ingenuity or creativity; or simply the fluctuations in performance that are an inherent and unavoidable part of humans (because the Reliability of Cognition is imperfect!).

Of greater interest is the performance variability that can be related to human erroneous action. As described in Chapter 1, the erroneous actions can be of two main types: context dependent (**system induced**) and context independent (**residual**). In both cases a host of specific factors (Common Performance Modes, PSFs, etc.) may play a role. The main difference is that whereas the system induced erroneous actions can be reduced to a given level, the residual erroneous actions cannot be significantly reduced.

5.3 Communication and Interaction

People usually work together with other people, either in the sense that they form a group or a team, or in the sense that they depend on a specific action from another person to continue their own line of activity. The communication between people is important for providing needed information, feedback, sharing tasks, etc. The interaction is important as a way of sharing work, delegating responsibilities, implementing hierarchisation of plans, and in its own way providing the parallelism that may be required from an overall task analysis point of view (Leplat, 1991). Examples which illustrate the significance of communication and interaction are found, for instance, in the field of Air Traffic Control (Hollnagel & Cacciabue, 1993).

5.4 Requirements for Human Reliability Analysis Methods

If the results of a human reliability analysis are to have any appreciable external validity they must clearly match these requirements. It follows that the description of the tasks or actions must be detailed enough to enable an assessment of these aspects, and furthermore that the underlying model must be rich enough to generate such descriptions. The basis for making analyses of human reliability has often been taken in the so-called Step-Ladder Model (SLM) proposed by Rasmussen (1986). Although this model or description of idealised decision making represents a significant theoretical development it is unable to support a description which matches the list of requirements presented above, as shown by the discussion in Chapter 4. An adequate description must explicitly reflect the cyclical nature of human action (e.g. Neisser, 1976; Hollnagel & Cacciabue, 1991) as well as the way in which human action is determined by the subjective context. The COCOM qualifies as such a description.

A human reliability assessment method which does not take these requirements into account will be impoverished. Yet the requirements cannot by themselves serve to define the dimensions of a task description, nor as examples of Common Performance Modes - or for that matter as PSFs. Anything that is to serve as input categories for a reliability analysis must correspond to a potential, if not an actual, operational measure. The reason is that it is of little value to use a category (a factor, a dimension, an aspect, ...) of description unless it is possible in some way to determine whether or not the particularly quality being described is present, as well as the degree or extent to which it is present. It is for that reason that, for example, stress appears to be such a good descriptor. We know that it can be present in a situation, and we know - or at least hope - that it can be measured. It is for the same reason that "information interface" or "work methods" are bad descriptors. It is very hard to imagine how their presence can ever be ascertained in a reasonably objective way - unless they are defined with specific reference to a unique part of the interface or a single working method.

The problem is similar to the notion of phenotypes (manifestations) and genotypes (causes) in the description of human erroneous actions (Hollnagel, 1991a). The phenotypes refer to what can be observed, while the genotypes refer to what is usually inferred in terms of unobservable causes, mechanisms, and the like. Much of the work on human erroneous actions has been impaired by the failure to distinguish between the manifestation and the cause. The result has been a muddled and imprecise set of categories which, predictably, has led to a confusion of results. The danger of making the same mistake in the specification of the Common Performance Modes is tangible. In fact, the common use of Performance Shaping Factors provides ample evidence. Some descriptors (like stress) may sound very plausible although they in reality are very difficult to apply. Such descriptors are therefore ill-suited as a basis for human reliability analysis methods and as candidates for a characterisation of the context. It is with these warnings in mind that the Common Performance Modes are defined.

6. Common Performance Modes

As argued above, the description of the overall performance must be based on a coherent point of view. In the present case this point of view is referred to as the Common Performance Modes (CPM). The Common

Performance Modes are defined as the conditions under which the performance takes place. The conditions are those which apply to the "performance as a whole", rather than the possibly unique terms that may apply for individual steps of the performance. The determination of precisely how much "performance as a whole" includes will depend on the circumstances. In terms of the task analysis it will be the segments that naturally correspond to the achievement of a major goal. This definition may seem a bit circular; the important thing is, however, that the focus of analysis is significantly larger than the individual task steps that are normally used.

Based on the general experiences from work and performance analysis, the modes could include such things as:

(1) the previous development or history of the event,

(2) the psychological conditions (attention, knowledge intensity, etc.),

(3) the interface (general and specific appropriateness),

(4) the adequacy of operational support,

(5) the task context (single, multiple tasks),

(6) the organisational conditions (single, group),

(7) the process state (normal, contingency),

(8) etc.

In accordance with the argument in the previous section, one requirement to a Common Performance Mode is that it refers to an operational measure - or at least that it is possible to define a procedure by which its presence can be ascertained and its magnitude estimated. If we apply this criterion to the list above it is easy to see that only a few of the entries comply. It is, therefore, necessary to become more specific. This can be done by using the COCOM as a foundation. Referring to the description of the main COCOM parameters given in the previous chapter, the following CPMs can be defined.

6.1 Available Time

The time that is available to carry out the task is clearly a significant Common Performance Mode (e.g. Moray *et al.*, 1991). Time plays a role in many different ways. One example is the event horizon, defined either as the history size or the prediction length (cf. Chapter 4). Both of these are, however, subjective aspects which cannot be observed. So is the subjectively available time that was one of the main COCOM parameters. All of these may, in certain cases, be inferred from the observed performance, but such inference is usually *post hoc*, hence of little value as a CPM.

In contrast, available time can in many cases be observed in the sense that the actual time used can be compared to the expected or nominal time for the task. The latter can be derived from the results of a GMTA, and can also in itself serve as a first approximation to available time. From another point of view, available time corresponds to how well the task execution is synchronised to the process dynamics. In cases where time is lost, various dominant phenomena can be expected as, for example, time compression (Decortis & Cacciabue, 1988; Decortis & De Keyser, 1988). In most other cases reasonable assumptions can be made about time availability, either during observations, in post-accident analyses - or even at the stage of task analysis and task design.

6.2 Availability of Plans

Another important descriptor is whether the person has available appropriate plans for action. (I use the term plans, since it has already been used in the description of the COCOM. Other equivalent terms could be procedures, familiar patterns, heuristics, routines, etc.) The question of whether a person faces a situation where he knows what to do, or one where he is unprepared to act, is central in all attempts to account for human performance. It looms large in the terminology offered by Rasmussen (1986) in the notions of skill-based, rule-based, and knowledge-based performance. It has in some cases even be suggested that the purpose of interface design is to ensure that the person can work on, for instance, a rule-based level, hence apply the procedures he knows (e.g. Rasmussen & Vicente, 1987, or Yoshimura *et al.*, 1992). (One should be aware of the potential circularity in this argument. It might equally well be stated that the person should know the procedures that were required by any situation the interface could support.) It is clearly a prerequisite of adequate performance that the person knows what to do; this can come about either because the plan is known *a priori* or because it has been made to fit in the situation. If no plan is available, performance is doomed to be inefficient and inappropriate; the person may succeed by sheer luck, but this should not be assumed in an analysis. If a plan is available the person may potentially be able to handle the situation. The availability of a plan does, however, not guarantee that it is also successfully employed.

It may be undecidable whether or not appropriate plans are available to the person at any given point in time. One thing to consider is the distinction between whether the person **knows** the plans and whether he is able to **recall and use** them. It is quite conceivable that the person will be unable to make use of what he knows if he functions in an opportunistic control mode (cf. below). These matters may be beyond a clear decision even in a *post hoc* analysis of an event - although it can usually be determined whether the person actually used a plan. But it is still possible to make reasonable assumptions about this aspect when a task scenario is investigated. It is, furthermore, possible to use it as a basis for producing design requirements to the interface as well as to the organisational support, training, etc.

Another way of accounting for this descriptor is to consider the level of preparedness of the person to act. This can also be seen as an expression of the familiarity - or, conversely, the novelty - of the situation. If the situation is familiar it is reasonable to expect that the person knows what to do. If it is unfamiliar, the ability to cope may depend on the ability to make an appropriate plan. A concept such as familiarity (or novelty) is, however, of a subjective nature, hence ill suited as a descriptor. Since it is assumed that familiarity correlates strongly with the availability of plans, and since plans or templates are part of the model, the latter descriptor is preferred.

6.3 Number of Simultaneous Goals

A third descriptor is the number of simultaneous goals. This corresponds to the number of tasks that the person is engaged in pursuing at the same time, or to the number of lines of action that are being carried out. If there are two or more simultaneous goals it simply means that the person considers more than one goal when an action is chosen. The number of simultaneous goals is an important indicator of the task context, in terms of the task load. It can also be expected to correspond, if not actually to correlate, with the available time: the more goals the person is pursuing during the same time period, the more likely it is that time will be a limited resource - although that relation is not a necessary one. Regardless of how the human mind or human cognition is modelled it is an empirical fact that attention capacity is limited, hence that there is a limit to how many goals a person can pursue simultaneously. This descriptor is therefore an important one.

Although the number of simultaneous goals cannot be observed, it can be ascertained in several ways. Firstly, the person may himself provide the count - from introspection or retrospection. Secondly, in the case of an empirical analysis, a task description will clearly reveal how many lines of activity were active at the time. Thirdly, in the case of a design analysis or assessment, a task analysis - such as the Goals-Means Task Analysis - will also clearly show the number of simultaneous goals, either directly or by representing the results in an Operational Sequence Diagram (Kurke, 1961). (Additionally one may assume that specific events, such as the onset and detection of an alarm, may increase the number of goals.) This is in fact one of the outcomes of a task analysis which may be vital for the way in which the task allocation is decided. Of the three descriptors mentioned so far, the number of simultaneous goals is easily the one which can be determined with least uncertainty.

6.4 Mode of Execution

In Chapter 4 a distinction was made between two modes of execution: subsumed *versus* explicit. The former referred to the automatic, skilled (or skill-based) performance of a task, which did not involve attention controlled feedback. The latter referred to the execution of a task where each step was followed by the evaluation of the feedback before the execution was continued. The mode of execution is therefore clearly an essential descriptor of the performance, hence belongs to the set of Common Performance Modes.

The mode of execution can be ascertained by analysis of performance records which contain a detailed account of verbalisations and the interaction between crew members. It is, however, more important as a design parameter, in the sense that a given mode of execution may be required by the design. It may be necessary for the person to carry out the entire task as one smooth operation. But even if the task consists of a combination of task steps and tests, many of the task steps may be expected to be performed in the subsumed mode. Failure to do so may jeopardise the overall performance and/or introduce new opportunities for performance degradation. Even if the mode of execution cannot be unequivocally determined from an event log, it still remains an important descriptor of performance.

As an example, consider the step in the previous example where Auxiliary Feedwater (AFW) should be re-established. Because this task step is highly crucial and because it only occurs in unusual circumstances, it would be reasonable to require that it could be carried out as a subsumed action. But reality shows something else, as evidenced by this description from NUREG-1154 (pp. 3-5):

> "... it is vital for an operator to be able to quickly start the AFWS. There could have been a button labeled simply "AFWS - Push to Start." But instead, the operator had to do a mental exercise to first identify a signal in the SFRCS[2] that would indirectly start the AFW system, find the correct set of buttons from a selection of five identical sets located knee-high from the floor on the back panel, and then push them without being distracted by the numerous alarms and loud exchanges of information between operators."

This was clearly a task that had to be carried out in an explicit mode of execution. As a result, each step of the task, rather than the task as a whole, must be considered in relation to the context in order to assess the reliability of performing the task.

6.5 Process State

The process state is an obvious descriptor, even if the categories are restricted to normal *versus* contingency states. (More categories may be defined if an adequate state-space description of the process is provided.) It is very reasonable to expect that the reliability of the performance will be affected by the state of the process; if the process is in a contingency state - in contrast to being in a normal state - the reliability is assumed to be reduced, due, in part, to the concomitant changes in the other descriptors, e.g. available time, number of goals, available plans, etc.

The process state is, furthermore, a descriptor for which a measurement can rather easily be obtained. Many highly automated processes have explicit indications of the process state, for instance by means of annunciators or alarm displays. For processes which are still in the design state, the task analysis in relation to the nominal performance criteria (limit values) will easily defined which state the process is in.

---

[2] SFRCS = Steam and Feedwater Rupture Control System (author's note).

6.6 Adequacy of Man-Machine Interface and Operational Support

The adequacy of the man-machine interface and the operational support, including the adequacy of procedures, is obviously also an important descriptor. If the operator must work in conditions where adequate operational support is lacking, e.g. where the procedures are incomplete or insufficient, then the reliability of the performance is certainly going to be affected. Since this is something which characterises the task as a whole, it must be considered as one of the Common Performance Modes.

The adequacy of the operational support must, however, be specified further to be of any value. The specification is required in order to define an operational measure, as well as to be able to assess the impact. The adequacy is therefore only meaningful in relation to a specific task or a specific context. To take the example used above (the start of the AFWS), the adequacy of the man-machine interface and the procedures for starting the AFWS can be quite accurately characterised. (In the case reported in NUREG-1154 neither the interface, nor the procedures were adequate.) But the adequacy of this particular part of the system would not make much sense in relation to other tasks or goals. The adequacy of the man-machine interface and the operational support is therefore a descriptor that must be used with care. In some cases it may be appropriate, in other cases not.

(The adequacy of the man-machine interface and the operational support should definitely **not** be confused or equated with concepts such as user-friendliness or usability. These concepts are, firstly, not sufficiently specific in what they actually refer to and, secondly, impeded by a number of problems in their use, e.g. Booth, 1989, Chapter 5.)

6.7 Adequacy of Organisation

Just as the adequacy of operational support is an important Common Performance Mode, so is the adequacy of the organisation. By the adequacy of the organisation is meant the distribution of roles and responsibilities among the crew members, the availability of additional assistance, the communication system (command and control, in particular the way in which control may be delegated), the Safety Management System, the instructions and guidelines for externally oriented activities (i.e., external to the control room) or for overall contingency actions, the role of external agencies, etc.

The adequacy of the organisation is a conglomerate of many different things. It is consequently important that it is well-defined when it is used in an assessment of performance reliability. In that way it is subject to the same limitations that pertain to the adequacy of operational support (cf. above), although perhaps even more so. It is therefore a descriptor which may not always be used; the decision may depend upon whether the execution of the task can be done within the closed world of the control room or whether it involves intervention from other persons.

6.8 Other Common Performance Modes

The CPMs described above have only partly matched the parameters described for the COCOM in Chapter 4. Two of the parameters have not been included in the CPMs. One is "determination of outcome". The reason is simply that since this refers to cognition, i.e., to how the person can determine the outcome of the previous action, there is no obvious way in which this can be associated with an operational indicator or measurement. The other is the event horizon. As mentioned in the description of "Available Time", the event horizon is related to available time; if available time is inadequate, the event horizon will obviously be limited since evaluating the past and anticipating the future are activities that require time. The event horizon in itself is, however, a way of characterising the person's cognition rather than an operational indicator. Both "determination of outcome" and "event horizon" are inferred parameters which are necessary for modelling but inappropriate as a CPMs.

A possible additional descriptor is **control mode**. The control mode differs from the previous descriptor, mode of execution, by being derived from the underlying model rather than from general experience. This descriptor is therefore difficult to relate to operational measures, but it is considered because it is essential for the model.

The control mode was defined as one of {scrambled | opportunistic | tactical | strategic}. The description of the COCOM in Chapter 4 did not provide any details on how the control mode was determined by the model parameters. An initial approach to that can be found in Hollnagel (1992a). The control mode is important for the modelling of cognition in the COCOM, hence for the analytical approach to the Reliability of Cognition. In the case of using the COCOM as a simulation to produce synthetic performance data, such as it is done in the SRG project, the control mode is the main determiner of how an action is performed.

However, for the present introductory description we may leave the control mode as a possible, undeveloped contender for a descriptor. It clearly does not fulfil the requirement of being related to an operational measure, and is therefore not included in the present list.

The Common Performance Modes defined so far are shown in Figure 5-8. A tentative distribution has been made between CPMs that are assumed to increase reliability and CPMs that are assumed to reduce it. The control mode has been kept as an intervening descriptor, i.e., as something which is assumed to play a role in the mediation of the effects, but which cannot be considered at the same level as the other CPMs.

In contrast to the PSFs, which were an outcome of the decomposition approach, the CPMs differ in several ways. Firstly, there is not necessarily a linear (arithmetic) relation between the size of the operational measurement or indicator and the amount or level of the CPM. As an example, consider available time. The subjectively available time can be set in relation to various time measures, but the relationship may be quite complex. Secondly, the CPMs are definitely not independent of one another, and therefore they cannot be rated in relation to each other. As a simple example, available time clearly interacts with both number of goals and process state - and possibly with other CPMs as well. This further means that the joint impact of the CPMs cannot be found by means of a simple function or an aggregation (summation). In fact, while it may be possible to describe it qualitatively, it may not be an easy matter to translate that into appropriate quantitative terms - nor may it be necessary.

7. From Task Analysis to Common Performance Modes

The first part of describing the Dependent Differentiation method was to specify a set of Common Performance Modes and describe the principles of how they may be related to operational measures. This must be followed by a description of how the outcome of a task analysis can be used as a basis for identifying and characterising the relevant CPMs. The primary principle is that the CPMs should be used on the level of the tasks rather than on the level of single events. This may be done as follows.

```
┌─────────────────────────────────────────────────────┐
│  ░░░░░░░░ CPMs that may increase reliability ░░░░░ │
│  ┌───────────────────────┐  ┌──────────────────────┐│
│  │ Sufficient available  │  │ Few simultaneous     ││
│  │ time:                 │  │ goals                ││
│  └───────────────────────┘  └──────────────────────┘│
│  ┌───────────────────────┐  ┌──────────────────────┐│
│  │ Plans are available   │  │ Normal process state ││
│  └───────────────────────┘  └──────────────────────┘│
│  ┌───────────────────────┐  ┌──────────────────────┐│
│  │ Adequate MMI & support│  │ Adequate organisation││
│  └───────────────────────┘  └──────────────────────┘│
│  ┌─────────────────────────────────────────────┐    │
│  │ Execution mode: Actual = required           │    │
│  └─────────────────────────────────────────────┘    │
└─────────────────────────────────────────────────────┘
```

Figure 5-8: Common Performance Modes for assessing the reliability of cognition.

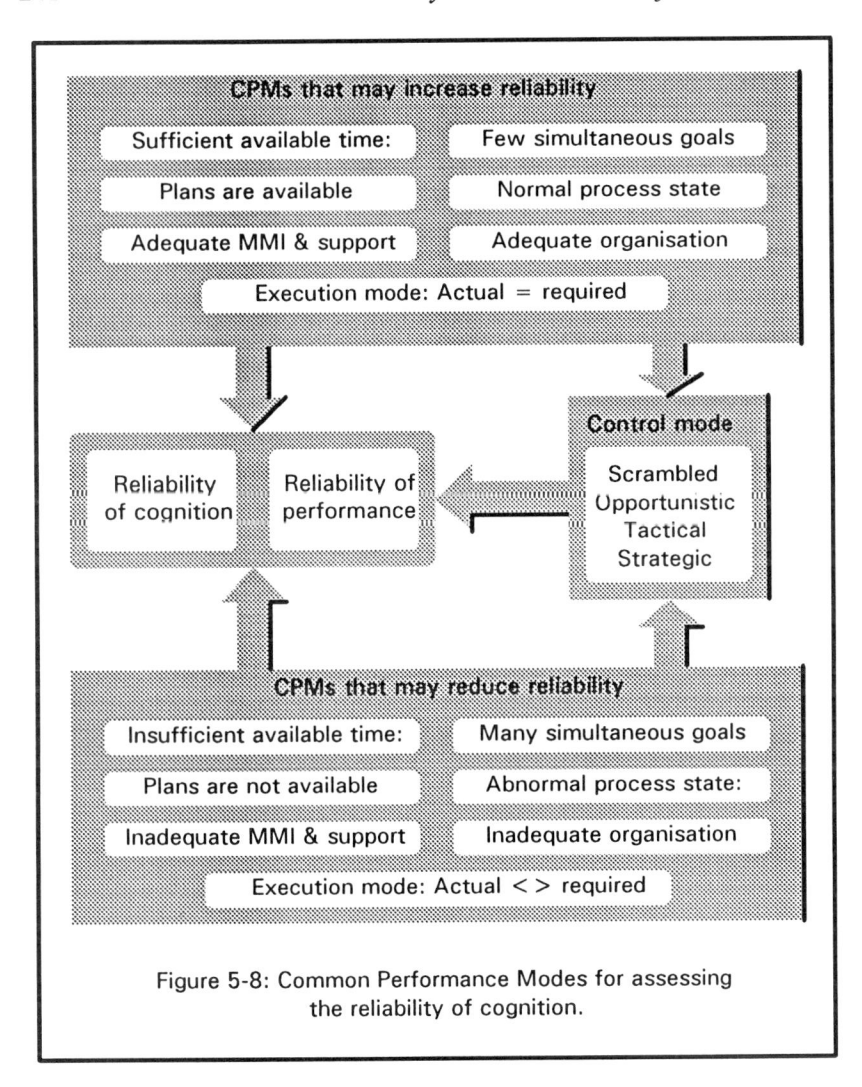

(Diagram shows "Reliability of cognition" and "Reliability of performance" boxes connected to "Control mode" with values: Scrambled, Opportunistic, Tactical, Strategic. Below is a "CPMs that may reduce reliability" section with: Insufficient available time, Many simultaneous goals, Plans are not available, Abnormal process state, Inadequate MMI & support, Inadequate organisation, Execution mode: Actual <> required.)

7.1 Stage 1: Task Analysis

For a given task, set of tasks (e.g. a scenario) or procedure, an analysis must be made using, for instance, a Goals-Means Task Analysis. (Although the GMTA is proposed as an appropriate tool, other task analysis methods may be applied as well.)

The Dependent Differentiation Method

The outcome of this task analysis is a characterisation of the task in terms of task steps and the specific performance conditions. These include ways in which task steps may depend on each other (pre-conditions, post-conditions) as well as the relation between the task goals and the means by which they can be achieved. This serves as a basis for characterising the overall performance conditions, for example, whether several activities must be carried out simultaneously, the time and resource limitations, the communication and information needs, dependencies between various persons, etc. From this description the relevant CPMs for the overall task can be identified.

The development of this overall picture may require repeated analyses of the scenario, e.g. as talk-through with the design team or with appropriate experts. The first pass will identify the fundamental structure of the task, but the dependencies between task steps can, of course, only be realised after the task steps themselves have been described. The analysis must therefore consist of at least two passes. The resulting **combination** of the CPMs will produce the first estimate of the performance mode for the task as a whole. (The term combination is used deliberately instead of, for instance, aggregation, which implies a quantitative procedure. The combination may be qualitative or quantitative, depending on the needs and possibilities.) This exercise should be carried out for the nominal task description as well as for likely variations of it.

7.2 Stage 2: Assessment of CPMs

Considering each of the identified CPMs in turn, an assessment should be made regarding its likely value or level in the task. This entails considering the task in the context of other tasks or as part of larger scenarios (single event, multiple events) - regardless of whether this has been included in the task analysis in the first place.

In some cases quite accurate judgments may be possible, e.g. for the adequacy of man-machine interface, available time, or operational support - because these CPMs will have a well defined reference in the context, hence a unique meaning. In other cases less precise statements must suffice. The importance of this stage is that each CPM is considered against an overall description of the task context without any reference to a specific task step or operation. It is consistent with the definition of a Common Performance Mode that it describes what is **common** for the task as a whole. The way in which the value or level is described may vary from case to case, according to the specific requirements of the analysis. It is, however, conceivable that in many cases a fuzzy description will be an appropriate intermediate format.

7.3 Stage 3: Identification of Critical Events

As a third stage, the critical events in the task should be identified and further analysed. This more detailed analysis could consider error modes, possible causes, failure modes and effects, etc. using some of the standard methods. This will not only help to identify areas where the Reliability of Cognition may be particularly important, but may also contribute to a refinement of the CPMs themselves. It is conceivable that the combined effect of the critical events may lead to a change in some of the CPMs. Conversely, an absence of critical events might also lead to a general improvement of the situation.

In particular, the occurrence of possible unwanted actions and their consequences ("error modes") should be assessed in terms of the probable impact they may have on the overall conditions as described by the CPMs. If this leads to substantial changes in the identified CPMs, either in their expected level or because a new CPM is introduced, the previous stages in the procedure must be repeated. The whole cycle should continue until a reasonably stable characterisation has been achieved of the overall task and of the performance conditions. The determination of what "reasonably stable" means cannot be made in absolute terms. A practical rule-of-thumb is that if a renewed analysis does not lead to significant changes in the CPMs (in their level or their number), then the analysis has been completed. There may be cases where external constraints such as time or money do not permit the luxury of repeating the analysis. If so, it is the responsibility of the analyst to ensure that there are no major deficiencies in the outcome.

It is this third stage which constitutes the dependent differentiation. The meaning of the term is that the differentiation only takes place if the preceding analyses indicate the need for it. The differentiation is therefore no longer the basic guiding principle for the analysis, as in the decomposition approach, but is only needed to achieve a sufficient level of detail of the results: the level of differentiation is not determined by the nature or granularity of the "elementary" actions or steps, but by the effect on the outcome. When a differentiation of task steps no longer improves the results, i.e., when it does not provide additional useful details, the procedure should be terminated.

7.4 Stage 4: Quantification

In order for the outcome of the analysis to be used as a part of the typical PSA/PRA, it is necessary that the qualitative descriptions of the task and of the expected operator performance is turned into a quantitative one. The reason for this is that most analytical methods require quantitative input, as discussed in the preceding chapters. The quantification can be done of the basis of, for example, fuzzy descriptions and entails two stages. Firstly, the overall reliability of the performance - of the task as a whole - should be estimated based on the information about the CPMs. (The term "estimated" is used deliberately. It is not possible to calculate the reliability because the process is one of translating from a qualitative to a quantitative representation. Within the quantitative representation results can be transformed according to the usual rules of arithmetic, scaling, etc.) Secondly, the likelihood of specific error modes and specific consequences should be quantified, referring to the outcome of the dependent differentiation in the third stage.

(The use of fuzzy descriptors is only meant as a suggestion and does not present an ideal solution. Many people feel uncomfortable about the notion of fuzzy set theory, although the reasons may sometimes be hard to divine. If that is a problem any other systematic approach may be used instead.)

The outcome of the first three stages is in essence a qualitative analysis and description of the task(s). This is in itself sufficient to indicate whether the performance can be expected to go smoothly or whether unwanted consequences may occur. It is in particular sufficient to show whether one can assume that the performance of the operator will be reliable, i.e., whether or not it will be a significant factor of the overall system reliability. As suggested above, the description on this level may be given in fuzzy terms. The advantage of using fuzzy descriptors is that they are meaningful to the analyst, although not unique. In contrast, numbers are unique but not meaningful. Fuzzy descriptions can later be transformed to quantitative descriptions using various rules or algorithms. The advantage of the fuzzy description is that the second stage can be modified while retaining the overall qualitative description conveyed by the first stage - the fuzzy description. If, in contrast, the result was immediately expressed in numbers, any change would correspond to a re-analysis.

7.5 Utilisation of the Dependent Differentiation Method

The final outcome of the DDM is a description of the task in its context which takes into account the Reliability of Cognition and the reliability of MMI. This description makes it possible both to identify or predict the situations where problems may arise and describe the conditions that may either be the cause of the problems or have a significant effect on how the situations develop, thereby complying with the goals of human reliability research. The overall purpose of the analysis is not to provide a numerical estimate of the reliability of performance *per se*, but rather to find out whether the reliability of performance can be expected to be satisfactory and, if not, to specify how the situation can be improved. The steps in the Dependent Differentiation Method are shown in Figure 5-9.

The outcome of the DDM indicates how complex the task is by showing how actions can be embedded in other actions, how large the cognitive load is likely to be, and thereby also how likely it will be that a sub-goal is not completed or that the backtracking goes wrong. In other words, it shows how human reliability can be affected. The outcome also automatically provides a context for the lower level actions. This is important to determine, for example, whether or not misinterpretation of information is likely (for instance, using frequency or recency heuristics). It also suggest how contexts may unintentionally be substituted for each other, e.g. that the person may forget either the current context and use the context from the next higher level, or incorrectly retain the current context when he returns to the previous level.

Finally, the outcome of the DDM shows the individual executable steps, and can therefore be used to assess the effects of failures of these. If, for instance, the operator makes the observation after moving to position B instead of position A, we can estimate the consequences by considering the context in which it occurs. In other words, single actions (such as making an observation) are no longer considered by themselves but only as parts of a context. They can therefore be given a far more precise meaning and interpretation. In the present example, making an observation may be part of a diagnosis which again may be part of restoring normal operation. This is therefore different from making an observation as part of, for instance, checking system status, and both the causes and the consequences of making an incorrect observation will be different.

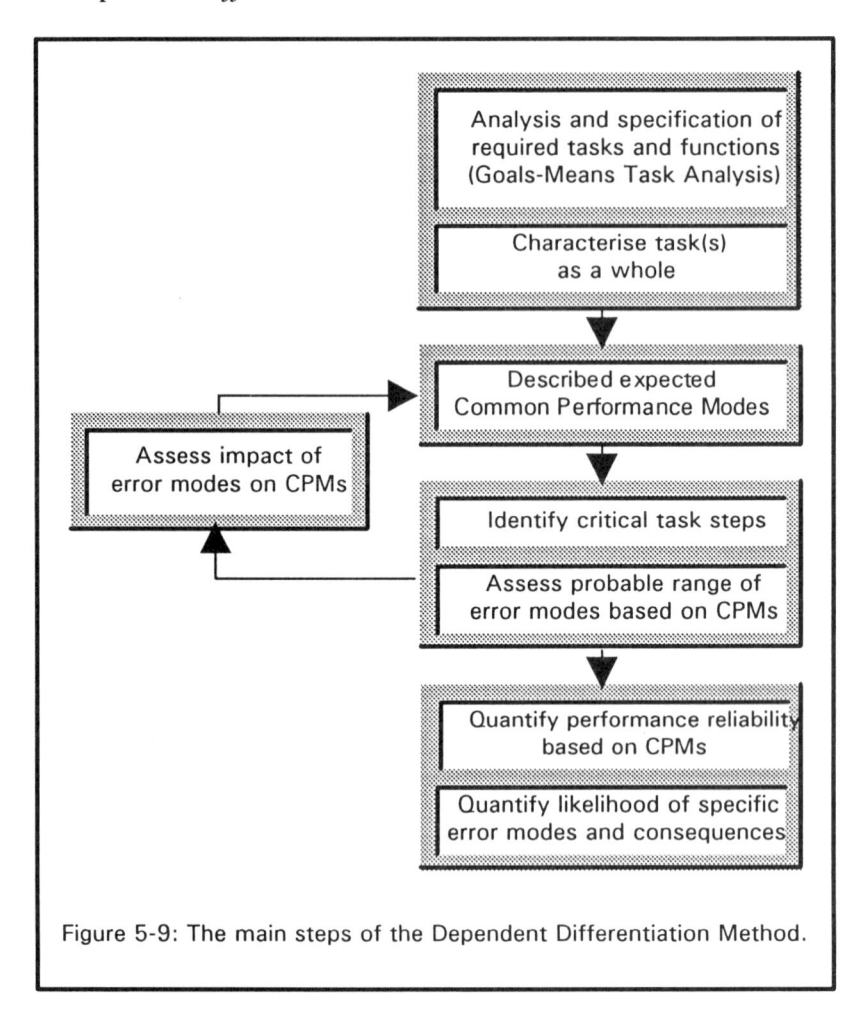

Figure 5-9: The main steps of the Dependent Differentiation Method.

8. Using the Dependent Differentiation Method

In order to illustrate how a Dependent Differentiation Method may work in practice, the previous "Feed And Bleed" example will be used.

8.1 Stage 1: Task Analysis

The first stage is to perform a task analysis to define the task steps as well as the overall performance conditions. Part of this has already been

Using the Dependent Differentiation Method

achieved in the description of the Goals-Means Task Analysis which, applied to the generalised FAB scenario, gave the following result (cf. Figure 5-7):

Goal: <Core cooling has been established>

    Task: Establish core cooling.

    Pre-condition: <AFW has been re-established> **OR** <FAB has been established>

    Goal: <AFW has been re-established>

        Task: Restart AFW from control room.

        Pre-condition: <AFW is in working condition>

        Goal: <AFW is in working condition>

            Task: Send assistant operators to restore the AFW system.

    Goal: <FAB has been established>

        Task: Establish FAB.

        Pre-condition: <FAB criterion is met> **AND NOT** <AFW has been re-established>

        Goal: <FAB criterion is met>

            Task: Monitor FAB criterion. "The criterion is that steam generator level is below a critical limit."

            Pre-condition: <High Pressure Safety Injection has been turned on>

            Task: Manually actuate all HPI pumps.

            Pre-condition: <PORVs have been opened>

            Task: Open all pressuriser PORVs

The GMTA shows that the FAB task is successful if FAB is established **or** if AFW is re-established. However, a pre-condition for trying to establish FAB is that AFW has **not** been re-established. This means that the following sequence should be kept:

The Dependent Differentiation Method   248

(1) **Try to re-establish AFW.**

This is an unconditional step which represents the first response to the incident, which is that neither the main feedwater system nor the AFW are available to cool the steam generators.

(2) **If this fails, then prepare to establish FAB.**[3]

This is a conditional step. Whereas the attempt to re-establish AFW could be an exclusive task step (in the sense of being carried out on its own), the preparation to establish FAB must be performed in parallel with the continued attempts to re-establish AFW.

(3) **In the meantime, continue the attempt to re-establish AFW.**

This step must be performed in parallel to the preparations for establishing FAB, and in particular to the monitoring of the FAB criterion. Since actually using FAB is a very last resort, the continued attempts to re-establish AFW become highly important. They will be carried out under increasing time pressure, as the FAB criterion comes closer to being met.

(4) **If the FAB criterion is met *and* AFW has not yet been re-established, then try to establish FAB.**

The FAB must usually be established within a relatively short interval after the criterion has been met, in order to prevent damage to the reactor core. Since the FAB procedure is very rarely executed, it can be assumed to be quite unfamiliar to the operators - even if they have practiced it in a simulator. Although this is, in principle, an exclusive task step it is far from being simple, and it must furthermore be carried out under considerable pressure - in terms of time as well as expectancy of success.

The FAB procedure is clearly not a simple task, but has a number of complicating elements even in this simple rendering. In practice one may suspect a number of confounding conditions. The operators may, for instance, carry out the FAB procedure to the letter or vacillate between using FAB as a precaution or as a last resort. The difference between

---

[3] This is not actually a step that is specified in the procedures. There are, however, a number of things that may need to be done before FAB can be established and for the sake of the example this step has therefore been included.

doing one or the other is important because it determines how much time is available and how many resources can be allocated to other tasks/solutions. The actual carrying out of the tasks, reestablishing AFW and establishing FAB may also be hampered by technical problems. In the only case that has been documented so far no less than twelve malfunctions occurred within a period of 30 minutes - representing a mixture of technical and operational factors (NUREG-1154). It is, of course, impossible to say whether this is representative of the FAB procedure, but the record of accidents and incidents in nuclear power plants does nourish the suspicion that extreme situations are usually further confounded by inadequate system functionality.

8.2 Stage 2: Common Performance Modes

Following the GMTA and the overall characterisation of the task, the second stage is to consider the possible Common Performance Modes. Looking at each of the CPMs defined above in turn, the result is as follows:

8.2.1 Available Time

The FAB task is one where the available time is limited in several ways:

(a) Firstly, the AFW must be re-established as soon as possible; if the first attempt fails, further attempts have an even more severe time limitation.

(b) Secondly, establishing the FAB must be initiated when the FAB criterion is met, i.e., at a precise point in time or within a short time interval.

(c) Thirdly, establishing the FAB must be concluded in the first attempt. Even after the FAB criterion has been met, the condition of the reactor continues to deteriorate and it is crucial that sufficient cooling is established before the reactor core is damaged.

In summary, available time is clearly an important CPM for this task. Available time is limited for the task as a whole, and the limitation may become acute for specific task steps.

The Dependent Differentiation Method 250

8.2.2 Availability of Plans

The FAB is included in the set of Emergency Operating Procedures for PWRs. There will therefore be a plan or a prescribed procedure available. It is furthermore a scenario which can be assumed to be well-known in the community of reactor operating crews - not least due to a few well-published cases where it might have been used. The fundamental understanding of what should be done can therefore be expected to be very good, although crews may be unfamiliar with the details of the procedure. In summary, the relevant plans are available. This CPM therefore should not present a problem.

8.2.3 Number of Simultaneous Goals

As the above discussion of the GMTA has shown, there are at least two simultaneous goals in the FAB task. These are to re-establish AFW and to prepare for FAB, followed by monitoring the FAB criterion. Although the task is carried out by a crew rather than by a single operator, the existence of multiple goals means that there is a need to maintain common awareness of the goals and how close they are to being achieved. At the point in time when the FAB criterion is close to being met there is a specific competition between re-establishing AFW and establishing FAB; the former is a more attractive solution and the decision to abandon this goal may be hard to make.

In conclusion, the FAB task contains at least two simultaneous goals throughout most of the task.

8.2.4 Mode of Execution

The mode of execution must clearly be explicit. The FAB task is very rarely performed and therefore requires the full attention of the operators. Very few task steps can be assumed to be subsumed, i.e., to be performed as routine tasks.

8.2.5 Process State

The process state is evidently an abnormal condition. This is the case throughout the FAB task - well before the FAB procedure is initiated and probably long after it has been concluded.

8.2.6 Adequacy of Man-Machine Interface and Operational Support

The adequacy of the man-machine interface and the operational support is less easy to determine. Judging from the case referred to above (NUREG-1154), this CPM seems to be less than fully adequate. A more precise characterisation will, however, depend on the specific plant and the specific organisation. The adequacy of the man-machine interface and the operational support will therefore not be considered here.

8.2.7 Adequacy of Organisation

The adequacy of the organisation is also a CPM which can only be characterised precisely when the details of the task environment (the crew, the plant, the organisation) are sufficiently known. It will therefore not be considered for this example.

8.2.8 Summary of Common Performance Modes for FAB

The preceding analysis can be summarised as providing the following qualitative description:

(1) **Available time:** Available time is either limited or very limited.

(2) **Availability of plans:** Appropriate plans are available, although they cannot be expected to be familiar to the crew.

(3) **Number of simultaneous goals:** There are two, or more, simultaneous goals. Furthermore the two goals may partly be conflicting.

(4) **Mode of execution:** Execution mode is predominantly explicit. Full attention is demanded through most of the task.

(5) **Process state:** The process state is clearly abnormal.

(6) **Adequacy of man-machine interface and operational support:** Not considered. This CPM cannot be characterised due to lack of precise information.

(7) **Adequacy of organisation:** Not considered. This CPM cannot be characterised due to lack of precise information.

The Dependent Differentiation Method 252

The outcome of the second stage is overall a picture of a task which is very demanding, both in terms of precision and timeliness, which may have multiple and conflicting goals, and which must be performed under abnormal process conditions. The Common Performance Modes are therefore not conducive to high performance reliability. The successful achievement of the task goals depends to a considerable degree on the operators' cognition - in addition to more manual skills - and the Reliability of Cognition is therefore an essential element when the overall probability of success or failure is to be assessed.

8.3 Stage 3: Identification of Critical Task Steps

The third stage in the Dependent Differentiation Method is concerned with the identification and further analysis of potentially critical task steps. One way of approaching this is to look for task steps which, if they fail, will lead to failure of the overall task. This can be done, for example, by describing the FAB scenario by means of an Operator Action Event Tree, which might look as in Figure 5-10. The event tree shows the sequence of the four main task steps: re-establish AFW, prepare for FAB, initiate FAB, and establish FAB. The event tree also shows the different outcomes, two of which are labelled "success" while the other three are labelled "failure".

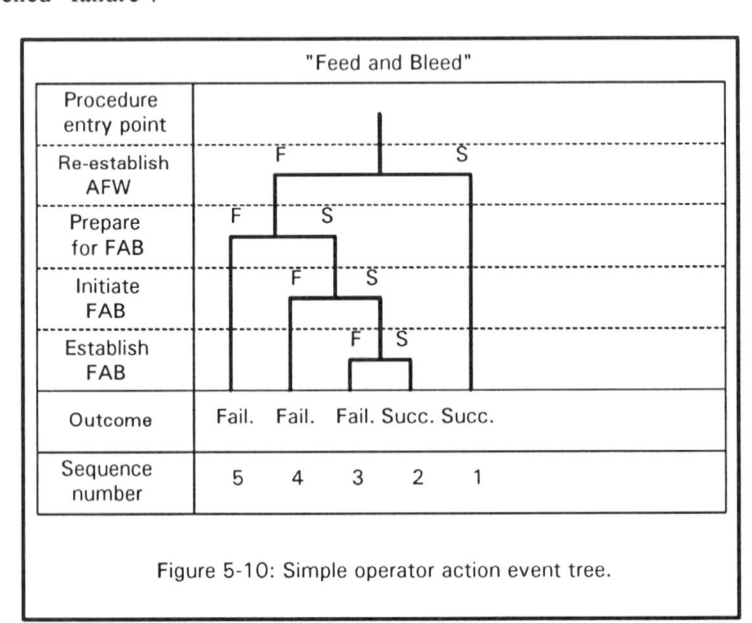

"Feed and Bleed"	
Procedure entry point	
Re-establish AFW	F S
Prepare for FAB	F S
Initiate FAB	F S
Establish FAB	F S
Outcome	Fail. Fail. Fail. Succ. Succ.
Sequence number	5 4 3 2 1

Figure 5-10: Simple operator action event tree.

If the event tree in Figure 5-10 is compared to the description of the task given above, it is obvious that the event tree representation is incomplete. The existence of two simultaneous sub-tasks, as well as the dependence of one sub-task upon the outcome of another, are not obvious from Figure 5-10. This may either be the fault of the underlying analysis or of the form of representation. Since the GMTA analysis in this case has found both the parallelism and the dependency, it is reasonable to suspect the event tree representation. (If, however, a different type of analysis had been used, that might have contributed to the incompleteness of the event tree representation.) We may, for instance, try to include the fact that re-establishing of AFW continues throughout the task, thereby producing something like the Operator Action Event Tree shown in Figure 5-11.

The simple event tree of Figure 5-10 has here been extended by introducing a second attempt of re-establishing AFW after preparing for FAB both if the preparations succeed and if they fail. In the first case, re-establishing AFW will render the actual use of FAB unnecessary. In the second case, re-establishing AFW is the only possible way of preventing a failure. Even the extended event tree is, however, incomplete compared with the outcome of the GMTA. There are several reasons for this:

(1) **The event tree representation cannot easily describe task steps**

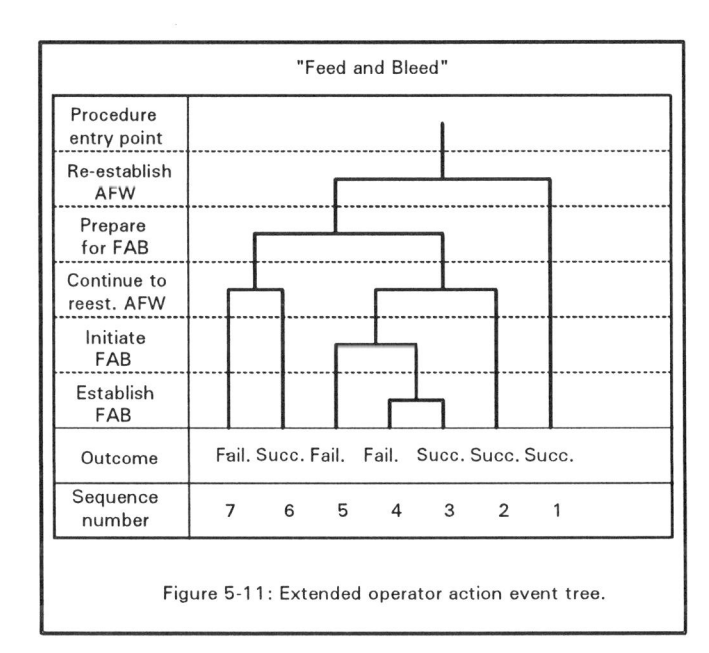

Figure 5-11: Extended operator action event tree.

that occur in parallel. A task step has to be place in a sequence of other task steps. If the task step is only put in one place the result is an oversimplification, e.g. as in Figure 5-10. If the task step is put in several places, the result may easily become very clumsy and difficult to use. Imagine, for instance, that "Reestablish AFW" was inserted after "Initiate FAB" as well as after "Establish FAB" in Figure 5-11. Then think of what would happen if the task had been shown with more steps than here. This problem is not particular to the Operator Action Event Trees, but occurs with any kind of event tree representation. The danger is that the realisation of this difficulty may easily lead the analyst to simplify the preceding analysis.

(2) **The event tree representation is not well suited to showing conditional task steps**, unless the condition refers to the immediately preceding step. This is simply due to the very sparse descriptions that can be included in an event tree representation. Neither is the event tree well suited to showing possible recovery from a failed operation. Other types of event trees, such as THERP, have this facility - but only to a limited extent. This becomes obvious as soon as one tries to make a THERP representation of the GMTA from the FAB scenario.

(3) **The event tree representation only gives information about individual task steps**, but not about the overall performance. This is, perhaps, the most severe limitation. It has the consequence that the analysis concentrates on each step in isolation without considering the conditions under which it occurs. This is a consequence of the decomposition approach; it is an inherent part of any analysis using this approach, but it is particularly pronounced in this particular type of representation.

Despite these shortcomings of the event tree representation it is sufficient to identify the main critical task steps in the present example. (The main critical events are, of course, identified when the operator action event tree is constructed, rather than by the tree itself. But the event tree representation makes it easy to recognise these events.) Both Figure 5-10 and Figure 5-11 shows that there are four main critical task steps: re-establish AFW, prepare for FAB, initiate FAB, and establish

FAB. For the sake of demonstrating the Dependent Differentiation Method, only one of these will be described in detail: Initiate FAB.

Table 5-1: Failure Modes for Perception/Observation.

PERCEPTION/OBSERVA-TION			
General Effect	Specific Effect	General Cause	Specific Cause
Failure to notice signal or alarm		Absent from place	Physiological needs
		Attention failure	Recent failures Time compression Work overload
Incorrect/incomplete recognition of value	Reading wrong indicator	Operation mode misjudged Reading wrong value	Lack of training
		Selecting wrong indicator	Work overload

The task step "initiate FAB" is essentially the decision to go ahead with the FAB. The decision is crucial because the process at this point has reached a condition where there are no other possible ways of recovery. A failure to act thus means that a system failure is unavoidable. The decision, in turn, depends on noting that the FAB criterion has been met and determining that the AFW has not been re-established. (The former is a relatively straightforward decision, while the latter is less clear-cut and involves considerable subjective judgment.) These are assumed to be predominantly perceptual processes, hence can be described by the set of general failure modes for perception/observation presented by Hollnagel & Cacciabue (1991), as shown in Table 5-1.

The failure to make the right decision when required may, however, otherwise be classified as a decision failure using, for example, the set of categories in Table 5-2.

It is possible to extend and refine the analysis of this single task step even further; in principle there will be very few limits to the number of causes or combinations of causes and conditions that can be imagined (cf. the discussion in Chapter 3). In relation to the Dependent Differentiation

Table 5-2: Failure Modes for Planning/Choice.

PLANNING AND CHOICE			
General Effect	Specific Effect	General Cause	Specific Cause
No choice made		Decision paralysis (Shock, Fear, ...)	Physiological condition/needs Recent failures Time compression Work overload
Incorrect choice of alternative	Use of wrong decision rule	Satisficing Recognition primed choice	
	Incomplete matching of alternatives	Pre-condition not considered Side-effect not considered Sub-goal not considered	Time compression Work overload
...	...	...	...

Method it is, however, more important to note that even if this task step fails it will not have much effect on the analysis made so far, i.e., there is no need to repeat the analysis. The failure to initiate FAB may certainly have serious consequences for the condition of the process, and may deteriorate the situation by increasing the pressure to find a solution - although there does not seem to be any obvious solution. The task description, however, does not go much beyond this step. Any subsequent actions to deal with the possible worsened plant state will have to be considered in their own right; the performance conditions for those should certainly include the consequences that may arise from failing to establish FAB. For the present, the dependent differentiation of the FAB task need not be extended.

8.4 Stage 4: Quantification of Analysis Results

The quantification of the results is usually considered the indispensable final stage in the analysis. The need to express the results in numerical terms comes from the way in which reliability analyses of technological systems traditionally have been carried out (cf. the discussion in Chapter 3). On second thoughts, the demand for numerical results is not a pursuit of numbers *per se*, but rather a need for results expressed in a precise and - apparently - unambiguous way.

Despite the predilection to prefer quantitative results over qualitative results, there is also a clear recognition that the qualitative aspects of the analyses are at least as important as the quantitative aspects (Poucet, 1989; Swain, 1990). The final stage is therefore not indispensable; quantification does not improve the **quality of the analysis**, but it may considerably improve the **usefulness of the results**. It is therefore necessary to consider how the outcome of the Dependent Differentiation Method can be quantified.

In the currently used methods, such as SHARP or THERP, the quantification is done for each step in the task. Consider, for instance, the event tree in Figure 5-10. The first step, re-establish AFW, may be seen as carrying out a procedure. It is therefore considered to be justified to use the Human Error Probability (HEP) data for procedural mistakes, and in this way assign a value to the probability of failing to accomplish this first step. As another example, consider the step of initiating FAB. As discussed above, this may be seen as a decision preceded by the detection of whether a certain measurement has reached a certain value. If this is treated as a case of perception, data from exercises or perception models may be used to assign a value to the probability of failing to accomplish this step. Altogether this will yield an expression like:

$$HEP_{FAB} = (P_1 + P_2 + P_3 + P_4)$$

where

> P_1 is the conditional probability that the operators fail to re-establish AFW.
>
> P_2 is the conditional probability that the operators fail to prepare for FAB.
>
> P_3 is the conditional probability that the operators fail to initiate FAB.
>
> P_4 is the conditional probability that the operators fail to establish FAB.

The outcome may either be a single number or a range. In both cases the interpretation of this quantitative result is crucial. The numerical indication in itself is close to being meaningless. In order to be properly applied it is necessary to know what the context is. It would therefore clearly be desirable if the outcome of the analysis maintained some indication of the context of the task.

The outcome of the first three stages of the Dependent Differentiation Method was a description of the task from the GMTA and a description of the relevant CPMs. The quantification must be based on the CPMs, but it is important that the overall task description is not neglected. The quantification is not to be a direct transformation, but rather a thoroughgoing interpretation and therefore needs the task description.

One way of approaching this interpretation is to view the characterisation of the CPMs as expressing the membership in a fuzzy set. This will provide a convenient and meaningful way of transforming qualitative descriptions into quantitative ones. For the sake of illustration only, a set of terms may be proposed as shown in Table 5-3. For each

Table 5-3: Possible Value Categories for Common Performance Modes.

Common Performance Mode	Possible descriptive terms		
Available time	Very limited	Limited	Adequate
Availability of plans	None	Unfamiliar	Familiar
Number of goals	Many	Two	One
Mode of execution	Mixed	Explicit	Subsumed
Process state	Abnormal (unknown)	Abnormal (known)	Normal
Adequacy of MMI/OS	Unknown	Low	High
Adequacy of organisation	Unknown	Low	High

CPM is shown a number of terms which apply to that CPM. Thus, available time might be described as varying from "zero" to "infinite" with the typical values in between being "very limited", "limited", and "adequate". Similarly, mode of execution may be described as "explicit", "subsumed", and "mixed".

(Lest there should be any misunderstanding, I want to emphasise that the terms shown here are completely hypothetical and proposed only as an illustration of the principle. If the Dependent Differentiation Method is to be adopted for real use, a considerable amount of empirical work is required to specify a set of terms and the set of rules that is needed to interpret them.) The set of terms must be complemented by a set of rules which can be used to transform the CPM terms into a conclusion - which, by the way, will be a fuzzy set itself. The general form of a rule would be:

If $(x$ is $L)$ **and**

$(y$ is $S)$ **and**

...

then

z is H "

A concrete example of a very desirable situation could be:

If (*available time* **is** *adequate*) **and**

(*availability of plans* **is** *routine*) **and**

(*number of goals* **is** *one*) **and**

(...) **and**

(*adequacy of organisation* **is** *high*)

then

reliability **is** very high.

If the same set of terms is applied to the case of the FAB task, the qualitative result of the Dependent Differentiation Analysis could be:

(1) **Available time:** Limited or very limited.

(2) **Availability of plans:** Unfamiliar.

(3) **Number of simultaneous goals:** Two or more.

(4) **Mode of execution:** Explicit.

(5) **Process state:** Abnormal (known).

(6) **Adequacy of man-machine interface and operational support:** Unknown.

(7) **Adequacy of organisation:** Unknown.

A rule to apply in that situation could be:

If (available time **is** limited) **and**

(availability of plans **is** unfamiliar) **and**

(number of goals **is** two) **and**

(mode of execution **is** explicit) **and**

(process state **is** abnormal [known])

then

reliability **is** very low.

The outcome refers to the membership function of the fuzzy set of "very low", which may look like the example shown in Figure 5-12. (This, again, is a purely hypothetical function.) This membership function can be used to transform the conclusion of the rule into a numerical value. In the example shown in Figure 5-12, the degree of membership value 0.7 corresponds to a high probability of failure - which is the same as saying that reliability is low.

8.4.1 Quantification Issues

In order for this kind of transformation to work it is necessary that rules are defined to cover all possible combinations of membership functions for the CPMs. This is certainly a considerable job to undertake, although definitely much smaller than trying to find numerical values for all possible task steps (of all possible tasks!) or find a comprehensive set of Human Error Probabilities. The use of fuzzy sets also makes it considerably easier

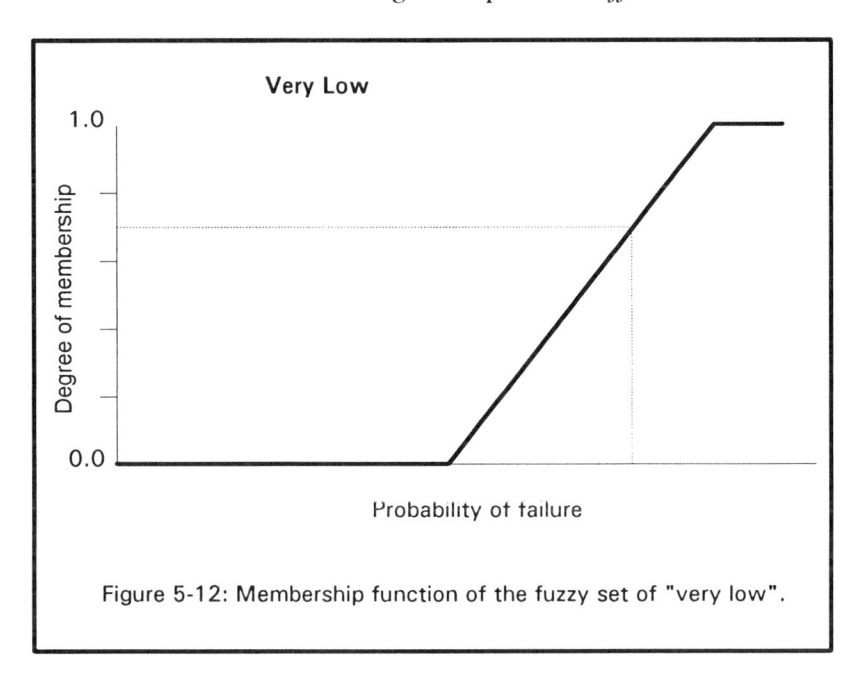

Figure 5-12: Membership function of the fuzzy set of "very low".

to accommodate new information, since it basically means a change in the individual membership functions. The relative simplicity of this approach does, however, not mean that it is inherently more correct than the decomposition approach referred to above. In both cases the validity of the results depends on the availability of sufficient empirical data. It is not unreasonable to hope that the use of fuzzy sets may make the construction of the necessary empirical database more manageable. But neither can it be ruled out that other approaches may do the same. The use of fuzzy sets here is basically meant to illustrate one way in which the qualitative outcome of the Dependent Differentiation Method can be turned into quantitative results, without losing the information about the context in which the task is carried out.

The transformation procedure described above provides a measure of the reliability of task performance as a whole, rather than a measure of the probability of failure for a specific task step. The reason for this is to be found in the purpose of making the analysis in the first place. The objective is not to find a precise probability for the success or failure of a specific action. It is rather to find out whether one can expect the performance as a whole to be reliable, and to find out if there are specific steps or parts which are particularly prone to produce unwanted consequences. This knowledge can then be used to change the design of the system, to introduce specific measures of compensation, to construct barriers and recovery options - in other words to make the whole system either more reliable or more forgiving.

It follows that in some cases it may be necessary to find the probability of failure of a specific task step. This can be done in a number of ways, which either can be found in the existing armory of human reliability assessment methods or in relatively unproved approaches, such as the fuzzy sets. In the example used here there was little need to do so, because the Dependent Differentiation Method did not find any particularly critical events. It is nevertheless important that these cases are not considered as isolated actions. This means, firstly, that the empirical data should not be taken from experimental studies which are based on the classical empirical paradigms; and secondly, that the failure probability of that particular action should be seen as yet another contribution to the reliability of the overall performance. **The critical task step will never occur by itself, but always as part of a context.** If that context does not enter into the analysis, the outcome is bound to be misleading.

One final argument for the Dependent Differentiation Method is that if a human reliability analysis is based on individual situations and/or individual events, the very fact they are individualised, small units makes for a large variety, i.e., that their generality necessarily is very limited. The description of the scenarios in terms of CPMs is less detailed, but in return there will probably be fewer different situations that need to be described. This corresponds to a description on the level of types or principles rather than on the level of exemplars. Such a description may be easier to generalise. There is, of course, always a trade-off between precision and generality. The value of both should be carefully assessed so that the right level of description can be found. It can be argued that the classical decomposition approach errs on the side of too many details, while the structural (Dependent Differentiation Method) approach errs on the side of generality. But that, at least, should make the problem easier to see.

9. The Dependent Differentiation Method and COCOM

In the beginning of this chapter it was argued that a basis for analysing the Reliability of Cognition could be found in the Contextual Control Model (COCOM). So far the COCOM has only been used in the sense that the emphasis has been on the context of the task, rather than on individual task elements. This hesitation is quite deliberate; it reflects a concern for becoming too dependent on a specific theory. If the coupling between a theory and a method is too tight, the danger is that the method must be abandoned if or when the theory fails. Staying at the general level or the general thrust of the theory may be enough to change the direction of the approach without becoming dependent on specific and hypothetical details. The hesitation is also due to the concern for having operational measures, which was discussed earlier on.

It is nevertheless quite possible to apply the COCOM to a much greater degree than has been done so far. It is in particular relevant in connection with the third stage of the method, the identification of critical task steps. In the above example use was made of established methods and the commonly accepted classification of error modes and error causes. It was noted then that the disadvantage was that there could be a practically endless combination of conditions and causes which could be used to explain how the critical event could come about. The seemingly open-ended search for a cause presents some practical difficulties; although it enables an analyst to come up with an explanation for practically any type of situation, it makes it very difficult when more precise and more limited characterisations are to be made - in particular when need of the final quantitative expressions is kept in mind.

The example used before was about the step "initiate FAB". It was shown how this could be considered either as a case of perception or as a case of decision making. For both possibilities a characteristic set of causes was pointed out - although this was by no means an exhaustive set. If, instead, the COCOM was used as the basis for analysing the task step "Initiate FAB", the concern would be about the **control mode** rather than about the possible causes. The present version of the COCOM makes use of four control modes called scrambled, opportunistic, tactical, and strategic (cf. the definitions given in Chapter 4). If the critical event is considered in these terms, the question is no longer what the possible causes may be, but what the likely control mode is going to be.

The analysis should try to determine which control mode would be present when the critical action needs to be carried out. It follows from the definitions of the four control modes - which again refer to the detailed assumptions embedded in the model - that a task step would be more likely to fail in achieving its goal in, say, the opportunistic mode than in the tactical mode. With reference to the preceding discussion of stage 4 of the Dependent Differentiation Method, it may even be suggested that this more detailed relationship can be described by fuzzy sets. Suffice it to say that if the probable control mode can be established or surmised, then the probable outcome of the critical step can be predicted. If the probable outcome of the task step can be predicted, then it is possible to assess whether it will have an impact on the CPMs thus far defined; and this is precisely what is needed by the Dependent Differentiation Method! There is no longer any need to be concerned about PSFs, because the control mode in itself is an expression of the joint effect of the PSFs. It is therefore feasible to predict the control mode based on the CPMs described so far, thereby establishing the coupling to the previous stages of the Dependent Differentiation Method.

The effect of using the COCOM as part of the more detailed analysis of critical events is thus that it abolishes the need to find specific causes, hence also the need to quantify those causes (i.e., establish numerical values for their probability). As argued throughout this book, this avoids the dependence on isolated actions and the whole need for data about those actions. By seeing the critical event in a context, and by describing (and quantifying) that context rather than describing (and quantifying) the critical action and its discrete causes, the dependence on the artificiality of the decomposition approach is dispensed with. This leads to a far more manageable analysis, and one which can be characterised on a task level rather than on the level of individual error modes and individual error causes.

9.1 Quantification of the Reliability of Cognition

The Dependent Differentiation Method has been presented, and an example given, as a way of producing a qualitative and quantitative description of the reliability of performance - or as a way of saying something sensible about the probability that unwanted consequences will obtain. It has not been presented as a method to describe and quantify the Reliability of Cognition.

The reason for this should be obvious by now. The main purpose of the method is to avoid the dependency on specific causes or specific factors and rather base the analysis on an understanding of the context. It is consistent with this that no attempt has been made to quantify the Reliability of Cognition *per se*. The attempt to quantify the Reliability of Cognition in itself would go against the spirit of the method. Cognition does not occur by itself nor does it express itself directly in action. Cognition is a component of action - and in many cases a very important one - but it can only be properly understood and properly assessed when seen as a part of that action. The emphasis of the Reliability of Cognition, in this book and in the COCOM, is fully taken into account by the Dependent Differentiation Method as such. The Reliability of Cognition is an important argument in favour of finding an alternative to the decomposition approach. If it only led to the inclusion of cognition as yet another component, the whole enterprise would have been in vain. The answer to the question about the quantification of the Reliability of Cognition is therefore simply, that this does not make sense as a separate goal. The reliability of performance depends on the Reliability of Cognition - among many other things. But the final result cannot be found by aggregating the results from sub-analyses.

10. Summary

In a typical reliability analysis involving human performance the analyst will focus on a critical event and the ways in which it can go wrong (the error modes), determine, for example, the causal factors (the error causes), and then go on to identify the Performance Shaping Factors (PSF) - for that event alone! The decomposition approach to human reliability analysis often ends with the details of individual actions plus a set of Performance Shaping Factors. But human performance is obviously conditional upon the common modes of performing; if, for instance, noise level is high, this condition will be common for a large number of events. This can be expressed by using the term Common Performance Modes (CPM). Tasks should not be analysed into elementary components that can be studied in isolation, neither should the operator be treated as a component. Human actions should not be investigated in isolation from other actions or from the task context.

Some PSFs can - usefully - be interpreted as general conditions for the performance. Examples of these are micro-gravity, temperature and other physical variables that are independent because they are not affected by how the operator performs. In that sense there are a number of the "classical" PSFs which will remain valid. However, it is still a simplification to consider them solely as numerical factors or weights. Their importance lies in the way in which they contribute to the actual working conditions - hence being a kind of CPMs - rather than as modifiers to specific quantitative results.

CPMs do not only influence error causes, but are causes in themselves.[4] The difference is that CPMs, as "causes", have an impact on the situation as a whole rather than a specific effect for a specific type or range of operations. The analysis of the reliability of MMI must therefore look to the general performance conditions first, and only then to the specific influences. The reason is that the general performance conditions may be more important and exert a proportionally larger influence, so that specific conditions can be left out with minimal impact to the result. One thing that must be remembered, however, is that there is a reciprocal influence, in the sense that the CPMs are determined by the outcome of specific actions as well as the working conditions as a whole. In other words, there are some actions which will result in a marked deterioration (or improvement) of the working conditions, hence on the CPMs. This reciprocal relationship must be carefully accounted for in any model that claims to serve as a basis for human reliability analysis and the Reliability of Cognition.

Once these modes (factors) have been defined their probable value must be derived from a characterisation of the situation as described by the task analysis. This leads to a set of rules or guidelines that can be used to determine a probable value for each mode. It must then be considered whether the ensuing description may lead to a change in the conditions. If so, the stage is repeated anew. When no more changes are foreseen or expected, the likely control mode (and changes in control modes) are identified. Once this has been done, the detailed assessment can be started. It is clearly only for the more reliable control modes that estimates of individual actions should be attempted. Of particular interest is the change in control modes, i.e., how and when they may occur, since such changes will be accompanied by changes in the probability level. The further conditions where the changes may take place need to be elucidated.

This creates a need for finding values of actions in different control modes, or the likely impact (chance of success) of a given control mode. The consequence for system design is to assist in maintaining performance in the higher control modes and to avoid degradation. This takes the

---

[4] It is a little puzzling that the decomposition approach has not adopted the same interpretation, since some of the PSFs might well serve in that capacity. There will thus be a significant overlap in the names of the PSFs and the CPMs, although the interpretation may differ.

emphasis away from the single isolated action, except possibly as a triggering condition.

The analysis of the reliability of performance requires a description of tasks and actions which can account for the effects of continuity, parallelism, hierarchisation, performance variability, communication, and interaction. This cannot be achieved by the existing procedural prototype models, but requires an alternative approach. The Contextual Control Model (COCOM) was proposed as a candidate.

Tasks can be described by a simple goal structure and refinement principle such as the Goals-Means Task Analysis, which enables qualitative statements about human reliability to be made. This also provides the framework for a more detailed - and potentially quantitative - assessment of human reliability, which includes full recognition of the Reliability of Cognition.

Human reliability in general, and cognitive reliability in particular, denote an important aspect of human performance. Human reliability is part of the way in which we understand human performance. It should therefore be treated as a concept, not as a number. Based on this analysis specific solutions can be proposed which may either prevent such situations from occurring or reduce their impact. Examples are:

(1) better design of systems and their parts (interfaces, procedures, task allocation),

(2) functional requirements for knowledge-based support and assistance,

(3) clearly focused training programmes, and

(4) specific technological solutions, such as error tolerant interfaces or performance monitoring.

It remains the task of cognitive engineering to devise ways in which these solutions can be turned into concrete implementations.

6.

Discussion

1. Analysing Human Performance

The problem with applying conventional methods to the analysis of human performance and human reliability is that the human, considered as a system, does not have a fixed and well-known configuration or structure. In some situations humans behave very much like machines. This is the case for highly skilled actions which are carried out by an "inner cognitive machine" without the need for attention and conscious control - and, indeed, without the need for cognition in the sense of thinking and reasoning. For these cases it may be warranted to use relatively simple analysis methods, since the reliability of cognition *per se* is of little importance. But for the rest of the cases, we cannot afford to ignore cognition and therefore cannot use methods that do not properly deal with cognition. This is the case regardless of whether the methods are relatively simple fault trees and operator action trees or more complex structures such as Sneak Path Analysis or Failure Mode and Effects Analysis.

Approaches to human reliability analysis that are based on the decomposition principle fail not only because humans are not machines but also because there is an essential difference between what tasks are and how they are described. Task descriptions - in the form of event trees or as procedural prototype models - are idealised sequences of steps. Tasks as they are carried out, or as they are perceived by the person, are rather incompletely ordered segments of actions. (This, in turn, may partly be attributed to the fact that people are not information processing machines!) It can be argued that the task descriptions have so far been inadequate and unable to capture the real nature of human work. The decomposition principle has encouraged - or even enforced - a specific way of task description, and this formalism has been self-sustaining. It has, however, led human reliability analysis into a *cul-de-sac*.

On the level of controlled, overt action the functioning of the human is clearly not causally deterministic and it is both practically and principally impossible to predict specific (re)actions. This applies on the level of covert action (cognition) as well. The view of the human as a deterministic machine or an information processing system should preferably be restricted to functions which do not involve or depend upon the workings of the human mind, i.e., functions that are purely physiological or biological or highly practiced skills. Accordingly, it does not make much sense in performance analysis to refer to possible combinations of principally deterministic sequences or events. This reservation is valid not only for human reliability analysis, but goes equally well for other attempts to understand human nature through the principle of decomposition. Here the absence of a deterministic "regulation" will make any combination of possible sequences or kind of cross-coupling possible. The only restrictions are the psychological ones.

In Chapter 4 it was argued that the main weakness of procedural prototype theories is the assumption that there is a pre-defined, highly likely sequence or ordering of actions, and that actual performance can be described with reference to that - and in particular as deviations from that. This assumption is, however, not tenable, and with that falls the basis for conventional decomposition approaches - since the aggregation that is part of them implies that a certain ordering is meaningful. Instead the analyst is faced with a set of combinations which in practice is exceedingly large, and in principle is unlimited. Unless it is possible to find some reason to exclude part of them, **all** of the combinations must be considered. The situation is made even worse if the analysis is extended by a search for causes that include cognitive functions and factors. Without a strong guiding principle cognitive functions can be combined willy-nilly to produce a seemingly limitless number of plausible explanations.

The most evident feature of human performance is probably that it is goal-oriented or purposeful and carried out according to plans. If that feature can be sustained as a theoretical assumption, the analysis of human reliability might be based on the plans and the deviations from them. Yet it is quite likely, as the contextual control model makes clear, that the plans are incorrectly executed - which in effects means that performance is no longer properly organised. So whereas the existence of plans can be used as a supporting ingredient of the analysis, they do not relieve the analyst from the obligation to consider less structured cases as well. Other candidates for reducing the number of combinations that must be considered are, for instance, the Law of Requisite Variety or the notion of contextual control. Only one thing is certain: that the very principle of decomposition is invalidated by the main uncertainty and infinity of human actions - overt, covert, and cognitive.

1.1 Accident Analysis and Performance Prediction

The study of human performance and the concern for the influence of human reliability can be focused in two different directions - accident analysis and performance prediction. As the names imply, the former is concerned with analysing events that have happened in order to find the probable causes, and perhaps even find the root cause. The latter is concerned with describing the probable ways in which an event may develop, in particular the ways in which the variability of human performance can influence how a system may evolve. In both cases the purpose should be to identify the possible weaknesses of the system and to develop designs or design modifications to overcome these.

There is, however, a difference between accident analysis and performance prediction which has serious consequences for the use of models and methods. That difference is the existence of a context (see Figure 6-1).

In the case of accident analysis the endeavour is to understand better something that has already happened. We therefore know precisely (more or less, at least) what took place, what the conditions were, which parts of the system failed (if any), what the consequences were, etc. The whole field of accident investigation, in nuclear power, aerospace, transport, medicine, chemical process industries, etc. is a testimony to that. Because the context is known, it is possible to follow the event step-by-step in the direction backwards from the observed consequence to the probable cause(s). It is therefore relatively easy to use models of cognition; the control problem neatly disappears because the context defines how actions were controlled. The many things that **could have happened** at a given stage or the many things that **could have been** the cause of an event are all reduced to one - by virtue of knowing what actually did happen. In this way the context provides the information needed to remove any ambiguities from the models. For the very same reasons the method can be a gradual decomposition or fractionation of the

Discussion

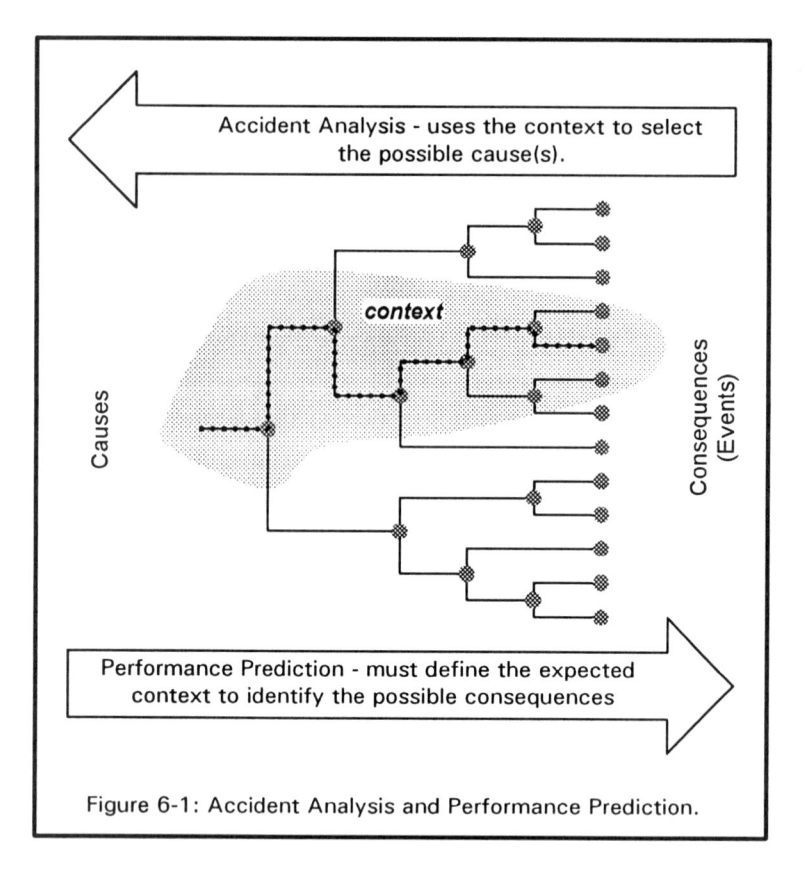

Figure 6-1: Accident Analysis and Performance Prediction.

whole into its parts. The many possible interactions between parts will, in the specific case, have been reduced to the interactions that actually did exist at the time.

In the case of performance prediction the aim is to find out what could possibly happen under given conditions, e.g. if a component fails, if there is a lack of time to act, or if a person misunderstands a procedure. In a sense, the art of performance prediction is **to specify the context that is most likely to exist**, rather than to specify which actions will occur. If the goal is taken to be predictions of specific actions, the problem is precisely that each step will present so many possibilities that the total quickly becomes unmanageable. A good theory or a good model can, of course, predict **any** accident or event, if only enough possibilities are considered.

In principle the task is simple; all one has to do is to retrace the path from consequence to cause, but this time start with the cause (cf. Figure

6-1). The path, however, only exists for a given context. If the conditions had been slightly different, the path might have looked completely different (e.g. Cacciabue *et al.* 1992). Thus event trees, such as those developed by THERP or those shown in Figures 5-10 and 5-11, are misleading because the sequence of events cannot be assumed to remain the same under all conditions. This means that a prerequisite for performance prediction is really to develop or create a probable context. It is only because the prevalent analytical methods require the identification of specific performance steps, that there is a need to predict performance in such detail. The use of a contextual method, such as the Dependent Differentiation Method, would mean that performance predictions should focus on the Common Performance Modes rather than on specific actions. This, in turn, has as a consequence that the purpose of performance prediction is to estimate the probable performance conditions rather than to predict specific events!

2. Consequences of the Contextual Control View

The contextual control view has important consequences for attempts to model cognition and for theories of cognition and action. The most obvious is, perhaps, that several issues or problems disappear or become superfluous.

For instance, the attempts to explain the shifts between levels of procedural prototypes, notably in terms of the categories skill-based, rule-based, and knowledge-based (e.g. Hannaman & Worledge, 1988) are no longer needed. The shifts are an artifact of the underlying model and therefore disappear when the model is forsaken. In its place comes the changes in the level of control, exemplified by the four control modes; these change are, however, explicitly addressed by the model of contextual cognition. Similarly, it is no longer necessary to design systems such that performance takes place on a specific theory-prescribed level (e.g. Rasmussen & Vicente, 1987). Instead one should design systems such that appropriate control can be maintained.

In fact, the whole issue of MMI design changes: the point is no longer to support specific procedure forms or performance types, but rather to support the understanding of the context. This has serious implications for man-machine studies, interface design, training, and work situation design - on both the theoretical and the practical levels. The real problem is now to describe how actions are organised, with the

Discussion 274

interpreted context being the main factor. Rather than trying to support specific modes of functioning or of decision making one should try to provide a representation of the context or the situation which is undivided and easily understandable. The contextual approach to understanding and analysis thus emphasises the structuralistic or holistic nature of cognition, and forces the view to a relatively high level of description. Cognition is not just situated; rather, the situation is cognition.

2.1 Interface Design

The consequence of the contextual control view for interface design is that the starting point must be an analysis in terms of the scenario as a whole and of the main determining factors, rather than an analysis in terms of constituent functions. The concern of the design changes from being the function of an instrument, measurement or device *per se*, to become how the instrument or device functions in the characteristic context. This, of course, introduces a trade-off between designing for anticipated scenarios and designing for unexpected events. But that trade-off is inevitable in any design.

The very notion of individual instruments and devices does, of course, go a bit against the spirit of the contextual control view - although it does not support the opposite approach. The designer clearly cannot disregard the fact that the available information is segmented in specific ways and that there are certain physical, technical, and organisational restrictions on which configurations are possible. But rather than succumb to a bottom-up, component-based design a viable alternative is to define information groups conforming to **representative contexts**. That alternative is, in a sense, similar to the notion of ecological interface design, although derived from an entirely different basis. The possible disadvantage is that this solution may disturb the constancy or cognitive momentum because the representative contexts sometimes may not match the actual conditions.

The first priority for the design of interfaces and working environments must, of course, be the needs of the task and the avoidance of possible risks as revealed by risk and reliability analyses - but contextual ones. The consequence of using contextual theories and models is that the design is no longer driven by the details of a theory (i.e., in a normative sense) but instead must reflect the needs of the application. At most, the contextual theory may emphasise the control functions and the availability of essential competencies (which in turn may be supported individually). Methods such as the Goals-Means Task Analysis can identify the functions that are needed to accomplish a given task as well as the control that must go along with it. In many ways the total result is therefore more a change in emphasis than a change in details. One could argue that detailed theories of micro-cognition (Cacciabue & Hollnagel, 1993) anyway were only bad as a basis for interface design, and that the few attempts that have been made to follow a specific theory have not been successful.

2.2   Other Issues

Interface design is just one out of several issues where there are consequences of the contextual control view. Other candidates are procedure design, operation support, "intelligent" support systems, training, work organisation, etc. It is, however, not possible to go into these at this point, since it would lead to a completely different book.

In relation to the Reliability of Cognition the consequences of the contextual control view can be summarised by repeating what was said in Chapter 2: man viewed as a behaving system is not simple; on the contrary, the complexity and variety of the environment requires equally complex cognition in order for a person to be able to cope. This means that the observed complexity of performance cannot be explained except by the use of models and theories that adequately capture the same kind of complexity. The complexity can apparently be brought to disappear if the analysis is confined to individual actions or steps, or if the organisation of actions is described by a simple principle, such as procedural prototypes. The consequence of that is, however, that the problem effectively changes because the whole cannot be reconstituted by a simple aggregation of the parts - or even by adding the effect of performance shaping factors. The results from analysing the simplified problem are therefore, not surprisingly, inappropriate for the solution of the real problem. The solution is to acknowledge the complexity of human performance and try to maintain it throughout the descriptions and analyses. This can in my belief best be achieved by explicitly trying to model and describe the influence of the context on how human's control their performance.

Discussion 276

3. Hypernatural Environments

If the context affects how actions are organised and controlled - regardless of the mode of actions - then it makes sense to try to provide and sustain an appropriate context. The corresponding principle has been called the design of hypernatural environments (Hollnagel, 1988). The notion of natural environments is closely tied to the notion of the reduction of unwanted consequences - a natural environment will reduce the number of unwanted consequences, just as an "unnatural" environment will increase them, cf. the notion of clumsy automation (Wiener, 1977; Wiener & Nagel, 1988). The basic tenet of hypernatural environments is that people perform adequately under natural circumstances, i.e., when sufficient adaptation of the work to people has taken place - rather than the other way around. The corollary is that if natural work environments can be created, then reliability will be high. Conversely, if even better environments - hence the term hypernatural - can be designed, then reliability will be further enhanced.

The problem of establishing efficient man-machine interaction can be described as the problem of how to maintain system performance within a specified envelope. The purpose of a system design is to enable the person to work efficiently by giving him the necessary facilities to obtain information about the situation and to use the available control options. The success of the design depends on the designer's ability to anticipate the possible situations and to invent means by which to deal with them. The tacit assumption is that the person is capable of handling the situation, i.e., that he has the cognitive resources to provide the required control. But to ensure that this potential can be realised the designer must provided the person with the proper facilities to **retrieve** the information needed, to **transform** it into a presentation form which is adequate for the problem at hand, to **combine** it on the presentation surface so that it is well suited to the current needs, and to **effectuate** the intended control actions. Because it is impossible to anticipate all situations, the person must be able to control these aspect of information usage. The actual interface must therefore have a somewhat flexible way of functioning which makes it possible for the person to achieve his goals - even during unanticipated situations.

In order to achieve an efficient coupling between humans and machines it is necessary that some kind of adaptation takes place. It can be said, with considerable justification, that the reason why most systems function satisfactorily is that the person is able to adapt to the design. But the human ability to adapt should not lead designers astray; the situation should preferably be that the machine must adapt to the human, rather than the other way around. In order to achieve this it is, however, necessary that one knows how to perform this adaptation. Here the following cases can be distinguished: (1) adaptation through design; (2) adaptation during performance; and (3) adaptation through management (Hollnagel, 1992c). Each of these will be described in more detail in the following.

3.1 Adaptation through Design

The basis for adaptation through design is the ability to predict the consequences of specific design decisions. If a designer wants to achieve a certain effect, he must be able to foresee what will happen under specific circumstances, and the art of design is to increase the constraints governing the design so that only one solution or alternative is left; in this respect design is a clear example of the problems of performance predictions. If the design criteria mainly relate to technical aspects of the system, then the outcome may easily be one that puts undue requirements to the person's ability to adapt. It follows that the dominant design criteria should include some that address the man-machine interaction and the characteristic features of human performance, thereby hopefully making it easier for the person to adjust to the system.

In order to achieve this goal it is necessary to have sufficient knowledge about how the person will respond. Such descriptions are commonly referred to as "user models". In some cases such user models are used prescriptively as a basis for promoting a specific form of performance through the design; in other cases user models are applied more sparingly, as condensed descriptions of how one can expect a person to respond for a given set of circumstances (e.g. Smith *et al.*, 1992). In either case the fundamental requirement is an adequate model description of a person's capabilities and sufficiently powerful design guidelines derived from this model.

Models are generally static, i.e., verbal or graphical descriptions which can be used to estimate specific (expected) reactions to specific input conditions. The very complexity of system design, however, means that static models are inadequate. The obvious alternative is to use animated or dynamic models. Although this is significantly more difficult, it is also a worthwhile undertaking because it forces the designer to make

Discussion

all assumptions explicit. A dynamic model of the user must in computational terms describe how a person will respond to a given development, whether that response is covert or overt; nothing can be taken for granted unless it has been clearly specified. An example of a solution of this kind can be found in the SRG (System Response Generator) project (Hollnagel *et al.*, 1992) while a more general discussion of uses of dynamic modelling or simulation is found in Cacciabue & Hollnagel (1993).

3.2 Adaptation during Performance

Even with the best of models, and abundant resources set aside for system development, it may be impossible to ensure sufficient adaptation through design. This it not just because knowledge about the user is necessarily incomplete, but rather because it is impossible to predict all the conditions that may arise. Even in a completely deterministic world the sheer number of possibilities may easily exhaust the resources that are available for design purposes. The alternative solutions to design completeness are (1) to rely on human adaptation and ingenuity, (2) to achieve some kind of adaptation during performance, and (3) to achieve adaptation through management. The first of these solutions, while frequently applied, is clearly not desirable. To some extent this solution has been a necessary evil because the other options have been limited. Adaptation through management may not be feasible unless the proper company spirit is present. This leaves adaptation during performance as the main option

Adaptation during performance may require technical solutions that are very difficult to specify and implement. The practical work on adaptive man-machine interaction was negligible until Artificial Intelligence started to apply inference techniques to plan recognition (e.g. Kautz & Allen, 1986). Coupled with notable advances in the development and use of knowledge-based systems, it suddenly became possible to build convincing demonstrations of adaptive man-machine interfaces (e.g. Peddie *et al.*, 1990; Hollnagel, 1990). Unfortunately, it does not make the designer's task any simpler because artificial adaptation must be designed in detail. Basically the system must change its performance such that it matches the needs of the person - as far as they can be ascertained in a dynamic fashion. Although the use of AI technology, such as plan recognition and pattern identification, has overcome the need to define simple performance measurements, the limitations are nevertheless more or less the same: that the designer has to anticipate the possible situations.

The designer therefore must consider not only, as before, the passive design of the interface but also the design of the active adaptive facilities. This increases the demands for modelling of the user and for maintaining an overview of the design, for example.

3.3 Adaptation through Management

The design of an efficient man-machine interaction does not occur in a vacuum. It must take into account the organisation where the work takes place, hence also the way this organisation is managed because the way in which the management is carried out may contribute to (or detract from) the efficiency of the man-machine interaction.

Unlike adaptation through design and adaptation during performance, adaptation through management cannot always be planned. In spite of that management can serve to adapt the working environment to the specific tasks. The tasks are determined by a combination of the situation specific goals, the facilities for interaction and support, the local organisation of work, etc. Management can effectively counter-balance design oversights and compensate for structural or functional changes in the system that may occur after the design has been completed. Management can also provide the short-term adaptations that may be needed until a part of the system can be re-designed.

Adaptation through management requires the continuous monitoring of effects, as well as data collection and analysis, and occurs continuously although with some delay. It is able to cope with large deviations even if they are completely unanticipated and is effectively a kind of adaptation by continuous redesign, hence working over and above what active adaptation may achieve. Adaptation through design occurs at long intervals when the system is designed/re-designed. Adaptation during performance occurs continuously and rapidly but can only compensate for smaller deviations - essentially only for those that have been anticipated.

3.4 Adaptation and Reliability

The consequence of the contextual control view is that performance is seen as controlled by a sequence of cognitive goals (cf. Bainbridge, 1991), rather than by a plan or as following procedural prototypes. Therefore, the design ought to support the understanding and maintenance of the relevant goals and the construction and retrieval of the actions that are pertinent to achieve the goals. The assumption is that performance reliability will improve if the design (i.e., the degree of adaptation) is improved. A badly designed system, where the onus of adaptation is left to the person and only little support is provided for that, cannot be expected to produce very reliable performance. The reliability of cognition *per se* is fundamentally the same, but there will be a demand for different types of cognition. In addition to accomplishing the primary task the person must typically also keep track of several things, weighing alternatives, sampling unnecessary information (unnecessary if he knew what he was supposed to be doing, but necessary now that he is not really certain), etc. The person is therefore charged with tasks that should have been avoided through a proper design, and the influence on performance as a whole - in particular the reliability of performance - is bound to be detrimental.

The contextual control view is best suited to support adaptation through design. The COCOM and the Dependent Differentiation Method are both useful as ways to improve the knowledge about the user, to build dynamic user models (as in the SRG project, cf. the Appendix), and to identify where and how specific design changes should be made.

4. Performing a Human Reliability Analysis

It is frequently acknowledged that the practical problems in making analyses of human reliability or human performance are almost prohibitive if the analysis is done in a step-by-step fashion, i.e., if done sequentially and manually. This is so even if computers are introduced to handle some of the routine administrative tasks. Thus a straightforward THERP analysis of a simple sequence with only ten steps requires a tree of $2^{10} = 1.024$ branches. If there were 20 steps the complete tree would have 1.048.576 branches. Little wonder then that the preferred solution to this problem is to look at only a selected subset of the task or application. This selection often is based on *a priori* criteria and may therefore be potentially detrimental as the criteria may exclude events or branches that in the event turn out to be essential. A "second generation" method, such as the Dependent Differentiation Method described in Chapter 5, provides a different solution to the problem by gradually making the selection part of the analysis. The overall endeavour must be to minimise the demands the analysis makes in terms of time and effort; in addition the method should use terminology and concepts that reasonably can be assumed to be comprehensible without an extensive background, e.g. in cognitive ergonomics.

The practical solution to the problems in making a human reliability analysis is to enrol the services of information processing systems - not just as a clerical or administrative facilitation but as a true information processor. This can be done either as a help to an investigation of all possible or probable combinations and conditions, or as a simulation based on a pre-defined set of assumptions about the scenario, the process, the user(s), and the way in which they interact. The latter is by far the more interesting use of computers. Even a very basic application of information processing techniques makes it possible to expand the scope of existing analysis methods, in particular to include, for example, state changes and temporal relations. An example of that is found in Cacciabue et al. (1992). Here a simplified Approach To Landing procedure was analysed using both THERP (Swain & Guttmann, 1983) and an integrated simulation concept known as the System Response Generator (e.g. Hollnagel et al., 1992). The THERP analysis of one case considered 32 out of the 64 possible branches and resulted in finding 12 success branches and 20 failure branches. The computerised analysis of the same case generated 38 sequences of which only seven were success branches. The discrepancy was due to the fact that the analysis could consider the changes in system states that occurred over time (essentially the position of a simulated airplane). The THERP analysis considered only one predefined sequence of events, while the SRG analysis generated three different sequences of the same events where the ordering depended on how the system simulation developed. The difference in results from the two cases is rather striking, and inevitably raises the question of which of the two outcomes is the (more) correct one! The difference is important even if it is "only" in terms of the qualitative analysis.

Static methods and the decomposition principle go hand in hand. This is probably because the limited techniques - manual analyses - make it impossible to handle the complexity. The need to write down elements and confine the analysis to paper, enforces the decomposition approach. Attempts of different approaches were therefore doomed to limited success because of a lack of a medium for simulation. Decomposition was for a long time the only practical way of handling the complexity of working situations. But the tradition of using decomposition has continued even when it was no longer necessary.

4.1 Model Verification

The use of simulation as a technique in human reliability assessment and probabilistic safety assessment is attractive because the relations and dependencies described by means of the simulation can be very complex, and certainly an order of magnitude more complex than when the analysis is done by hand. However, the simulation must reflect empirical facts, i.e., it must amount to a **verifiable** model of the target system, whether it is the process or the user. This is particularly important, and particularly difficult, in the latter case because there is no *a priori* correct set of observations to rely upon. For that reason expert judgment and expert assessments are still needed as an important contribution to analyses of the Reliability of Cognition, just as they are to human reliability assessment and probabilistic safety assessment. The use is, however, different from that described in Chapter 3. The objective is no longer to produce estimates of point probabilities, but rather to generate estimates of dependencies and relations - both in terms of their nature (quality) and their extent or degree (quantity).

It is a *sine qua non* for any type of science, and for any use of models, that a proper correspondence between empirical facts and model assumptions is established and maintained. The risk in simulation-based analyses is that the tool or the simulator becomes a closed world and that the analysis takes place inside that world without considering the assumptions that are made or the conditions that are implied. It becomes an analysis within the paradigm that does not question the validity of the paradigm, nor even recognise that the problem may exist. It is therefore incumbent upon the analyst to remain calm and detached, and realise that the ultimate value of the tools that are used depends upon whether they match the reality he is trying to model and describe.

4.2 Computerisation of the Dependent Differentiation Method

In the Dependent Differentiation Method the initial Goals-Means Task Analysis provides a definition of the task steps. The Goals-Means Task Analysis is an obvious candidate for computerisation both because it develops the basic structure rather than starts from it, and because it is a recursive approach. Recursion is easily handled by computers, but only very badly so by humans. Following the initial task analysis the next step is to consider the context or the conditions (performance modes) under which the task steps are to be carried out. The properties of this context can be derived both from a consideration of the COCOM (since this is the implementation environment) and from general and specific experience about human action.

Once the Common Performance Modes have been identified their probable impact must be derived from a characterisation of the context as described by the task analysis. This is done by finding the critical task steps and their dominant error modes. At present, this step must be carried out manually, but it is an obvious candidate for computerisation. It must then be considered whether the ensuing description leads to a change in the performance conditions. If so, the steps are repeated. When no more changes are foreseen or expected, the likely control modes are identified. Only when this has been done can the detailed assessment be started. It is clearly only for the more pervasive control modes that estimates of individual actions should be attempted. Of particular interest are the **changes in control modes**, i.e., how and when they may occur, since such changes may lead to changes in the probability level, hence in the quantification of the Reliability of Cognition. The further conditions where the changes may take place need to be elucidated.

In order to use the Dependent Differentiation Method it is necessary to find "values" of actions in different control modes and the likely impact (chance of success) for a given control mode. The consequence for system design is to assist in moving to higher control modes and to avoid degradation of performance. This takes the emphasis away from the single isolated action, except possibly as an activating condition.

It might seem that the Dependent Differentiation Method is not that much different from a conventional reliability analysis method, such as e.g. SHARP (cf. Figure 2-1). The similarity is due to the fact that both methods are described in the same way, i.e., as steps in a sequence. This is, however, because of the very nature of methods: a method is an ordered description of how something should be done. The similarity of description must therefore not mislead anyone to think that the contents of the two methods is the same. It is, indeed, unfortunate that a description of a method - precisely because it entails a series of steps - may cause a disposition to use the decomposition principle. This is nevertheless not a necessity; it can be avoided, as the DDM has shown.

4.3 Application of DDM Outcomes

The outcome of the analysis, the description of the Reliability of Cognition, can be used in two ways. Firstly, to identify those activities or functions which are difficult to perform, i.e., where the reliability is expected to be low. This can then serve as an input to a task design/work design with the aim to avoid such situations. Here the conclusions from a specific theory, e.g. the contextual control model of cognition, are combined with the outcome of a task and reliability analysis to generate design input (or re-design input, as the case may be).

A second use is to identify the important Common Performance Modes. The assumption is that the erroneous action might not have occurred if the Common Performance Modes had been different (provided, of course, the design was otherwise appropriate). The human reliability analysis is in particular a good basis for finding the decisive aspects of organisational impact. In other words, the reliability concept (of safety, of performance) must be extended to the environment and organisation as a whole. Simply focusing on Reliability of Cognition is missing the context and thereby missing the point of the analysis.

A properly designed human reliability analysis will clearly use more than one method. The situation can be characterised as in Figure 6-2. The three main phases of the analysis are the **qualitative analysis**, the **sensitivity analysis** (differentiation), and the final **quantification**. Figure 6-2 also suggests how various analysis methods fit into this view. The Goals-Means Task Analysis, for instance, covers the first phase, the qualitative analysis, while the Dependent Differentiation Method can be used for both the first and the second phases. The quantification is typically based on human error rates or human error probabilities. The relative size and importance of the three phases varies from case to case; the emphasis is very often put on the quantification. But it is clearly important that the two previous phases are allowed to play their part in full. The first two steps provide the substance of the human reliability assessment; the last phase translates it into a specific form. It is quite conceivable that the third phase - the quantification - may become less important or even superfluous. A good qualitative description may in many cases provide all the information that is needed, for instance to make a change in the design.

5. Comparing Analysis Methods

The Human Factors Reliability Benchmark Exercise (Poucet, 1989), which was described in Chapter 3, clearly showed that different methods for human reliability assessment gave different results - even if the

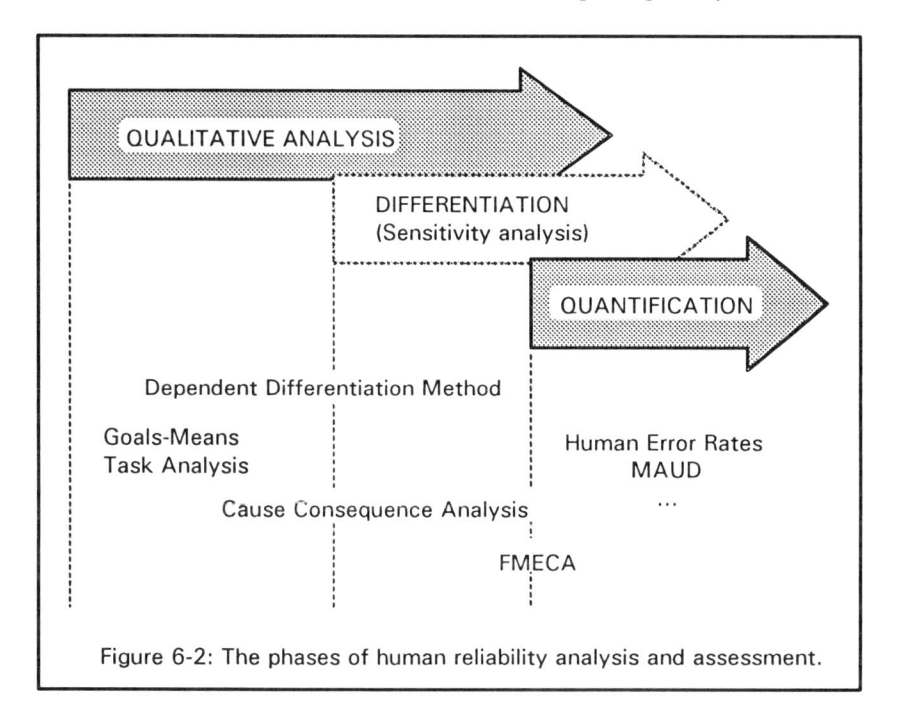

Figure 6-2: The phases of human reliability analysis and assessment.

methods were of the same type. The study made by Cacciabue *et al.* (1992) also showed that different approaches to the analysis of interactive systems (as a precursor to human reliability analysis) gave quite different results. As described in Chapter 3, several attempts have been made to compare and evaluate the various candidate methods, although without any unambiguous conclusions. It is obviously an important question whether it is possible to compare different methods, and in particular whether it is possible to say that, for example, a "second generation" method (Dougherty, 1990), such as the Dependent Differentiation Method, is better - or worse - than a typical "first generation" method, such as THERP.

There are various ways in which such a comparison can be approached. It clearly hinges upon the definition of the criterion that is applied, i.e., a definition of what "better" means. The surveys mentioned in Chapter 3 all used specific criteria such as, for example, usefulness, acceptability, and practicality. Another solution would be to invoke the notions of reliability and validity: if one method is more reliable than another and if the results produced are more valid, then that method clearly is to be preferred. A third solution could be the degree of "social

Discussion 286

acceptance", i.e., that there was less scope for disagreement or alternative interpretations of the results, and less uncertainty in the raw data used (if such indeed are needed). A fourth solution could be to consider how much one could achieve at the various stages of the analysis, e.g., how much advantage the quantitative analysis really gave *vis-a-vis* the goal of improving system safety, compared with the preceding qualitative analysis. Yet another solution would be to consider how well a method could support the common reactions to accidents; these include establishing new barriers, redesign, and elimination (cf. Chapter 1, Figure 1-6). If the chosen response, for example, is to change part of the existing procedures, then it is important that the reliability analysis not only shows that there is something wrong with the procedures, but also more expressly can contribute to specify which changes should be made.

5.1 Completeness, Consistency, and Decidability

A somewhat different way of approaching the comparison of methods is to use the three criteria of completeness, consistency, and decidability.[1] These criteria are useful because they not only address the method but also the concepts (assumptions, theories, and models) that lie underneath it. In relation to human reliability analysis the three criteria can be given the following meaning:

(1) **Completeness**: For every statement about human reliability it must be possible to determine whether it is correct or incorrect. This means that it must be possible, by using the method and the underlying theory, to address every conceivable event or combination of events (and conditions) and decide whether or not they are in accord with the method.

Completeness is related to the **scope** of the method. It is fairly obvious that most of the known methods have limited scope and the presumption is that the Dependent Differentiation Method has a larger scope. It is, however, not analytically possible to prove that the scope of a method is sufficiently large; only experience can give an indication. But when new methods are developed, care can be taken to maintain as broad a scope as possible.

---

[1] The criteria were used in the discussion of the foundation of mathematics in the 1930s.

(2) **Consistency**: It must not be possible to arrive at an incorrect result by using the method. This requirement is clearly violated by the main decomposition methods because they can produce results that are mutually incompatible, cf. the Human Factors Reliability Benchmark Exercise. It is difficult to say about any existing method whether it complies strictly with this requirement because there are few cases where a correct result is known. (A possible exception is post-accident analyses where the outcome is known because it **has** happened.) A variation of this criterion would therefore be that if the same method is used repeatedly or by different analysts, then it should not produce results that are too discordant.

(3) **Decidability**: It must be possible to find a method which, for every assertion, produces a correct outcome. This means that it must be possible to find, to define or to derive an estimate or description for any meaningful situation or set of conditions. In other words, there must not be any cases that cannot be covered by the theory or method.

This does not necessarily mean that it must be possible to produce a probability estimate for each event or task step, but only that it must be possible to characterise it in a way which is in accordance with and contributes to the purpose of the analysis. There must therefore not be any kind of event or situation which cannot be covered by the method. A good qualitative analysis, such as Goals-Means Task Analysis, will always meet this criterion, whereas methods which depend on the use of pre-defined categories or check-lists may sometimes fail.

The three criteria can be used both to compare two methods and to characterise a specific method. If we look at the Dependent Differentiation Method, we find that completeness can always be achieved on the qualitative level since the method starts with a Goals-Means Task Analysis. The question of consistency is more difficult to answer because a proper reference is missing. There are few cases where the correct results are known. The question of consistency is a little easier if it is viewed on the level of qualitative results. Here a criterion could be the level of agreement with known and accepted empirical material. Finally, the Dependent Differentiation Method can meet the requirement of decidability because the general nature of the COCOM and the inclusion of the Goals-Means Task Analysis puts very few, if any, restrictions on the cases that can be considered.

The real improvement of the Dependent Differentiation Method lies in the adherence to the contextual control view, and in the recognition that performance is determined by context and not by inner mechanisms of cognition. Any method which recognises this and systematically tries to encompass it will be better than a method based on the decomposition principle. The improvements are due to the ability to see the influence of dominant factors (whether or not they are called Common Performance Modes) and treat it appropriately rather than reducing it to specific influences (PSF) on individual actions. The second advantage of the DDM is that it better supports a dynamic approach - although it is not in itself dynamic. The contextual control view, however, lends itself to dynamic analysis methods, simply because it addresses a complexity that very easily gets lost when it is subjected to static means.

A further advantage is that the analyst is no longer forced to consider individual actions. The very nature of a static decomposition forces a narrowing of the view, a pointing to elements; if contextual factors are considered it is only as they may affect that element. In this way the whole context is irretrievably lost.

6. Attention and the Reliability of Cognition

Although neither the Contextual Control Model nor the Dependent Differentiation Method directly made use of attention, they both indirectly refer to it, for example, through the notion of execution mode. It is also clear that the concept and description of the various control modes, as well as the very idea of different degrees of control, must involve attention in one way or another. Attention is not only an essential concept for the scientific investigation and understanding of human performance, but also for the practical study and understanding of human reliability. The fact that a person to a very large extent is able (1) to select what he will take notice of or work with, as well as (2) how much effort he will put into it, are both ascribed to attention. Attention is also the crucial distinction between controlled (willed) and automatic (skilled) performance, as captured, for instance, by the popular classification of performance as being either skill-based, rule-based, or knowledge-based (Rasmussen, 1986). Attention is, moreover, a fundamental concept in learning, monitoring, communication, decision making, etc.

"Attention is a central concept in human behavior. It is a general, global aspect of human cognition, intimately connected to one's state of self-awareness, of consciousness. We see that highly learned, automatic activities may require little or no conscious attention to be performed. These highly automated activities can be disrupted, and when that happens they require conscious control, often to the detriment of their normal performance."
(Norman, 1976, p. 70)

In relation to human reliability and the Reliability of Cognition an initial assumption could be that reliability will be high if attention either is constantly high or if it is absent - which is also a form of constancy. In the first case the argument is that performance will benefit from the undivided attention of the person at every step; deviations from the intended development will be quickly detected and compliance with the goal can therefore be expected to be maximum. In the second case the argument is that performance is dominated by well-learned automatic activities, and that it consequently will be highly efficient and reliable (assuming, of course, that the goal is the right one). According to these assumptions it is therefore only when attention is incomplete or when it fluctuates or varies that reliability deteriorates.

The problem of attention is, however, not just the question of whether or not attention is available. There are clearly two possible scenarios for how a situation can develop: in the first scenario attention is required but is only partly available or not available at all; in the second scenario attention is not really required to carry out the task but is nevertheless applied, for instance because the person has nothing else to do. In both cases the result may be unfavourable. In the first case the actions do not receive all the attention they need; this may affect all aspects of performance: signals or information may be missed or incompletely perceived, interpretation and evaluation may be careless or incomplete; and execution may be prone to errors (being open loop rather than closed loop). In the second case the reason is that because the person attends to how the actions are carried out he forces an "unpacking" of automatic performance which obstructs an otherwise smooth execution and which may bring it out of synchronisation with the process. (If this seems far fetched, just think of what happens when you try to explain to someone how you perform a skilled action, such as tying a shoe lace.)

Discussion   290

The question is therefore in which situations attention will fluctuate and what the factors are which influence that. The impact of attention is an inherent part of the classical performance shaping factors, for instance in the concepts of workload (or mental workload), stress, etc. But this should be improved by providing a taxonomy of situations where attention is the crucial factor. If one can characterise events and situations with regard to the need for attention, and furthermore characterise them with regard to whether attention can be expected to be available, then this will provide a first approximation to a model for the Reliability of Cognition, hence a method for human reliability analysis.

The purpose of the following sections is to highlight some essential aspects of attention that are important for assessing the Reliability of Cognition. The reader is assumed already to be familiar with the elementary aspects of attention as it is clasically treated in cognitive psychology (e.g. Kahneman, 1973; Norman, 1976).

6.1 The Limits of Attention

It is a generally accepted truism that human attention is of limited capacity; this is often expressed by saying that a human being can only attend to one thing at a time. A more precise way of saying that is by noting that any type of task demands attention. If the task is easy it demands little attention; if the task is difficult it demands much attention. (There is a potentially dangerous circularity in these definitions, since the difficulty of a task may be defined by how much attention it needs. It is therefore necessary to be able to define either "attention" or "difficulty of task" independently. For the sake of the discussion here I will simply assume that this is possible and not go further into that.)

If the demand on attention is greater than the capacity, then performance will falter or may even fail completely. On the other hand, success is not guaranteed simply by having more capacity than demanded, cf. the two situations mentioned earlier. The detailed explanation of why this is so is the concern of capacity theories of attention (e.g. Kahneman, 1973). A simple rendering of the relation between capacity and demand is shown in Figure 6-3. The figure indicates how capacity (= supply) is increased as the demand increases, until a certain limit when the maximum is approached. When demand is low, total capacity is more than enough to cope with the task; there may even be spare capacity. If the demand to attention continues to increase, the result is that the spare capacity is

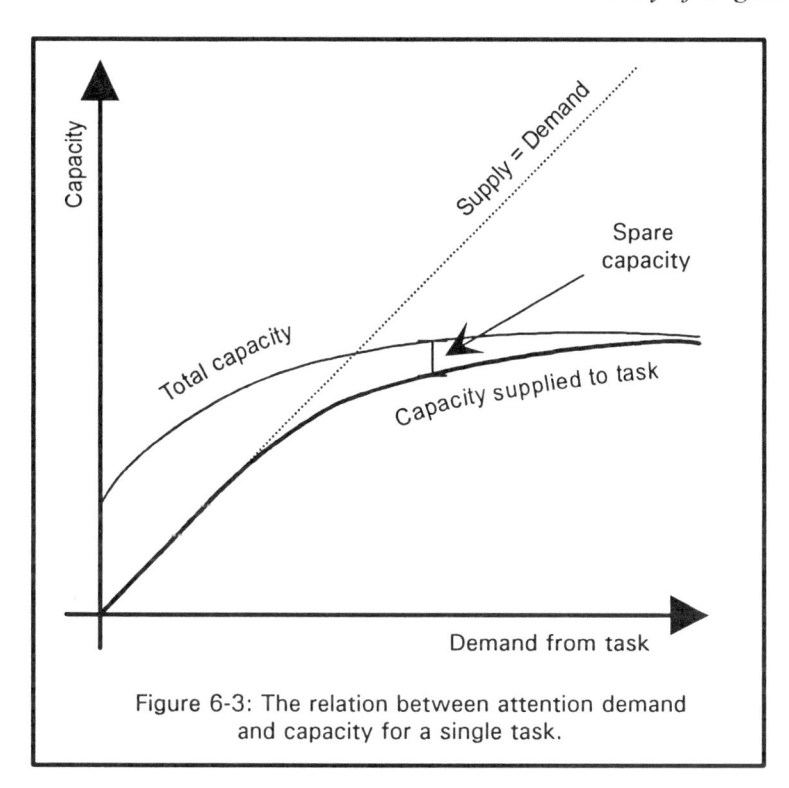

Figure 6-3: The relation between attention demand and capacity for a single task.

reduced until the point is reached when there is no spare capacity left and where the demand by far exceeds the supply.

Most tasks seem to demand so much attention that it is impossible to be engaged in two tasks at the same time. Humans cope with this dilemma in several different ways:

(1) Tasks can be organised so that attention can be shifted and shared between them without impairing overall performance.

(2) The person can learn to do things without paying constant attention to them, i.e., frequently executed tasks or activities may become automatic.

(3) The nature of the task can be changed, i.e., by being less rigorous with regard to discrimination of perception or precision of action (e.g. the case of information input overload described by Miller, 1960).

Discussion 292

It is reasonable to expect that if the attention demands of a task do not exceed the current capacity, then it is possible to perform the task adequately - subject, of course, to the spontaneous variations in capacity that may always occur. Yet this formulation really begs the question of what attention demand and attention capacity are. The answers to these questions are at the basis of the Reliability of Cognition. Unfortunately, they are not easy to give.

6.1.1 Attention Capacity

Although attention is one of the phenomena that have been studied from the very early days of psychology, definitions are not easy to find. Thus neither Norman (1976) or Kahneman (1973) provide a definition of attention. Attention has typically been defined as:

> "the active selection of, and emphasis on, one component of a
> complex experience, and the narrowing of the range of objects
> to which the organism is responding."
> (English & English, 1958, p. 49)

This definition highlights the two main features of attention, that it is **selective** and that it can **vary in intensity**. It is a well established fact that attention is limited both in relation to selection and intensity, although the limits may be hard to define. Under normal circumstances there is a clear limit to how many things a person can attend to or discriminate between at the same time;[2] there are also limits for how long a person can maintain the focus of attention on the same object or the same task (e.g. Kahneman, 1973). It is nevertheless not difficult to find situations where the "normal" capacity is exceeded (Neisser, 1982).

Although experience tells us that there are limits to our mental capacity (expressed in terms of attention span, short-term memory, vigilance, switching speed, difference thresholds, etc.) there are also considerable variations between different persons in the same situation as well as between different situations for the same person. For the purpose of reliability analyses the precise values of attention capacity are of less

---

[2] In the heyday of information processing psychology this became known as the "magical number seven, plus or minus two" (Miller, 1956). Although the phenomenon is real, the notion of a "magical" number as a limit of capacity is an oversimplification, hence misleading.

importance than the following two facts: (1) that capacity is limited and (2) that capacity varies. Even though people may perform memorable feats it is in practice possible to assign a reasonable upper bound to attention capacity, hence use this as a basis for analysis and design.

6.1.2   Attention Demand

The problem of **attention demand** is much harder to solve. Whereas the capacity mainly reflects the constitution of human cognition and, ultimately, the way our brains work, the demand is related to the state of the environment. The demand to attention is thus a function of the task and the working conditions. It makes a big difference whether the person has to attend to one task or many, whether the system is in a normal or disturbed state, whether the work environment (physical, physiological, psychological) is supportive or disturbing, whether resources are adequate or scarce, whether it is day or night, etc.

All this is commonly amassed under the ubiquitous concept of **workload** - a concept which is often used but rarely agreed upon (Wickens, 1990). Workload is often treated as if it could be described by a single dimension and measured to a precise value. But workload or demand must, of course, be seen in relation to capacity. It is therefore not the absolute size of the demand, but the size of the demand relative to the available capacity which is important. Accordingly, it seems better to talk about a **workload/capacity ratio** (cf. Hollnagel & Cacciabue, 1991). In order to be practically useful, the workload/capacity ratio must be defined operationally with reference to some measurable parameters; several of these are also part of the Common performance Modes:

(1) **Number of simultaneous tasks**. The assumed relation would be that a high number of tasks corresponds to a large value of the workload/capacity ratio (increasing workload without increasing capacity). Although the number of tasks strictly speaking is a work demand, it is safe to assume that it has an impact on the person's subjective workload.

(2) **Time available for each task**. The assumed relation would be that the less time there is available, the larger the workload/capacity ratio will be.

(3) **Time passed since a major disturbance or alarm**. The assumption could be that the more time that passes without a

solution having been found, the larger the workload/capacity ratio will be. The assumption is that even without a definite time limit, the failure to find a solution will lead to an increase in workload and/or a decrease in capacity.

Note that this relation is the opposite of what is suggested by the notion of the Time-Reliability Correlation as put forward by Hall *et al.*, 1982. The workload/capacity ratio and the Time-Reliability Correlation are also incompatible in other ways although they superficially seem to address the same phenomenon.

(4) **Other factors** which might be important are: circadian rhythm, time since start of work (e.g. for flight crews), time zone change, ambient conditions (temperature, noise, etc.), individual conditions, social conditions, etc. Just as for performance shaping factors the problem is not to think of possible factors which may have an effect. The problem is rather to identify those factors which have the largest effects, and to account for the ways in which they may interact.

6.2 Consequences for Design

The consequences of the limited attention capacity, and the limited cognitive capacity in general, are usually either (1) that one must automate everything that the human cannot attend to, (2) that the capacity is artificially increased by introducing operational support systems, or (3) that the demands must be reduced until they fit the capacity. The first option frequently means that automation substitutes for human performance and that the person is reduced to do the tasks that either are too trivial to bother about or too difficult to automate. The second option is theoretically attractive but difficult, because it may lead to the introduction of a cognitive prosthesis (Reason, 1988) which in the end may make situation worse. The third option is, perhaps, the better solution, but it needs to be expressed differently. One should not just reduce the demands but rather **reformulate** them.

6.2.1 Coping with Multiple Tasks

Accept, for a moment, the conservative assumption that a person can only attend to one task at a time - and furthermore that this can only be done for a limited duration. This, however, does not say anything about what the task may be and what level of attention it may demand. If a situation requires that the person must attend to two tasks at the same time, then one solution might be to combine the two tasks into one - rather than to leave one for the person and the other for automation or to split the tasks in some other way (e.g. shift them in time) (e.g. Millot & Mandiau, 1992). This, of course, presupposes that it is possible to combine the tasks. They should not simply be added together, but rather be **integrated**. If there is no physical or functional interdependence between the two tasks, then it may be difficult to transform them into one. In the case of working with a physical process this would only happen if the systems were physically and functionally decoupled - *de facto* and not just by design. It is therefore usually possible, in principle, to define an integrated task - although it may sometimes be difficult to do so at design time for more than a limited number of cases.

If a new, combined task can be defined, then it may be assumed that the task can be handled by the person because it is a **single** task. The attention demanded by two, or more, tasks is not simply the sum of the demands from either task; in addition the person must also spend some effort in sharing his attention between the tasks (monitoring, planning, scheduling, etc.). The proper way of solving the problem of workload/capacity mismatch is not to reduce the demands by eliminating a task, but instead to produce a reduction of demands or workload by transforming the tasks into a simpler one. That may eventually lead to a reduction of unwanted consequences as well. This also points to a principle for the design of support systems, such as expert systems, decision support systems, etc. They should not take over any of the functionality or the tasks unless they can do them completely, thereby relieving the person of the need to follow what is going on - and that would in itself require a complete decoupling. Instead, the support systems should assist the integration of tasks and thereby improve the working conditions for the person.

The reformulation of a task is not necessarily the same as a simplification of the interface. The interface can be made simpler without changing the underlying tasks, e.g. by removing the information that is deemed unnecessary. Such simplifications, however, run the risk of being counter-productive since they may result in the loss of fine grained monitoring and control. If that is the case, then the situation is actually made worse for the person, who is faced with the same task but without adequate support. The reformulation of a task should change the way actions are organised and should therefore clearly manifest itself in terms, for example, of a different outcome on the level of a GMTA.

7. COCOM and the Reliability of Cognition

Despite what I hope is the obvious correctness of the cognitive viewpoint, the fact remains that practically all models and methods are based on the S-O-R paradigm in one form or another - even if they are cleverly disguised as information processing models. All current methods also subscribe to the mechanical view in one way or another, as demonstrated by Swain (1989), who explicitly uses the compliance with the notions of skills, rules, and knowledge as a criterion for rating human reliability analysis methods. The methods that fail to do so fail for the wrong reason, usually resorting to even simpler models and notions, for instance the Time-Reliability Correlation or the notion of Human Cognitive Reliability. Thus, although the idea of the person as a machine is fundamentally wrong, it is so prevalent that it is necessary to examine the consequences of that view a bit further.

7.1 Attention and Performance Reliability

Although attention is a central concept we may assume that it is not attention in itself, or the lack therefore, which leads to unwanted consequences or erroneous actions. In human performance attention is only the guiding or controlling element; the actual substance of human action is something else: the low-level, but highly stable procedures or habitual ways of doing things, such as the heuristics described by Tversky & Kahneman (1975) or Kahneman *et al.* (1982), or phenomena like similarity matching and frequency gambling as described by Reason (1990). (I use the term "substance of human action" for lack of something better.) Similarly, the substance consists of the processes that perform storing and searching in memory, matching of problems to solutions, generalisation from specific to general cases (and, conversely, specialisation from general to particular cases), intention formation, etc.

In all control modes the actual execution of actions is skilled, i.e., automatic and without attention. The basic actions, the skills, may be more or less organised or aggregated. It is the level of aggregation that corresponds to the higher levels of control - just as in the skill-based, rule-based, knowledge-based trichotomy (Rasmussen, 1986). Even highly controlled activities - such as strategic, explorative, or goal-based behaviour - rely on lower level skills actually to achieve anything. The skills themselves are uncontrolled but the degree of control of the aggregation of skills will vary. The notion of a skill (in the sense of the basic level of competence) is furthermore relative to the situation and to the desired level of analysis. For any given level of analysis one can always question the definition of skills, and choose to analyse them further. There are possibly some skills that are relatively universal in terms of the general population. They may, however, be far below the level of skill (the template) that is used in a specific analysis.

From the reliability analysis point of view it is reasonable to assume that completely unattended tasks make do without the higher level heuristics such as representativity or frequency gambling. In a skill, for instance, there is no need for frequency gambling or other heuristics: the actions are executed automatically and within the skill there can be no uncertainty in, for example, what should be recalled or what should be associated as a response. It is in the nature of the skill that every step is pre-determined. Reliability only suffers if the skill is mixed with attention guided events in an uncontrolled way, for instance when we try to "unpack" it.

There are, however, not many true skills, in the sense of completely autonomous responses. In the majority of cases actions are better described as a combination of skills and controlled action, i.e., something which requires attention. An everyday example is breaking eggs, for instance as a preparation for baking a cake. The sequence can be described as follows:

Step	Specific Description	Generic Description
1	Break egg	[Break object-X]
2	Separate albumen (white) in a bowl	[Put something from object-X into container-A]
3	Put yolk in another bowl	[Put something from object-X into container-B]
4	Throw the empty shells into the bin	[Put something from object-X into container-C]

The elementary actions in their generic form (i.e., breaking an object, dropping something into a container) are all skills, but an aggregated action is usually not a skill. The sequence of the actions

Discussion

therefore has to be controlled; if insufficient attention is paid to the control, the sequence will go wrong. Typically, I may end up by reversing the last two steps, hence dropping the yolk in the bin (container C) and the shells in the bowl (container B). This reversal is serious because while the shells can be recovered, the yolk usually cannot. The reversal may occur, for example, when the overall task is considered to be of a trivial nature, and insufficient attention therefore is allocated to it. When a task is considered to be unimportant the person may try to carry it out simultaneously with, and possibly also in competition with, other tasks, e.g. planning the next steps of baking the cake, listening to the radio, talking to someone in the next room, daydreaming, getting ideas for a talk, etc. The result of insufficient attention is insufficient control, hence unwanted consequences. The Reliability of Cognition is in this case low because control is relinquished; the control mode may nominally be tactical, but the overall priorities are inappropriate. The person pays insufficient attention to how the actions are executed and tactical control is therefore not fully achieved.

7.2 Attention, Specificity, and Control

Carrying out a subsumed action can be described as if an intention is set up to achieve a goal or target. A classical example is how to reach a specific destination when driving by car. The task is, however, usually not expressed as achieving a particular goal (e.g. "point-X reached") but rather as the more general procedure (e.g. "keep moving and follow cues to the current goal"). (This representation is more general because "point X" can be seen as a value of the parameter "current goal".) In other words, I will assume that there is a difference between the formulated intention, which refers to conditions or states of the world, and the "translation" of this into an operational description for the cognitive system which refers to elementary and generic functions.

This "translation" can be used to explain the occurrence of some erroneous actions. (In the commonly used classifications it will be the erroneous actions known as slips.) Assume, for instance, that the task is considered trivial; attention will consequently only be partial, and the performance will be controlled by generic functions such as "follow cue X" rather than specific intentions such as "go to point-X". At some time a cue may occur which is frequent (in relative terms), hence strong. If at that moment attention is low, then the appearance of a strong cue will activate the response associated with the cue. This may be different from the response which would lead to the goal: it will be the proper response to the cue in question, but under the circumstances the cue should not have been considered. (The classical example is going to an unusual destination along a familiar road and making a wrong turn.)

The explanation is that the selection or noting of cues is automatic and the result of simple but strong learning mechanisms or principles. If we look for a specific cue, e.g. "look for cue-X and then turn to the right" and attend to that, then we will succeed. But if attention is reduced, then the control changes to "look for cue which indicates a turn", i.e., a less specific condition which probably will lead to the erroneous action. A general explanation for erroneous actions can therefore be found in the relation between attention and specificity. If the task requires that specificity is maintained, attention must be maintained. Conversely, if attention is reduced for one reason or another, specificity is also reduced. The notion of the strength of a cue is therefore relative to the notion of specificity; and it is the relative rather than the absolute strength which is important.

7.2.1 An Example

Consider the following simple example. I get up in the morning and start to make tea. In order to do that I fill the electric kettle with water and switch it on. At the same time I fill the tea pot with hot water to warm it up. Some minutes later, when the water in the kettle is boiling, I prepare to make the tea. This involves three steps:

STEP	Specific Description	Generic Description
1	Empty the pot of hot water	Pour contents of container-A into container-B
2	Put the tea leaves in the pot	Put object-X into container-A
3	Fill the pot with boiling water	Pour contents of container-C into container-A

The possibility of making a mistake is due to the similarity between step 1 and step 3; this similarity is obvious from the generic description. To complicate matters even further, both containers have about the same weight and the same contents (hot water). If sufficient discrimination is not made, it is entirely possible to grab the kettle and pour its contents into the sink. Whenever that happens (and it does occasionally), I normally realise it before it is too late - but not before I have started the action.

The point is not that it is difficult to distinguish between the pot and the kettle. The point is rather that I do not pay sufficient attention to what I am doing (particularly because it is early in the morning). The problem arises because there are two items (pot, kettle) which fit the description of step 1 even in its specific form: "empty container-X for hot water". Both the pot and the kettle are containers which contain hot water and both may therefore match the activating conditions; if attention is low, then whatever happens to come first will be used.

The generality of this is the following. Each of the steps in a plan, seen either as actions or as sub-goals, can be described in terms of a set of pre-conditions, an action, and a set of post-conditions (i.e., the outcome). If the existing conditions match the pre-conditions, the action will be carried out. Therefore, if the pre-conditions can be fulfilled in more ways than one, and if proper attention is not maintained during that "test", then the possibility exists for an erroneous action leading to an unwanted consequence.

7.3 Attention and Control

Most of the things we do, and most of the objects we interact with, have multiple functions and can be used in multiple ways. This is something problem solving psychology excels in demonstrating (e.g. Duncker, 1945). In other words, if we do not actively search and choose but rather passively examine whatever presents itself, or if insufficient attention is paid to an activity, then there will now and again be cases where the wrong object is used. Making tea is a case in point, but using the wrong button or switch, reading the wrong measurement, etc. are other common cases. The general rule seems to be that the less attention there is, the easier it is to find a match (i.e., to fulfil the pre-conditions), hence to initiate the corresponding action.

In some cases things go wrong even if we do pay attention. That happens, for instance, if the objects are sufficiently similar, as when we try to open the wrong car in the parking lot. Here the difference may be the position of the car, but position is not always a salient cue; when I try to find my car I look for an object of a certain shape and colour, rather than for an object in a specific position.

When attention wavers or when attention is insufficient things may go wrong and unwanted consequences may occur. The possible effect of a lack of attention is, however, contingent upon the possibility that something can go wrong. Thus, if I was not in the habit of warming the tea pot as a part of making the tea, then there would be no possibility of a mistake between step 1 and step 3 - because there would not be a step 1. (There are, of course, other possibilities for incorrect performance; I might, for instance, forget to do step 2 or step 3. Expanding on that will, however, take us far into the issue of genotypes of erroneous actions, hence it is not pursued here.) Whether such a possibility exists can be determined by a proper analysis of the situation, e.g. in terms of a Goals-Means Task Analysis.

Although attention clearly is of central importance for the reliability of human action, it has not been included either in the parameters of the COCOM, nor among the Common Performance Modes. The reason for that is simply that attention is a derived rather than a primary measurement or parameter. The endeavour has been to produce a description and a method which as far as possible could be expressed in operational terms or clearly observable indicators. The assessment of the possible effects of a lack of attention should therefore be based partly on a "measure" like the workload/capacity ratio and partly on a description of the possibilities for unwanted consequences as suggested above. Both of these can be qualitatively assessed as part of the Dependent Differentiation Method, hence used to support a more detailed analysis of steps that are believed to be critical. (Provided meaningful HEPs can be found, this outcome can also be expressed in a quantitative form.) The following general rules can be suggested as guiding principles for that more detailed part of a human reliability analysis:

(1) If the subjectively expected seriousness of the unwanted consequences (or the subjective probability of unwanted consequences) is low, then attention will be low - unless there are other conditions, such as generally high arousal, that have the opposite effect.

    (a) If attention is **low** and if similarity is **high**, then it is **very likely** that the pre-conditions will be fulfilled by more than one object. If the action is initiated it is therefore **very likely** that unwanted consequences will occur.

    (b) If attention is **low** and if similarity is **low**, then it is **possible** that the pre-conditions will be fulfilled by more than one object. If the action is initiated it is therefore **likely** that unwanted consequences will occur.

(2) If the subjectively expected seriousness of the unwanted consequences (or the subjective probability of unwanted consequences) is high, then attention will be high.

    (a) If attention is **high** and if similarity is **low**, then it is **very unlikely** that the pre-conditions will be fulfilled by more than one object. If the action is initiated it is therefore **very unlikely** that unwanted consequences will occur.

    (b) If attention is **high** and if similarity is **high**, then it is **possible** that the pre-conditions will be fulfilled by more than one object. If the action is initiated it is therefore **likely** that unwanted consequences will occur.

Guiding principles of this nature may form the basis for a detailed situation analysis and assessment. They are, not surprisingly, consistent with the contents of the COCOM and can be seen as an enhancement of the descriptions of the tactical and opportunistic levels of control. In particular, the rules may be used in the Dependent Differentiation Method as part of the sensitivity analysis.

8. The Last Word

The need to include aspects of cognition in human reliability analysis and assessment is often argued as follows: (1) every now and then the person may do something wrong; (2) the wrong action may be preceded by a decision; (3) in order to understand the decision it is necessary to introduce a cognitive analysis. This line of argument is typical of the decomposition approach and encourages a piecemeal view. Yet cognitive aspects should not just to be added to a normal analysis in order to account for something that is missing. Rather, the whole approach must be based on the premise that - in Man-Machine Interaction and Man-Machine Systems - the very foundation for design and analysis must be an understanding of human cognition.

Our ways of doing science are dominated by the Western way of thinking, and has too often left us looking for traces of processes and specific factors that may or may not have an influence. One consequence of the contextual approach might be the realisation that the Western paradigm is unsuitable for a view which looks at the whole rather than the details. The notion of levels or control modes does, of course, in itself imply causality; this feature may be defended on the grounds that the performance we are concerned about is predicated by a system that is causally designed. That causal design forces the person to think causally about the system in order to control it - or at least we know that this will be an effective way of doing it (Lind & Larsen, 1992). But it is by no means certain that this solution is universal. In fact, for other, natural systems, it may be more appropriate to assess them on a different level, hence look to different methods.

As noted in Section 1.1, the problem with the use of cognitive models has been that they create too many opportunities for failure, i.e., that there are too many combinations to consider. The reason for this is clearly that the primary approach rests on the principle of decomposition; this produces a set of basic elements or causes which are then combined willy-nilly. The decomposition is partly done for the specific procedure or task that is being considered, and partly done *a priori* for the check-lists of causes that are used, for example. It should be obvious that a very detailed check-list of causes can produce a very large number of explanations, unless the causes are irreducible ("root" causes). The selection of possible causes can only be delimited if a context is found or established; the analysis must consequently be performed in a top-down (context driven) fashion rather than in a bottom-up (cause/element driven) fashion.

The so-called "first generation" human reliability analysis methods work by listing potential error modes and identifying the recovery/detection possibilities for each potential error mode. The alternative offered by the COCOM is that incorrect or erroneous actions are seen as part of specific control modes, hence as being of specific types. The context is provided by a very small set of control modes, rather than by a conglomerate of specific actions which - at least in terms of the analysis - are isolated from the modes. This means that there is no need to conjecture errors from the bottom up by "blind" combinations. The COCOM, together with the Dependent Differentiation Method, provides a top-down view which supplies a natural filtering, and puts the possible incorrect actions into a context. To the extent that decomposition is needed it should serve the purpose of improving the understanding of the whole - the composition - rather than of finding and isolating the parts.

Discussion 304

Understanding the whole can only be achieved when we know about the parts as well as the way they work together. **But we cannot understand the parts by looking at the parts by themselves**.

An important question is whether an approach like the Dependent Differentiation Method is better than the traditional ones. It is certainly different, but difference for difference's sake should be avoided. Above I have discussed the relative merits from a specific view - expressed by the criteria of completeness, consistency, and decidability. The most important conclusion is probably that the Dependent Differentiation Method and COCOM both emphasise the qualitative analysis. Adhering to the notion that the purpose of human reliability analysis and probabilistic safety assessment is to improve system design and prevent unwanted consequences, the quantification may not always be necessary. Quantification **without** an underlying qualitative analysis is useless. If the qualitative analysis is **inadequate**, the ensuing quantification becomes a senseless exercise. It adds numbers, which themselves are without meaning, to structures which are ambiguous. At the very least one must demand that the qualitative analysis is clear and complete/consistent.

The approach described in this book does not solve the problem of quantification, which is mainly due to the general lack of sufficient and appropriate empirical data. Even though the role of the human error probability is diminished, the need for data and specific details cannot be abolished. Yet it is an advantage to be more clear about what the data are needed for. Knowing what to look for makes it is so much easier to find! I hope that this book has contributed to that clarity.

Appendix: Introduction to the System Response Generator

The System Response Generator (SRG) project is a three year (1991 - 1994) research and development project sponsored by the Commission of the European Communities as part of the programme on **Science and Technology for Environmental Protection** (STEP), DG XII, Directorate-General for Science, Research and Development. The project is carried out by a consortium consisting of Computer Resources International A/S (Denmark), Aerospatiale Protection Systemes (France), and the Institute for Safety Engineering and Informatics, CEC Joint Research Centre at Ispra (Italy).

1. The Practice of Safety and Reliability Analyses

Technological systems today are highly dependent on the interaction between humans and machines. The systematic study of how this interaction can be analysed, described and designed is therefore of primary importance. The main objectives of Probabilistic Safety Analyses (PSA) are to reduce the probability that an untoward event occurs and to minimise the consequences of uncontrolled developments of accident conditions. Practical experience has repeatedly demonstrated that improper Man-Machine Interaction (MMI) may be the cause of unwanted consequences. Thus three problems face practitioners in the field of MMI for industrial processes:

(1) how to design the MMI to meet the functional requirements, particularly with regard to safety and efficiency,

(2) how to account for the impact of human action on the process (planned and unplanned interventions), and

(3) how to evaluate and/or validate the efficacy of the proposed design *vis-a-vis* the projected human performance.

It is consequently necessary to analyse and understand how the MMI takes place and how it may affect the development of the system.

1.1 Point-to-Point Analyses

Current safety and risk analyses use a number of different techniques. Common to them all are the concepts of a cause (a source) and a consequence (a target). The purpose of safety and risk analyses is to find the possible links between causes and consequences, in order to devise ways in which the activation of these links can be avoided. One can consider four types of situations:

(1) **One cause, one consequence.** Here the specific relations between a single cause and a single consequence are sought. The analysis can either start from the consequence or start from the cause. Examples are root cause analysis, which tries to identify the root cause of an event, and consequence propagation, which tries to find the specific consequence of a cause.

(2) **One cause, many consequences.** This approach is the investigation of the possible outcomes of a single cause. Typical techniques are Failure Mode and Effects Analysis (FMEA/FMECA), or Sneak Analysis (SA). One-to-many approaches are mainly analytical, starting with a single event and trying to compute the possible outcomes.

(3) **Many causes, one consequence.** This type of analysis looks for the possible causes for a specific consequence. Examples are Fault Tree Analysis (FTA), and Common Mode Failure analysis. The rationale is that once an (actual or hypothetical) event has occurred, there is a well-defined consequence, and the analysis can trace backwards to the causes. A more general type of analysis is Functional Analysis

(FA) which is a goal-driven approach to identify the basic functional elements of the system

(4) **Many causes, many consequences.** This is the investigation of the relations between multiple causes and multiple consequences. Examples are Hazard and Operability Analysis (HAZOP), Cause-Consequence Analysis (CCA), and Action Error Mode Analysis (AEMA). All of these start by a single event (the seed) and try to identify the possible causes as well as the possible consequences.

The common feature of these **point-to-point** analyses is that they are based on a limited number of pre-defined causes or consequences. This enables the analyses to focus on relevant subsets of the domain. The selection is necessary because it is computationally impossible to consider all possible events even for very small systems. Yet it is also a limitation because the quality of the outcome of the analysis is restricted by the completeness or scope of the selection. The set of initiating events and conditions must be both comprehensive and representative in order for the results to be practically applicable, and these conditions may be hard to establish.

1.2 Static and Dynamic Analyses

Point-to-point analyses are **static** analyses and therefore limited by the simplifications invoked by the **intermediate representation** and the use of a **fixed set** of causes and consequences.

Since system descriptions exist in a variety of formats, the first step of an analysis is usually to develop an intermediate (often diagrammatic) representation. But a static analysis cannot be more precise or complete than the intermediate representation. The intermediate representation is basically a snapshot or a frozen state description (an **operating mode**) which represents a single state of the system. Tracing the propagation of events through a snapshot is limited by the conditions contained in the snapshot. If these are insufficient, another snapshot must be made - which suffers from the same limitations. Any event or consequence that introduces a change in the conditions captured by the intermediate representation forces the analysis to be started anew - unless it can be guaranteed that the change only has a local effect. Although that assumption is sometimes made, it is usually not warranted.

Using a fixed set of causes and consequences creates the problem of enumeration. This can be overcome by using a formal description of the system, because this defines **conditions** rather than **examples**. If the conditions can be realised in some sense, for instance by using a systematic interpretation method, they correspond to different instantiations, hence to a number or range of different examples. The analysis therefore only requires the description of the principles (conditions) for specific cases rather the cases themselves. A formal representation coupled with a suitable interpretation method allows the introduction of dynamic aspects, ranging from considering multiple states and state transitions to full scope simulations. A dynamic analysis method can treat sets of potential causes without spelling them out. It can encompass configurations which are missed by the static analyses and can capture the effects of changes in parameters (e.g. from side-effects and post-conditions), even including the effect of the different timing conditions (delays, etc.). A dynamic analysis will, of course, be limited by the granularity and correctness of the condition descriptions, which thus define its validity. But it will still be many times richer than a static analysis.

In order to be manageable a dynamic analysis must start from a limited set of initial conditions - either causes or consequences. In the case of the System Response Generator (SRG) the starting point for the analysis is a scenario which describes the initial conditions of the system as well as the expected failure conditions of specific components - but **not** the set of consequences! The dynamic method does not try to find a many-to-many mapping between causes and consequences, but rather **generates** the set of possible consequences from the given set of causes. The analysis uses a formal description of the system dynamics, e.g. in the guise of simulators. This makes it possible to account for how the system evolves with time, how MMI may have an effect, and how cross-couplings may appear.

Because the analyst can state the general principles for how outcomes occur, rather than having to invent or imagine specific events, the dynamic analysis method overcomes the limitations in enumerating a large number of specific states and in keeping track of the combinations.

2. The System Response Generator

The overall purpose of the SRG project is to develop and implement a software tool which can be used to analyse the influence of human decision making and action on the way in which incidents in complex systems evolve. The SRG will serve to:

(1) **identify potential problem areas**, i.e., the aspects of the task and the MMI where problems are likely to occur,

(2) **evaluate the effects of specific modifications** to the system (e.g. of procedures, information presentation, control options, etc.), and

(3) **provide quantitative data** as input to a more formal Probabilistic Safety Analysis or Human Reliability Analysis.

The SRG is primarily intended to be used during system design; it provides a tool for systematically examining the possible ways in which a scenario can develop, depending on how the operator interacts with the system. The SRG can also be applied for the *post hoc* analysis of accidents which have occurred; in this case the SRG will be used to reproduce an accident in order to elicit additional information from experts and explore possible alternative sequences of development (near-misses). The SRG is designed to specify **how** an event may develop and **what** the consequences may be rather than **when** or **how often** it may occur. The SRG is an analytical instrument for probing deep rather than for looking at many different cases, and is thus more qualitative than quantitative.

The concern of the SRG is with the MMI, specifically the way in which misunderstandings and incorrectly executed human actions can change how an accident evolves. The fundamental innovation of the SRG is the consistent method for dynamically generating responses by combining simulations of both the technical process and the human operator, although the models themselves are not part of the SRG (cf. Figure A-1). Use of the SRG will be facilitated by an interface which offers the user the ability to define scenarios, "tune" operator and process models, and produce a comprehensive record of the results.

The advantage of the SRG is that a specific configuration will recreate the main characteristics of the application under study. This flexibility makes it possible to explore a wide range of topics as well as a number of "local" versions of a topic, without the need of creating each time a new and costly *ad hoc* model. The SRG will be instrumental in clarifying precisely how the configuration of the physical system and the interface may sometimes transform human erroneous actions into

Appendix

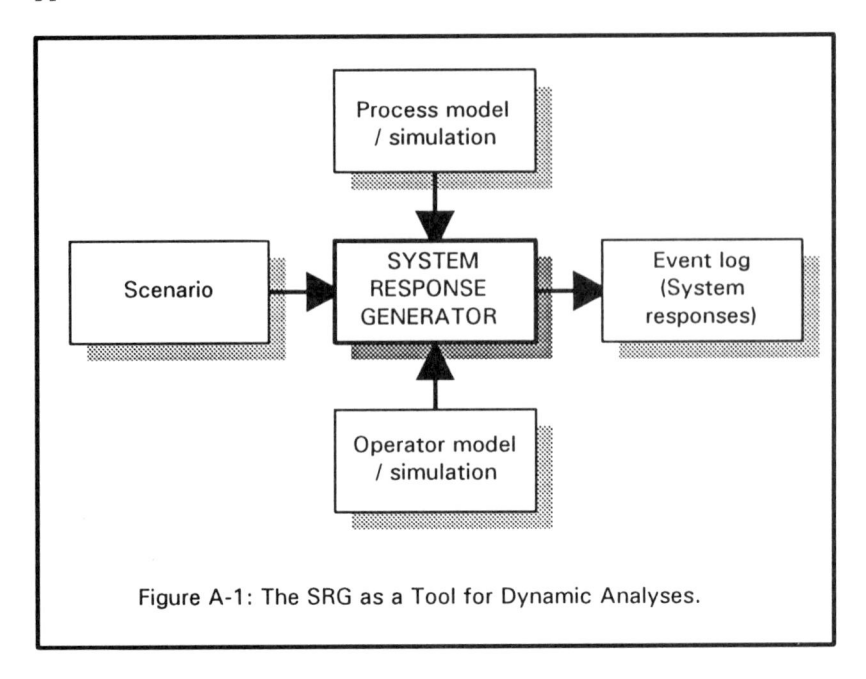

```
          ┌─────────────────┐
          │  Process model  │
          │   / simulation  │
          └────────┬────────┘
                   │
                   ▼
┌──────────┐  ┌──────────┐  ┌──────────┐
│          │  │  SYSTEM  │  │ Event log│
│ Scenario │─▶│ RESPONSE │─▶│ (System  │
│          │  │ GENERATOR│  │responses)│
└──────────┘  └─────▲────┘  └──────────┘
                    │
          ┌─────────┴───────┐
          │  Operator model │
          │   / simulation  │
          └─────────────────┘
```

Figure A-1: The SRG as a Tool for Dynamic Analyses.

accidents. On the basis of this knowledge it will be possible to devise (and even computationally to test) practical counter-measures aimed at improving human reliability under operational conditions.

2.1 The Generation of System Responses

The main functional principle of the SRG is that a basic event cycle is repeated until the terminating conditions for the scenario have been met. The SRG does not work with low probability events; thus the event driver is not based on a stochastic or probabilistic principle, but rather examines the consequences (responses) of specific events and conditions, which then later can be used as input for more extended calculations.

The scenario consists of a set of pre-defined event frames. Each event frame is described in terms of pre-conditions and post-conditions, and the selection of the "next" event frame is determined by whether there is a frame where the pre-conditions match the current conditions (i.e., the conditions existing at the end of the previous cycle). Once an event frame has been completed it is disabled, simply because the pre-conditions no longer match. The description of the conditions takes into account time, i.e., they reflect the dynamics of the process (contrary to standard production rule systems).

Each event frame contains a number of messages that are sent to the SRG modules when the event frame has been activated. The messages of an event frame may refer to control input for the process or an aggregated set of elementary actions, e.g. as a plan. The activation of the plan may trigger a number of cycles of the event frame, until the conditions for another event frame are met - or until some general stop criterion is reached (e.g. time). This means that an event frame that appears one time in the scenario description can appear several times in the Event Log.

The functional principle of the SRG is shown in Figure A-2.

2.2 SRG Modules

There are four main modules involved in determining how the accident evolves dynamically, and which thus constitute the basic event cycle as described above.

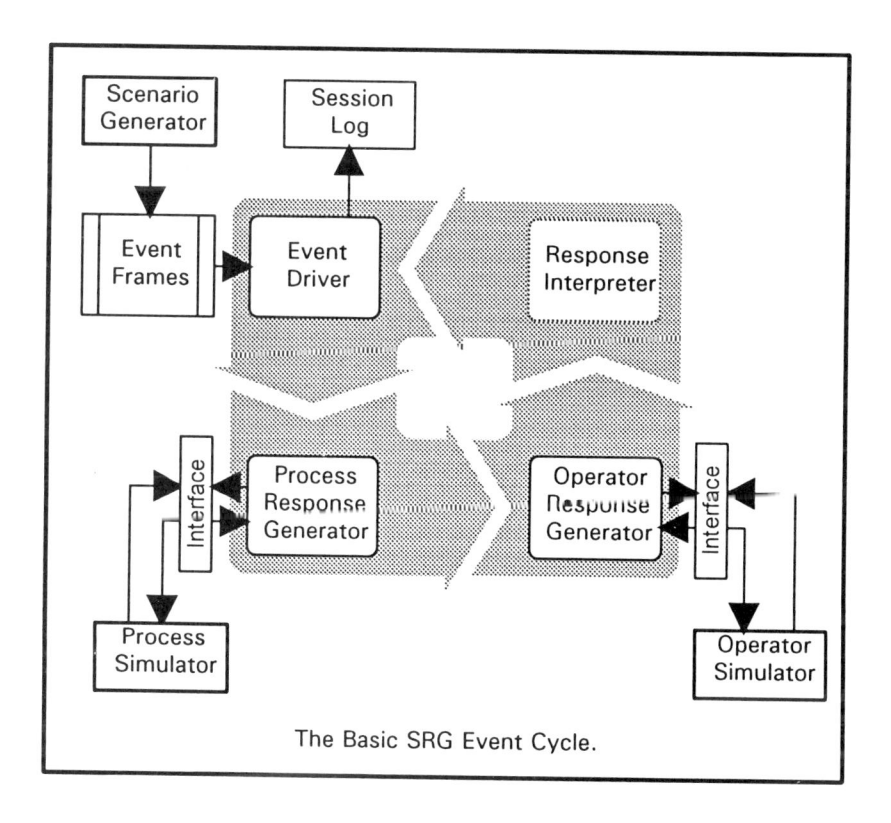

The Basic SRG Event Cycle.

Appendix

(1) The **Event Driver** controls the progress of the SRG from break-point to break-point. The event driver starts the chosen scenario and triggers each step according to the scenario specifications and the current input from the other modules, in particular the control input from the **Response Interpreter**. The inputs are scenario conditions, break-points, and analysed response (control input). The outputs are the next break-point specification and the event frame.

(2) The **Process Response Generator** provides the response of the process to a given event frame and control input, using the resources of an external process simulator which is interrogated through the interface. The inputs are process expertise and event frame conditions. The output is event frame information (process response).

(3) The **Operator Response Generator** assesses the output from the process response generator by providing the outcome of the operator's perception/discrimination/reaction/response, again using the resources of an external operator model or simulation. The inputs are event frame information and operator expertise. The output is the response.

(4) The **Response Interpreter** examines the output from the operator response generator and determines the implications of the operator's response for the event driver. The input is the response. The output is the event control information.

The basic event cycle is repeated until the desired output from the SRG has been produced or until other conditions for terminating the SRG session have been met. The session is defined by means of the **Scenario Generator** which provides the necessary details for a given scenario. The output from the scenario generator is information to control the event driver (the empty event frames) as well as initial parameter values to prime the process and operator response generators. Finally, a graphical interface provides the basis for using the SRG as an analysis tool.

2.3 Operator and Process Modelling

The purpose of operator modelling in the SRG is to emulate the response of an operator under specific conditions. The range of responses (outputs) from the model must therefore go from correct response to incorrect or erroneous responses. To make it manageable for the SRG it is assumed that performance can be modelled by three different levels:

(1) On **Level-0** the performance is flawless and correct. This corresponds to an operator with correct knowledge, infinite capacity, and infinite attention. The input maps directly onto the output and the nominal or prescribed relations between input and output variables are reproduced in any suitable form. Level-0 serves the purpose of testing the consistency and correctness of, for instance, procedure design and interface support.

(2) On **Level-1** the performance is expressed by **quantitative modelling** in terms of continuous functions, e.g. a probability function. This corresponds to an operator who is influenced by the working conditions, and the independent variables will all have to do with external conditions, such as interface design, structure of procedures, work demands (leading to work load), noise, etc. Level-1 includes a large variety of models, such as probabilistic models, supervisory control models, filter models, differential equations, Markov models, etc. Their common feature is that they do not make any pretence of incorporating psychological (cognitive) concepts or functions, but achieve their purpose by means of continuous, mathematical functions (e.g. signal processing models).

(3) On **Level-2** the performance can be expressed by **symbolic modelling** in terms of rules or heuristics. This corresponds to an operator who is influenced by the perceived conditions or internal states (cognitive states), e.g. assumptions, intentions, biases, preferences, habits, etc. The model is expressed in terms of symbol manipulation, typically as an information processing model.

Although the SRG is provided with a default operator model the system is designed to work with an arbitrary operator model. In order to accomplish that, the interface between the operator model and the SRG has been clearly defined and the operator model functions as an external module, in the same way as the process model does.

Process modelling is achieved by means of available process simulators for the application in question. The SRG provides a well-defined interface protocol for establishing the interaction between the SRG and the process simulator, as well as a method to define the

Appendix

additional process knowledge which is needed for a scenario to run. The development of a specific process simulator is outside the scope of the SRG.

The main test case chosen for the development of the SRG is the flying of a simulated commercial aircraft. The four basic scenarios are: Climbing To Cruise Altitude, Normal Descent, Intermediate Approach, and Emergency Descent. The scenarios have been evaluated and it has been found that they are sufficiently complex to serve as a basis for developing and testing the SRG. For each of these scenarios a number of variations can be defined (e.g. rapid descent and programmed descent) as well as numerous possibilities of combinations of human error and equipment malfunction. As part of the SRG project a detailed pilot model will be developed, using the general modelling principles described above.

References

Allen, J. (1983). Maintaining knowledge about temporal intervals. *Communications of the ACM, 26*, 832-843.

Amalberti, R. & Deblon, F. (1991). Cognitive modeling of figther aircraft process control: A step towards an intelligent onboard assistance system. *International Journal of Man-Machine Studies*, in press.

Amendola, A. (1985). *Results of the reliability benchmark exercise and the future CEC-JRC programme.* Proceedings of ANS/ENS International Topical Meeting on Probabilistic Safety Methods and Applications, San Francisco, February 24th - March 1st.

Anderson, J. R. (1980). *Cognitive psychology and its implications.* San Francisco: Freeman.

Annett, J., Duncan, K. D., Stammers, R. B. & Gray, M. J. (1971). *Task analysis.* Training Information Paper No. 6. H. M. S. O.

Apostolakis, G. E. (1990). Editorial. *Journal of Reliability Engineering and System Safety, 29*, 281.

Apostolakis, G. E., Kafka, P. & Mancini, G. (Eds.), (1988). *Accident Sequence Modeling.* London: Elsevier Applied Science.

Arbib, M. A. (1964). *Brains, machines and mathematics.* New York: McGraw-Hill.

Ashby, W. R. (1956). *An introduction to cybernetics.* London: Methuen & Co.

Attneave, F. (1959). *Applications of information theory to psychology: A summary of basic concepts, methods, and results.* New York: Holt, Rinehart & Winston.

Ayel, M. & Laurent, J.-P. (Eds.). (1991). *Validation, verification and test of knowledge-based systems.* Chichester, UK: Wiley.

Bagnara, S., Di Martino, C., Lisanti, B., Mancini, G. & Rizzo, A. (1989). *A human error taxonomy based on cognitive engineering and on social and occupational psychology* (EUR 12624 EN). Ispra, Italy: Commission of the European Communities.

Bainbridge, L. (1991). Mental models in cognitive skill: The example of industrial process operation. In A. Rutherford & Y. Rogers (Eds.), *Models in the mind.* London: Academic Press.

References

Beare, A. N., Gaddy, C. D., Parry, G. W. & Singh, A. (1991). An approach for assessment of the reliability of cognitive response for nuclear power plant operating crews. In G. Apostolakis (Ed.), *Probabilistic Safety Assessment and Management*. New York: Elsevier.

Bellamy, L. J. & Geyer, T. A. W. (1988). Addressing human factors issues in the safe design and operation of computer controlled process systems. In B. A. Sayers (Ed.), *Human factors and decision making: Their influence on safety and reliability*. Amsterdam: Elsevier Applied Science.

Boden, M. A. (1988). *Computer models of mind: Computational approaches in theoretical psychology*. Cambridge: Cambridge University Press.

Booth, P. (1989). *An introduction to human-computer interaction*. London: Lawrence Erlbaum Associates.

Bruner, J. S., Goodnow, J. J. & Austin, G. A. (1956). *A study of thinking*. New York: John Wiley.

Cacciabue, P. C. (1991). *Understanding and modelling man-machine interaction*. Principal Division Lecture, Proceedings of 11th SMiRT Conference, Tokyo, August 1991.

Cacciabue, P. C., Cojazzi, G., Hollnagel, E. & Mancini, S. (1992). Analysis and modelling of pilot-airplane interaction by an integrated simulation approach. *5th IFAC/IFIP/IFORS/IEA Symposium on Analysis, Design and Evaluation of Man-Machine Systems*, The Hague, Netherlands, June 9-11, 1992.

Cacciabue, P. C. & Hollnagel, E. (1993). *Simulation of cognition: Applications*. (In preparation.)

Christensen, J. M., Howard, J. M. & Stevens, B. S. (1981). Field experience in maintenance. In J. Rasmussen & W. B. Rouse (Eds.), *Human detection and diagnosis of system failures*. New York: Plenum Press.

Dawkins, R. (1988). *The blind watchmaker*. Harmondsworth: Penguin Books.

De Keyser, V. (1986). Technical assistance to the operator in case of incident: Some lines of thought. In E. Hollnagel, G. Mancini & D. D. Woods (Eds.), *Intelligent decision support in process environments*, pp. 229-253. Heidelberg: Springer-Verlag.

De Mey, M. (1982). *The cognitive paradigm*. Dordrecht: Reidel.

Decortis, F. & Cacciabue, P. C. (1988). Temporal dimensions in cognitive models. *4th IEEE Conference on Human Factors and Power Plants*, June 5-9, Monterey, CA.

Decortis, F. & De Keyser, V. (1988). Time: The Cinderella of man-machine interaction. *IFAC/IFIP/IEA/IFORS Conference on Man-Machine Systems*, Oulu, Finland, 14 - 16 June 1988.

Dougherty, E. (1990). Human reliability analysis - Where shouldst thou turn? *Journal of Reliability Engineering and System Safety, 29*(3).

Dubois, D. & Prade, H. (1989). Handling uncertainty in expert systems. Pitfalls, difficulties, remedies. In E. Hollnagel (Ed.), *The reliability of expert systems*. Chichester: Ellis Horwood Ltd.

Dörner, D. (1989). Managing a simple ecological system. *Proceedings of Second European Meeting on Cognitive Science Approaches To Process Control*. Siena, Italy, October 24-27.

Duncker, K. (1945). On problem solving. *Psychological Monographs, 58*(5), (Whole No. 270).

Embrey, D. E., Humphreys, P., Rosa, E. A., Kirwan, B. & Rea, K. (1984). *SLIM-MAUD. An approach to assessing human error probabilities using structured expert judgment* (NUREG/CR-3518). Washington, DC: USNRC.

English, H. B. & English, A. C. (1958). *A comprehensive dictionary of psychological and psychoanalytical terms*. London: Longmans.

Fischhoff, B. (1975). Hindsight ≠ foresight: The effect of outcome knowledge on judgment under uncertainty. *Journal of Experimental Psychology: Human Perception and Performance, 1*, 288-299.

Green, A. E. (1988). Human factors in industrial risk assessment - some early work. In L. P. Goodstein, H. B. Andersen & S. E. Olsen (Eds.), *Task, errors and mental models*. London: Taylor & Francis.

The Guardian, November 14th, 1990. "Boeing blames pilots for disasters."

Hagen, E. W. (Ed.), (1976). Human reliability analysis. *Nuclear Safety, 17*.

Hall, R. E., Fragola, J. R. & Wreathall, J. (1982). *Post-event human decision errors: Operator action trees/time reliability correlation* (NUREG/CR-3010). Washington, DC: USNRC.

Hamer, M. (1990). Lessons from a disastrous past. *New Scientist*, 22/29 December, 72-74.

Hannaman, G. W. & Spurgin, A. J. (1984). *Systematic human action reliability procedure (SHARP)*, EPRI NP-3583. Palo Alto, CA: Electric Power Research Institute.

Hannaman, G. W. & Worledge, D. H. (1988). Some developments in human reliability analysis approaches and tools. In G. E. Apostolakis, P. Kafka & G. Mancini (Eds.), *Accident sequence modeling: Human actions, system response, intelligent decision support*. London: Elsevier Applied Science, 1988.

Haugeland, J. (1985). *Artificial intelligence: The very idea*. Cambridge, MA: MIT Press.

Henley, J. & Kumamoto, H. (1981). *Reliability engineering and risk assessment*. New York: Prentice-Hall.

Hirschberg, S. (Ed.). (1990). *Dependencies, human interactions and uncertainties in probabilistic safety assessment* (RAS 470). Västerås, Sweden: ABB Atom AB.

Hollnagel, E. (1984). Inductive and deductive approaches to modelling of human decision making. *Psyke & Logos, 5(2)*, 288-301.

Hollnagel, E. (1988). Information and reasoning in intelligent decision support systems. In E. Hollnagel, G. Mancini & D. D. Woods (Eds.), *Cognitive engineering in complex dynamic worlds*. London: Academic Press.

Hollnagel, E. (Ed.). (1989). *The reliability of expert systems*. Chichester, UK: Ellis Horwood Ltd.

Hollnagel, E. (1990). *The design of error tolerant interfaces*. First International Symposium on Ground Data Systems For Spacecraft Control, Darmstadt, FRG, June 26-29. (ESA Special Publication SP-308.)

Hollnagel, E. (1991a). The phenotype of erroneous actions: Implications for HCI design. In G. Weir & J. Alty (Eds.), *Human-computer interaction and complex systems*, 73-12. London: Academic Press.

Hollnagel, E. (1991b). Cognitive Ergonomics And The Reliability Of Cognition. *Le Travail humain, 54(4)*, 305-321.

Hollnagel, E. (1992a). Modelling Of Cognition: Procedural Prototypes And Contextual Control. *Le Travail humain*, (submitted for special issue).

Hollnagel, E. (1992b). Coping, Coupling And Control: The Modelling Of Muddling Through. *Proceedings of 2nd Interdisciplinary Workshop on Mental Models*, Robinson College, Cambridge, UK. March 23-25, 1992.

Hollnagel, E. (1992c). The Art Of Efficient Man-Machine Interaction: Improving The Coupling Between Man And Machine. *3rd CADES Workshop*, St. Prix, France, April 15-16.

Hollnagel, E. (1992d). The design of fault tolerant systems: Prevention is better than cure. *Journal of Reliability Engineering and System Safety, 36*, 231-237.

Hollnagel, E. (1992e). *Requirements for dynamic modelling of man-machine interaction.* Post-ANS Seminar, November 4-5 1992, Kyoto, Japan.

Hollnagel, E. (1993). The phenotype of erroneous actions. *International Journal of Man-Machine Studies*, (in print).

Hollnagel, E. & Cacciabue, P. C. (1991). Cognitive Modelling in System Simulation. *Proceedings of Third European Conference on Cognitive Science Approaches to Process Control*, Cardiff, UK, September 2-6, 1991.

Hollnagel, E. & Cacciabue, P. C. (1993). *The limits of automation in ATC and aviation.* Report from a workshop on The limits of automation in ATC and aviation, Siena, Italy, 25-27 November 1992.

Hollnagel, E., Cacciabue, P. C. & Rouhet, J. -C. (1992). The use of an integrated system simulation for risk analysis and reliability assessment. *7th International Symposium on Loss Prevention*, Taormina, Italy, 4-8 May, 1992.

Hollnagel, E., Pedersen, O. M., & Rasmussen, J. (1981). *Notes on human performance analysis* (Risø-M-2285). Roskilde, Denmark: Risø National Laboratory, Electronics Department.

Hollnagel, E. & Woods, D. D. (1983). Cognitive systems engineering: New wine in new bottles. *International Journal of Man-Machine Studies, 18*, 583-600.

INPO (1984). *Analysis of root causes in 1983 significant event reports* (INPO 84-027). Atlanta, GA: Institute of Nuclear Power Operations.

INPO (1985). *Analysis of root causes in 1983 and 1984 significant event reports* (INPO 85-027). Atlanta, GA: Institute of Nuclear Power Operations.

Johannsen, G. (1990). Fahrzeugführung. In C. G. Hoyos & B. Zimolong (Eds.), *Ingenieurpsychologie*. Göttingen, FRG: Verlag für Psychologie.

Kahneman, D. (1973). *Attention and effort.* Englewood Cliffs, NJ: Prentice-Hall.

Kahneman, D., Slovic, P. & Tversky, A. (Eds.), (1982). *Judgment under uncertainty: Heuristics and biases*. New York: Cambridge University Press.

Kantowitz, B. H. & Fujita, Y. (1990). Cognitive theory, identifiability and human reliability analysis (HRA). *Journal of Reliability Engineering and System Safety, 29*, 317-328.

Kautz, H. & Allen, J. (1986). Generalized Plan Recognition. *Proceedings of AAAI-86*, Philadelphia.

Kurke, M. I. (1961). Operational sequence diagrams in system design. *Human Factors, 3*, 66-73.

Leplat, J. (1989). Error analysis, instrument and object of task analysis. *Ergonomics, 32*(7), 813-822.

Leplat, J. (1991). Organisation of activity in collective tasks. In J. Rasmussen, B. Brehmer & J. Leplat (Eds)., *Distributed decision making: Cognitive models for coorperative work*. London: Wiley.

LeVan, W. L. (1960). *Analysis of human error problem in the field*. Bell Aerosystems Company, Report No. 7-60-932004.

Leveson, N. G. (1991). Safety assessment and management applied to software. In G. Apostolakis (Ed.), *Probabilistic safety assessment and management* (Vol. 1). New York: Elsevier.

Lind, M. (1991). On the modelling of diagnostic tasks. *3rd European Conference on Cognitive Science Approaches to Process Control*, Cardiff, UK, September 2-6, 1991.

Lind, M. & Larsen, M. N. (1992). Planning and the intentionality of dynamic environments. *CADES Meeting on "Cognitive Simulation"*, JRC Ispra, Italy, April 9th, 1992.

Lindblom, C. E. (1959). The science of "muddling through". *Public Administration Quaterly, 19*, 79-88.

Mach, E. (1905). *Knowledge and error*. Dordrecht, Netherlands: D. Reidel. (English Translation, 1976).

Mancini, G. & Lederman, L. (Eds.). (1989). *Models and data requirements for human reliability analysis*, IAEA-TECDOC-499. Vienna: International Atomic Energy Agency.

Mandler, G. (1975). *Mind and emotion*. New York: Wiley.

Maruyama, M. (1963). The second cybernetics: Deviation-amplifying mutual processes. *American Scientist, 55*, 164-179.

Miller, G. A. (1956). The magical number seven, plus or minus two: Some limits on our capacity for processing information. *Psychological Review, 63*(2), 81-96.

Miller, G. A., Galanter, E. & Pribram, K. H. (1960). *Plans and the structure of behavior.* New York: Holt, Rinehart & Winston.

Miller, J. G. (1960). Information input overload and psychopathology. *American Journal of Psychiatry, 116*, 695-704.

Millot, P. & Mandiau, R. (1992). Men-machines cooperative organization: Formal and pragmatic implementation methodologies. *CADES Meeting on "Cognitive Simulation"*, JRC Ispra, Italy, April 9th, 1992.

Moray, N. (1990). Dougherty's dilemma and the one-sidedness of human reliability analysis (HRA). *Reliability Engineering and System Safety, 29*, 337-344.

Moray, N., Dessouky, M. I., Kijowski, B. A. & Adapathya, R. (1991). Strategic behavior, workload, and performance in task scheduling. *Human Factors, 33(6)*, 607-629.

Moray, N. & Lee, J. (1990). Operator intervention in automated systems. *Proceedings of Ninth European Annual Conference on Human Decision Making and Manual Control.* Ispra, Italy: CEC JRC.

Moray, N. & Rotenberg, I. (1989). Fault management in process control: Eye movement and action. *Ergonomics, 32(11)*, 1319-1342.

Moray, N. & Sabadosh, N. (1992). *Stresswl: An experimental environment for the study of strategic behavior in cockpit management.* University of Illinois at Urbana-Champaign: Engineering Psychology Research Laboratory.

Morick, H. (1971). Cartesian privilege and the strictly mental. *Philosophy & Phenomenological Research, 31(4)*, 546-551.

Muir, B. M. (1988). Trust between humans and machines, and the design of decision aids. In E. Hollnagel, G. Mancini & D. D. Woods (Eds.), *Cognitive engineering in complex dynamic worlds.* London: Academic Press.

Nagel, T. (1974). What is it like to be a bat? *Philosophical Review, 83*, 435-450.

Neisser, U. (1976). *Cognition and reality.* San Francisco: W. H. Freeman.

Neisser, U. (1982). *Memory observed.* San Francisco: W. H. Freeman.

Newell, A. (1980). Physical symbol systems. *Cognitive Science, 4*, 135-183.

Newell, A. (1990). *Unified theory of cognition.* Cambridge, MA: Harvard University Press.

Newell, A. & Simon, H. A. (1963). GPS, A program that simulates human thought. In E. A. Feigenbaum & J. Feldman (Eds.), *Computers and thought*. New York: McGraw-Hill.

Newell, A. & Simon, H. A. (1972). *Human problem solving*. Englewood Cliffs, NJ: Prentice-Hall.

Newell, A. & Simon, H. A. (1976). Computer science as empirical inquiry: Symbols and search. *Communications of the ACM, 19* (March).

Nisbett, R. E. & Wilson, T. D. (1977). Telling more than we can know: Verbal reports on mental processes. *Psychological Review, 74*, 231-259.

Norman, D. A. (1976). *Memory and attention* (2nd ed.). New York: Wiley.

Norman, D. A. (1986). Cognitive engineering. In D. A. Norman & S. W. Draper (Eds.), *User centered system design: New perspectives on human computer interaction*. Hillsdale, NJ: Lawrence Erlbaum.

NUREG-1154 (1985). *Loss of main and auxilliary feedwater event at the Davis-Besse plant on June 9, 1985*. Washington, DC: Nuclear Regulatory Commission.

Park, K. S. (1987). *Human reliability*. Amsterdam: Elsevier.

Peddie, H., Filz, A. Y., Arnott, J. L. & Newell, A. F. (1990). *Extraordinary computer human operation (ECHO)*. Presented at the 2nd Joint GAF/RAF/USAF Workshop on Electronic Crew Teamwork, Ingolstadt, Germany, 25-28 September, 1990.

Perrow, C. (1984). *Normal accidents: Living with High-Risk Technologies*. New York: Basic Books.

Porter, L. W., Lawler, E. E. III & Hackman, J. R. (1975). *Behavior in organizations*. New York: MacGraw-Hill.

Poucet, A. (1989). *Human Factors Reliability Benchmark Exercise - Synthesis Report* (EUR 12222 EN). Ispra (VA), Italy: CEC Joint Research Centre.

Poucet, A., Amendola, A. & Cacciabue, P. C. (1987). *Common cause failure reliability benchmark exercise* (EUR 11054 EN). Ispra, Italy: Joint Research Centre.

Prætorius, N. & Duncan, K. D. (1988). Verbal reports: A problem in research design. In L. P. Goodstein, H. B. Andersen & S. E. Olsen (Eds.), *Tasks, errors and mental models*. London: Taylor & Francis.

Rasmussen, J. (1973). *The role of the man-machine interface in systems reliability*. NATO Generic Conference, Liverpool, UK.

Rasmussen, J. (1974). *The human data processor as a system component. Bits and pieces of a model,* Risø-M-1722. Risø National Laboratory, Roskilde, Denmark.

Rasmussen, J. (1984). Strategies for state identification and diagnosis in supervisory control tasks, and design of computer based support systems. In W. B. Rouse & J. Rasmussen (Eds.), *Advances in man-machine systems research,* Vol. I. A. I. Press, Inc.

Rasmussen, J. (1985). Trends in human reliability analysis. *Ergonomics, 28(8),* 1185-1195.

Rasmussen, J. (1986). *Information processing and human-machine interaction: An approach to cognitive engineering.* New York: North-Holland.

Rasmussen, J. (1988). Human error mechanisms in complex work environments. In G. E. Apostolakis, P. Kafka, & G. Mancini (Eds.), *Accident sequence modelling: Human actions, system response, intelligent decision support.* London: Elsevier Applied Science.

Rasmussen, J., Duncan, K. & Leplat, J. (Eds.), (1987). *New technology and human error.* New York: Wiley.

Rasmussen, J. & Jensen, A. (1973). *A study of mental procedures in electronic troubleshooting (Risø-M-1582).* Roskilde, Denmark. Risø National Laboratory.

Rasmussen, J. & Vicente, K. J. (1987). *Cognitive control of human activities and errors: Implications for ecological interface design* (Risø-M-2660). Roskilde, Denmark: Risø National laboratory.

Reason, J. T. (1988). Modelling the basic error tendencies of human operators. In G. E. Apostolakis, P. Kafka, & G. Mancini (Eds.), *Accident sequence modelling. Human actions, system response, intelligent decision support.* London: Elsevier Applied Science.

Reason, J. T. (1990). *Human error.* Cambridge, UK: Cambridge University Press.

Reason, J. T. & Mycielska, K. (1982). *Absent-minded? The psychology of mental lapses and everyday errors.* Englewood Cliffs, NJ: Prentice-Hall.

Restle, F. & Greeno, J. G. (1970). *Introduction to mathematical psychology.* Reading, MA: Addison-Wesley.

Robinson, J. E., Deutsch, W. E. & Rogers, J. G. (1970). The field maintenance interface between human engineering and maintainability engineering. *Human Factors, 12,* 253-259.

Rosness, R., Hollnagel, E., Sten, T. & Taylor, J. R. (1992). *Human reliability assessment methodology for the European Space Agency* (STF75 F92020). Trondheim, Norway: SINTEF.

Sanderson, P. M. (1989). The human planning and scheduling role in advanced manufacturing systems: An emerging human factors domain. *Human Factors, 31*(6), 635-666.

Sanderson, P. M. & Harwood, K. (1988). The skills, rules and knowledge classification: A discussion of its emergence and nature. In L. P. Goodstein, H. B. Andersen & S. E. Olsen (Eds.), *Task, errors and mental models*. London: Taylor & Francis.

Schützenberger, M. P. (1954). A tentative classification of goal-seeking behaviours. *Journal of Mental Science, 100*, 97-102.

Searle, J. R. (1980). Minds, brains, and programs. *The Behavioral and Brain Sciences, 3*, 417-424.

Senders, J. W. & Moray, N. P. (1991). *Human error. Cause, prediction, and reduction*. Hillsdale, NJ: Lawrence Erlbaum.

Shannon, C. E. & Weaver, W. (1969). *The mathematical theory of communication*. Chicago, IL: University of Illinois Press.

Shapero, A., Cooper, I. J., Rappaport, M., Shaeffer, K. H. & Bates, B. J. (1960). *Human engineering testing and malfunction data collection in weapon system programs*. Wright Air Development Division Technical Report, 60-36. Dayton, OH: Wright-Patterson Air Force Base.

Simon, H. A. (1969). *The sciences of the artificial*. Cambridge, MA: MIT Press.

Singleton, W. T. (1974). *Man-machine systems*. Harmondsworth, UK: Penguin Books.

Smith, W., Hill, B., Long, J. & Whitefield, A. (1992). The planning and control of multiple task work: A study of secretarial office administration. In P. A. Booth & A. Sasse (Eds.), Mental models and everyday activities. *Proceedings of Second Interdisciplinary Workshop on Mental Models*, Cambridge, UK, March 23-25.

Spurgin, A. J. & Moieni, P. (1991). Interpretation of simulator data in the context of human reliability modeling. In G. Apostolakis (Ed.), *Probabilistic safety assessment and management*. New York: Elsevier.

Stassen, H. G. (1986). Decision demands and task requirements in work environments: What can be learned from human operator modelling. In E. Hollnagel, G. Mancini & D. D. Woods (Eds.), (1986). *Intelligent decision support in process environments*. Berlin: Springer-Verlag.

Stewart, I. (1989). *Does God play dice: The new mathematics of Chaos.* Harmondsworth, UK: Penguin.

Swain, A. D. (1987). *Accident sequence evaluation program: Human reliability analysis procedure,* NUREG/CR-4772. Washington, DC: U. S. Nuclear Regulatory Commission.

Swain, A. D. (1989). *Comparative evaluation of methods for human reliability analysis* (GRS-71). Garching, FRG: Gesellschaft für Reaktorsicherheit.

Swain, A. D. (1990). Human reliability analysis: Need, status, trends and limitations. *Journal of Reliability Engineering and System Safety, 29,* 301-313.

Swain, A. D. & Guttmann, H. E. (1983). Handbook of human reliability analysis with emphasis on nuclear power plant applications. NUREG/CR-1278. Sandia National Laboratories, NM.

Taylor, F. V. (1960). Four basic ideas in engineering psychology. *American Psychologist, 15,* 643-649.

Taylor, F. W. (1911). *The principles of scientific management.* New York: Harper.

Trager, Jr. T. A. (1985). *Case study report on loss of safety system function events.* AEOD/04. Washington, DC: US NRC.

Tversky, A. & Kahneman, D. (1975). Judgment under uncertainty: Heuristics and biases. In D. Wendt & C. Vlek (Eds.), *Utility, probability and human decision making.* Dordrecht, Holland: Reidel.

Wakefield, D. J. (1988). Application of human cognitive reliability model and confusion matrix in a probabilistic risk assessment. In G. E. Apostolakis, P. Kafka, & G. Mancini (Eds.), *Accident sequence modelling: Human actions, system response, intelligent decision support.* London: Elsevier Applied Science.

Weizenbaum, J. (1976). *Computer power and human reason: From judgment to calculation.* San Francisco: Freeman.

Welford, A. T. (1969). Ergonomics of automation. In D. H. Holding (Ed.), *Experimental psychology in industry.* Harmondsworth, UK: Penguin Books.

Wickens, C. D. (1990). Processing resources and attention. In D. Damos (Ed.), *Multiple task performance.* London: Taylor & Francis.

Wiener, E. L. (1977). Controlled flight into terrain accidents: System-induced errors. *Human Factors, 19,* 171-181.

Wiener, E. L. & Nagel, D. C. (Eds.) (1988). *Human factors in aviation.* San Diego: Academic Press.

Wilde, G. J. S. (1982). The theory of risk homeostasis: Implications for safety and health. *Risk Analysis, 2(4)*, 209-255.

Winston, P. H. (1984). *Artificial intelligence* (2nd ed.). Reading, MA: Addison-Wesley.

Woods, D. D. & Roth, E. M. (1990). Cognitive systems engineering. In M. Helander (Ed.), *Handbook of human-computer interaction*. Amsterdam: North-Holland.

Woods, D. D., Roth, E. M. & Pople, H. Jr. (1988). Modeling human intention formation for human reliability assessment. In G. E. Apostolakis, P. Kafka, & G. Mancini (Eds.), *Accident sequence modelling: Human actions, system response, intelligent decision support*. London: Elsevier Applied Science.

Index

Accident analysis, xxii; 271
act of God, 16
Action Error Mode Analysis (AEAM), 137; 307
activity set, 165; 166; 172; 201
Adapathya, R., 321
adaptation, 32; 33; 276 - 280
auxiliary feedwater (AFW), 220; 222 - 225; 237; 247 - 250; 252 - 255; 257
aggregation, 38; 88; 101; 108; 117; 162; 240; 242; 270; 275; 296; 297
Allen, J., 215; 278; 315; 320
Amalberti, R., 174; 180; 315
Amendola, A., 130; 315; 322
amplification, 7; 18; 84; 85
analysed event data, 115; 142
anatomy of an accident, 14; 46
Anderson, J. R., 62; 153; 315
Annett, J., 207; 209; 315
Apostolakis, G. E., xiv; xvii; 49; 50; 57; 125; 315; 316; 318; 320; 323 - 326
apparent complexity, 33; 34
Approach To Landing (ATL), 165; 281
Arbib, M. A., 146; 315
Arnott, J. L., 322

arousal, 177; 187; 301
artificial cognition, 85 - 87; 92
artificial intelligence, 92
Ashby, W. R., 37; 105; 120; 163; 315
ATC, 11
atomistic assumption, 94 - 97; 101; 141; 204; 205
attention capacity, 236; 292 - 294
attention demand, 292; 293
Attneave, F., 96; 315
Austin, G. A., 316
automation, xviii; 1; 4; 6; 18 - 20; 32; 62; 81 - 85; 92; 276; 294; 295
availability of plans, 179; 202; 235; 259; 260
available time, 20; 171 - 173; 177; 178; 182; 186; 188 - 190; 192 - 194; 202; 234; 236; 237; 239; 240; 242; 249; 259; 260
Ayel, M., 86; 315

Bagnara, S., 123; 315
Bainbridge, L., xxvi; 160; 161; 168; 279; 315
Bates, B. J., 324
Beare, A. N., 69; 124; 316

Index

Bellamy, L. J., 4; 316
Boden, M. A., 176; 316
Booth, P., xxvi; 238, 316; 324
Bruner, J. S., 169; 316

Cacciabue P. C., xxvi; 120; 141; 151; 165; 199; 231 - 234; 255; 273; 275; 278; 281; 285; 293; 316; 317; 319; 322
centralisation, 81; 82; 85; 92
Chaos theory, 107
chess, 190
Christensen, J. M., 4; 316
circadian rhythm, 40; 294
COCOM, xxi; xxii; 47; 48; 175; 184; 192 - 198; 202; 203; 205 - 207; 213; 229; 232 - 234; 239; 240; 262 - 265; 267; 280; 283; 287; 296; 301 - 304
cognitive approach, 63; 64; 125; 136
cognitive goal, 160
cognitive modelling, xiv; 53; 124; 125; 138; 139
cognitive myopia, 44
cognitive systems engineering, xiv; xvii; xxvi; 124; 149
cognitive viewpoint, xvii; xix; xx; xxv; 21; 22; 44; 46; 149; 150; 200; 201; 296
Cojazzi, G., 316
commission, 3; 60; 79; 91
common cause, 28; 44; 47; 130
common mode, 7; 39; 44; 45; 47; 306

competence description, 117; 142
completeness, 128; 229; 278; 286; 287; 304; 307
complexity of cognition, 34
component-driven, 54 - 56
computerisation, 82; 83; 85; 92;282; 283
conceptual description, 116; 117; 142
consciousness, 56; 289
consistency, 21; 57; 128; 130; 189; 198; 286; 287; 304; 313
contextual control, xx; xxi; xxii; 47; 152; 153; 161; 162; 164; 166; 176; 180; 194 - 196; 198; 201; 202; 270; 271; 273 - 275; 279; 280; 284; 288
continuity, 36; 187; 230; 267
control mode, 166; 169 - 171; 173; 174; 177; 180; 182; 184 -; 191; 193 - 195; 197; 198; 202; 230; 235; 239; 240; 263; 264; 266; 283; 298
Cooper, I. J., 324
core cooling, 220 - 224; 247
correct action, 71; 91
CPM, xxi; 45; 206; 233; 234; 240; 242; 243; 249; 250; 251; 259; 265
curve fitting, 145
cybernetics, 2; 105

Data analysis, 114 - 116; 229

data collection, 47; 112 - 115; 117; 119; 120; 123; 142; 143; 145; 279
Dawkins, R., 80; 316
De Keyser, V., 77; 169; 234; 316; 317
De Mey, M., 22; 316
Deblon, F., 174; 180; 315
decomposition approach, xix; 102; 115; 203; 240; 243; 254; 261; 262; 264; 265; 281; 302
decomposition principle, xix; xx; 15; 34; 37; 38; 43 - 48; 54; 57 - 60; 64; 80; 81; 87; 88; 90; 91; 95; 109; 122; 125; 127; 136; 140; 141; 204; 269; 281; 283; 288
Decortis, F., 77; 141; 234; 317
Democritus, 95
Dependent Differentiation Method (DDM), xxi; xxii;25; 48; 203; 206; 245; 246; 252; 255 - 259; 261 - 265; 273; 280; 282 - 288; 301 - 304
design principle, 141
Dessouky, M. I., 321
determinism, 157; 201
Deutsch, W. E., 323
Di Martino, C., 315
Dougherty, E., xiii; xvi; xvii; xxv; 50; 285; 317; 321
Dubois, D., 107; 317
Duncan, K. D., 66; 68; 315; 322; 323
Duncker, K., 300; 317
dynamic process supervision, 199

Dörner, D., 119; 317

Embrey, D. E., 41; 131; 317
empirical data, xx; 21; 39; 50; 56; 78; 110; 122; 123; 125; 129; 135; 138; 143; 206; 208; 261; 262; 304
enabling task, 214; 215
engineering approach, 50; 64
engineering quantification, 94; 97
engineering solution, 50; 54; 57; 90; 139
English, A. C., 61; 292; 320
English, H. B., 61; 292; 320
ergonomics, xiv; xvii; 1 - 3; 45; 51; 281
erroneous action, xix; 29 - 31; 67; 69; 70; 73; 75; 77; 91; 123; 126; 231; 284; 299; 300
event estimation, 90
event horizon, 168; 170; 180 - 182; 185; 187 - 190; 202; 234; 239
event tree, 34; 38; 116; 125; 252 - 254; 257
expert judgment, 109; 125 - 127; 136; 143; 282
explorative control, 169; 173; 175; 190
external validity, 231

Failure mode, 306
fallible machine, xx; 64; 65; 100; 112; 116; 148; 149
fault tolerant system, 19

fault tree, 37; 306
Feed and Bleed (FAB), xxv; 220; 222 - 225; 247 - 257; 259; 263
feedforward, 133; 173; 182; 190
Filz, A. Y., 322
Fischhoff, B., 190; 317
FMEA, 306
FMECA, 306
focus gambling, 169
Fragola, J. R., 317
Fujita, Y., xxvi; 94; 102; 104; 110; 320
fuzzy set, xxi; 244; 258 - 260

Gaddy, C. D., 316
Galanter, E., 157; 321
GEMS, 108; 137
genotype, 69; 70; 73; 91
Geyer, T. A. W., 316
Goals-Means Task Analysis (GMTA), xxi; 48; 153; 207; 210; 212; 213; 219; 220; 223; 224; 226; 227; 229; 234; 236; 241; 247; 249; 250; 253; 254; 258; 267; 275; 282; 284; 287; 288; 296; 301
Goodnow, J. J., 316
Gray, M. J., 315
Green, A. E., 12; 317
Greeno, J. G., 110; 323
Guttmann, H. E., 38; 40; 52; 131; 134; 281; 325

Hackman, J. R., 322
Hagen, E. W., 4; 70; 317
Hall, R. E., 207; 294; 317
Hamer, M., 9; 317
Hannaman, G. W., 54; 103; 132; 134; 273; 318
Harwood, K., 155; 324
Haugeland, J., 148; 318
HCR, 103; 104; 132; 137
Henley, J., 207; 318
HF-RBE, 129; 130; 132; 133
hierarchical task description, 213
hierarchisation, 230; 231; 267
Hill, B., 315; 322; 324
history size, 180; 190; 202; 234
holistic, 102; 108; 274
Hollnagel, E., xiv; 18; 29; 67; 68; 72; 86; 108; 113; 115; 116; 120; 121; 124; 149; 151; 153 - 155; 160; 164; 174; 178; 194; 199; 231; 232; 239; 255; 275 - 278; 281; 293; 316 - 319; 321; 324
homo economicus, 190
Howard, J. M., 316
human action, xiv; xxiv; 3; 7; 14; 29; 31; 39; 45; 56; 60; 67; 69; 78; 84; 89; 91; 93; 94; 108; 124; 136; 145; 151; 153; 168; 176; 181; 201; 206; 229; 231; 232; 283; 296; 301; 306
human cognition, xiv; xv; xvii - xx; xxii, 6; 23; 33; 37; 46; 47; 49; 50; 53; 56 - 58; 60; 64 - 66; 87; 91; 98; 102; 112; 145; 148; 149; 151; 160 - 162; 168; 198; 200; 205; 229; 236; 289; 293; 302

human error probability (HEP), 39; 257; 304
human error rate, 42
human factor, xviii; xix; 2
human factors engineering, 2; 18
human performance, xvi; xvii; xviii; xxiii; 3; 20; 31; 33; 40; 45 - 49; 58; 63 - 66; 79; 80; 81; 84; 85; 92; 94; 97; 104; 106; 108; 109; 111; 112; 115; 126; 127; 133; 138; 145; 150; 153; 161; 177; 198 - 200; 204; 205; 234; 265; 267; 269; 270; 271; 275; 277; 280; 288; 294; 296; 306
human reliability analysis/assessment (HRA), xiii - xvi; xviii - xxiii; 16; 21; 24 - 26; 32; 38; 41; 44; 46 - 48; 50; 51; 60; 62; 65; 67; 78; 79; 90; 92 - 94; 99; 102; 104; 109 - 112; 114; 115; 122; 129; 131 - 134; 136 - 138; 141 - 144; 200; 202; 203; 208; 229; 231; 232; 262; 265; 266; 269; 270; 281; 282; 284 - 286; 290; 296; 301 - 304; 309
Humphreys, P., 317

Inaccurate execution, 21; 23; 46
individual action, xix; 38; 44; 45; 47
information processing mechanism, 97; 148; 200
INPO, 4
insertion, 74; 77 - 79

interface design, 169; 173; 199; 234; 273 - 275; 313
intermediate data format, 115; 142
introspection, 56; 68; 91; 236
intrusion, 71; 72; 74; 75; 78; 79; 91

Jensen, A., 12; 323
Johannsen, G., 146; 319
jump backward, 91

Kafka, P., 125; 315; 318; 323; 325; 326
Kahneman, D., 290; 292; 296; 319; 320; 325
Kantowitz, B. H., 94; 102; 104; 110; 320
Kautz, H., 278; 320
Kijowski, B. A., 321
Kirwan, B., 317
knowledge mismatch, 21; 23; 46; 169
knowledge-based, 104; 109; 148; 158; 159; 195; 234; 267; 273; 278; 288; 296
Kumamoto, H., 207; 318
Kurke, M. I., 207; 236; 320

Larsen, M. N., 303; 320
latent failure, 14
Laurent, J.-P., 86; 315
Law of Requisite Variety, xxiii; 37; 47; 105; 106; 142; 163; 271
Lawler, E. E. III, 322

Index

Lederman, L., 136; 320
Lee, J., 51; 321
Leplat, J., xxvi; 209; 231; 320; 323
LeVan, W. L., 4; 320
level of risk, 8; 10; 11
Leveson, N. G., 15; 320
Lind, M., 156; 303; 320
Lindblom, C. E., 121; 320
Lisanti, B., 315
Long, J., 324

Mach, E., 65; 320
man-machine interaction (MMI), xiv; xxiii; xxiv; 8; 14; 24; 30; 31; 33; 37; 86; 98; 199; 245; 258; 266; 273; 276 - 279; 305; 306; 308; 309
man-machine system (MMS), xxiv; xxv; 3; 94; 199
Mancini, G., 125; 136; 315; 318; 320; 321; 323 -; 326
Mancini, S., 316
Mandiau, R., 295; 321
Mandler, G., 56; 320
manifestation, 58; 68; 69; 70; 79; 91; 232
Markov chain, 187
Maruyama, M., 7; 320
meaningful description, 26
mechanical model, 27
mechanisation, 84; 85; 92
mechanistic assumption, 96; 97; 100; 141; 204
mental capacity, 2; 292
meta-cognition, 121

micro-cognition, 275
micro-world, 119 - 121; 143
Miller, G. A., 96; 157; 291; 320
Miller, J. G., 171; 321
Millot, P., 295; 321
model of cognition, xv; xx; 79; 152; 154; 162; 176; 184; 203; 284; 38
model of competence, 162; 164; 166; 167; 201
model of control, 158; 162; 164; 199; 201
Moieni, P., 104; 133; 324
Moray, N. P., xvii; xxvi; 43; 51; 67; 110; 119; 175; 198; 234; 321; 324
Morick, H., 66; 321
MTBF, 61
Muir, B. M., 51; 321
Mycielska, K., 70; 323

Nagel, D. C., 276; 325
Nagel, T., 321
Neisser, U., 65; 97; 148; 151; 231; 292; 321
Newell, A., 80; 96; 97; 124; 148; 321
Newell, A. F., 322
Nisbett, R. E., 68; 322
Norman, D. A., 124; 289; 290; 292; 322
nuclear power plant, 103; 153

OATS, 109

omission, 3; 60; 69; 70; 72; 74; 75; 79; 91
operational support, 18; 233; 238; 239; 242; 251; 260; 294
opportunistic control, 169; 170; 171; 174; 186 - 189; 197; 198; 235
overshoot, 78

Parallelism, 230; 231; 253; 267
Park, K. S., 51; 146; 322
Parry, G. W., 316
Peddie, H., 278; 322
Pedersen, O. M., 319
perceived risk, 3; 9; 11
performance analysis, 208; 209; 233; 270
performance shaping factor (PSF), xix; 37; 40; 42; 265; 288
performance variability, 231; 267
Perrow, C., xvii; 3; 7; 322
phenotype, 69; 70; 73; 75 - 78; 91
plan recognition, 171; 278
planning, 2; 22; 35; 36; 82; 121; 141; 164; 168 - 171; 174; 175; 181; 185; 191; 295; 298
Pople, H. Jr., 326
Porter, L. W., 88; 322
post-condition, 218; 219; 228; 229
Poucet, A., 41; 130 - 132; 133; 166; 257; 284; 322
PRA, 107; 110; 134; 244
Prade, H., 107; 317

pre-condition, 157; 211 - 213; 215; 216; 218; 222 - 225; 228 - 230; 247
prediction, xxii; xxiii; 33; 117; 139; 140; 180 - 182; 185; 187; 190; 193; 202; 234; 271 - 273
premature action, 77
Pribram, K. H., 157; 321
privileged knowledge, 66
probabilistic safety assessment (PSA), 16; 18; 24; 49; 50; 67; 90; 93; 106 - 108; 111; 199; 203; 244; 282; 304; 305
procedural prototype, 97; 116; 152 - 157; 160; 162; 167; 170; 195; 201; 205; 267; 269; 270
processing capacity, 2; 45
Prætorius, N., 66; 68; 322
psychological mechanism, 27
pure task, 212; 213; 217
PWR, 220

Qualitative analysis, 25; 111; 132; 166; 200; 244; 281; 284; 286; 287; 304
quantification, xx; 15; 24 - 26; 48; 49; 54; 56; 60; 80; 90; 93; 94; 97; 102; 110; 111; 118; 124; 132; 133; 141; 143; 204; 244; 256 - 258; 265; 283; 284; 304
quantitative analysis, 45; 46; 286

Rappaport, M., 324

Index

Rasmussen, J., xvii; 4; 12; 32; 33; 64; 65; 70; 104; 148; 154; 155; 158; 159; 170; 195; 198; 231; 234; 273; 288; 296; 316; 319; 322; 323
raw data, xx; 68; 114; 115; 117; 142; 286
Rea, K., 317
Reason, J. T., xvii; 64; 68; 70; 108; 148; 149; 169; 181; 294; 296; 323
reliability analysis, xiii - xvi; xviii - xx; xxii; xxiii; 15; 16; 21; 24 - 26; 32; 38; 44; 46 - 48; 50; 51; 60; 62; 65; 78; 90; 93; 94; 99; 102; 104; 109 - 112; 114; 115; 122; 125; 129; 131 - 134; 136; 138; 141; 142; 144; 166; 200; 202; 208; 229; 231; 232; 262; 265; 266; 269; 270; 281; 283 - 286; 290; 296; 297; 301 - 304; 309
reliability benchmark, xx
reliable performance, 11; 36; 83; 280
repeatability, xix; 54; 61; 62; 91; 129
replacement, 6; 16; 19; 74; 77
Restle, F., 110; 323
reversal, 72; 75; 298
risk homeostasis, 3; 7; 9; 10; 11; 46; 84
Rizzo, A., 315
Robinson, J. E., 4; 318; 323
robustness, 53; 134; 135; 144
Rogers, J. G., 315; 323

root cause, xvi - xviii; 27; 28; 67; 271; 306
Rosa, E. A., 317
Rosness, R., 65; 138; 324
Rotenberg, I., 175; 321
Roth, E. M., 150; 326
Rouhet, J. -C., 319
rule-based, 104; 109; 148; 158; 159; 170; 174; 195; 234; 273; 288; 296

S-O-R, xx; 96; 97; 145 - 148; 150; 200; 296
Sabadosh, N., 119; 321
Sanderson, P. M., 155; 180; 324
satiation, 61; 62
schema, 174
Schützenberger, M. P., 35; 36; 192; 324
Scientific Management, 95; 80; 81; 92
scrambled control, 169; 185 - 188; 190; 191; 197; 230
Searle, J. R., 148; 324
Senders, J. W., xvii; 67; 324
sensitivity, 134; 135; 144; 284; 302
Shaeffer, K. H., 324
Shannon, C. E., 73; 324
Shapero, A., 4; 324
SHARP, 54; 137; 257; 283
similarity, xix; 61; 62; 63; 91; 119; 169; 170; 283; 296; 299; 301; 302
Simon, H. A., 33; 34; 64; 96; 124; 148; 322; 324

simulation, xiv; xxiii; 60; 122 - 125; 136; 138; 139; 240; 278; 281; 282; 312
simulator, 103; 122; 124; 129; 135; 143; 248; 282; 312 - 314
simultaneous goals, 177 - 179; 202; 235; 236; 250; 251; 260
Singh, A., 316
Singleton, W. T., xxiv; 324
skill-based, 104; 109; 148; 158; 159; 195; 234; 236; 273; 288; 296
slip, 70
Slovic, P., 320
Smith, W., 277; 324
software reliability, 15; 16; 86
specificity, 299
Spurgin, A. J., 54; 104; 132 - 134; 318; 324
Stammers, R. B., 315
Stassen, H. G., 146; 324
Sten, T., 324
step-ladder model (SLM), 148; 155; 156; 158; 195; 231
Stevens, B. S., 316
Stewart, I., 107; 325
strategic control, 168; 170; 171; 173; 179; 188 - 191; 197; 198; 230
stress, 26; 40 - 45; 194; 232; 290
sub-goal, 167; 211 - 213; 216; 245
subjectively available time, 171 - 173; 177; 178; 182; 186; 188; 190; 194; 202; 234; 240

Success Likelihood Index Methodology (SLIM), 41; 109; 131; 132; 137
Swain, A. D., 3; 4; 25; 31; 38; 40; 44; 52; 114; 131; 133; 134; 135; 138; 257; 281; 296; 325
system dependability, 9
system design, 18; 20; 23; 24; 44; 46; 117; 153; 208; 266; 276; 277; 283; 304; 306
System Response Generator (SRG), xxvi; 137; 196; 199; 240; 278; 280; 281; 305; 308 - 314

Tactical control, 170; 171; 173; 186; 188 - 191; 197; 198; 298
task allocation, 19; 208; 236; 267
task analysis, xxi; 25; 48; 125; 157; 206 - 209; 211 - 213; 217; 220; 226; 229; 231; 233; 234; 236; 238; 240 - 242; 246; 266; 282; 283
task description, 208; 209; 213; 220; 227; 232; 236; 242; 256; 258; 270
task design, 19; 234; 284
task element, 37
task load, 13; 40; 235
task representation, 209; 210; 220
task driven, 54; 55; 56
Taylor, F. V., 1; 325
Taylor, F. W., 80; 95; 325
Taylor, J. R., xxvi
Taylorism, 80

Index

team structure, 41; 42
technical system, 63
Technique for Human Error Rate Prediction (THERP), 38; 131; 132; 134; 137; 204; 254; 257; 273; 280; 281; 285
technological improvement, 9; 10
template set, 167; 172
time compression, 141; 175; 234
timing condition, 215
TOTE, 157; 158; 159; 182; 201
Trager, Jr. T. A., 4; 325
trial-and-error, 168; 185
trust, 51
Tversky, A., 296; 320; 325

Unexpected event, 12; 14
Unified Theory of Cognition, 80; 97
unwanted consequences, xvi; xix; xxii; xxiv; 3; 6; 7; 12; 18; 21; 22; 30 - 32; 45; 46; 49; 50; 51; 63; 65; 69; 70; 84; 91; 106; 109; 112; 122; 140; 244; 262; 264; 276; 295; 296; 298; 300 - 305

Validation, 86
validity, 68; 119; 121; 135; 160; 231; 261; 282; 285; 308
Very Simple Model of Cognition, 151
Vicente, K. J., 159; 234; 273; 323

Wakefield, D. J., 109; 325
Weaver, W., 73; 324
Weizenbaum, J., 148; 325
Welford, A. T., 81; 325
Whitefield, A., 324
Wickens, C. D., 293; 325
Wiener, E. L., 276; 325
Wilde, G. J. S., 8; 10; 326
Wilson, T. D., 68; 322
Winston, P. H., 153; 315; 321; 326
Woods, D. D., xiv; xxvi; 124; 149; 150; 199; 316; 318; 319; 321; 324; 326
Worledge, D. H., 103; 273; 318
Wreathall, J., 317